Undergraduate Texts in Mathematics

Undergraduate Texts in Mathematics

Undergraduate Texts in Mathematics are generally aimed at third- and fourth-year undergraduate mathematics students at North American universities. These texts strive to provide students and teachers with new perspectives and novel approaches. The books include motivation that guides the reader to an appreciation of interrelations among different aspects of the subject. They feature examples that illustrate key concepts as well as exercises that strengthen understanding.

More information about this series at http://www.springer.com/series/666

Matthias Beck • Sinai Robins

Computing the Continuous Discretely

Integer-Point Enumeration in Polyhedra

Second Edition

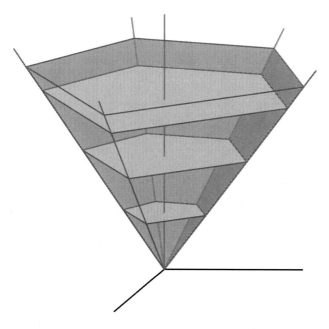

With illustrations by David Austin

 Springer

Matthias Beck
Department of Mathematics
San Francisco State University
San Francisco, CA, USA

Sinai Robins
Department of Mathematics
Brown University
Providence, RI, USA

ISSN 0172-6056   ISSN 2197-5604   (electronic)
Undergraduate Texts in Mathematics
ISBN 978-1-4939-2968-9   ISBN 978-1-4939-2969-6   (eBook)
DOI 10.1007/978-1-4939-2969-6

Library of Congress Control Number: 2015942640

Springer New York Heidelberg Dordrecht London

Printed on acid-free paper

Springer Science+Business Media LLC New York is part of Springer Science+Business Media (www.springer.com)

To Tendai   *To my mom, Michal Robins*

with all our love

Preface

The world is continuous, but the mind is discrete.

David Mumford

The mathematical interplay between polytopes and lattices comes to life when we study the relationships between the continuous volume of a polytope and its discrete volume. Since the humbling and positive reception of the first edition of this book, published in 2007, the field of integer-point enumeration in polyhedra has gained considerable momentum. Many fields of mathematics have begun to interact in even more surprising ways, and a beautifully simple unification among a multitude of classical problems continues to emerge when we use this combinatorial-geometric lens.

In this second edition, and with encouragement from many wonderful readers, we have added two new chapters: Chapter 9 introduces zonotopes, an extremely useful class of polytopes, and Chapter 10 explores some deep yet elegant relationships that are satisfied by the h-polynomials and h^*-polynomials. We have also added many new exercises, new and updated open problems, and graphics.

Examples of polytopes in three dimensions include crystals, boxes, tetrahedra, and any convex object whose faces are all flat. It is amusing to see how many problems in combinatorics, number theory, and many other mathematical areas can be recast in the language of polytopes that exist in some Euclidean space. Conversely, the versatile structure of polytopes gives us number-theoretic and combinatorial information that flows naturally from their geometry.

The *discrete volume* of a body $\mathcal{P}$ can be described intuitively as the number of grid points that lie inside $\mathcal{P}$, given a fixed grid in Euclidean space. The *continuous volume* of $\mathcal{P}$ has the usual intuitive meaning of volume that we attach to everyday objects we see in the real world.

Fig. 0.1 Continuous and discrete volume.

Indeed, the difference between the two realizations of volume can be thought of in physical terms as follows. On the one hand, the quantum-level grid imposed by the molecular structure of reality gives us a discrete notion of space and hence discrete volume. On the other hand, the Newtonian notion of continuous space gives us the continuous volume. We see things continuously at the Newtonian level, but in practice, we often compute things discretely at the quantum level. Mathematically, the grid we impose in space—corresponding to the grid formed by the atoms that make up an object—helps us compute the usual continuous volume in very surprising and charming ways, as we shall discover.

In order to see the continuous/discrete interplay come to life among the three fields of combinatorics, number theory, and geometry, we begin our focus with the simple-to-state *coin-exchange problem* of Frobenius. The beauty of this concrete problem is that it is easy to grasp, it provides a useful computational tool, and yet it has most of the ingredients of the deeper theories that are developed here.

In the first chapter, we give detailed formulas that arise naturally from the Frobenius coin-exchange problem in order to demonstrate the interconnections between the three fields mentioned above. The coin-exchange problem provides a scaffold for identifying the connections between these fields. In the ensuing chapters, we shed this scaffolding and focus on the interconnections themselves:

(1) Enumeration of integer points in polyhedra—combinatorics,
(2) Dedekind sums and finite Fourier series—number theory,
(3) Polygons and polytopes—geometry.

We place a strong emphasis on computational techniques and on computing volumes by counting integer points using various old and new ideas. Thus, the formulas we get should not only be pretty (which they are!) but also allow us to compute volumes efficiently using some nice functions. In the very rare instances of mathematical exposition when we have a formulation that is both "easy to write" and "quickly computable," we have found a mathematical nugget. We have endeavored to fill this book with such mathematical nuggets.

Much of the material in this book is developed by the reader in the more than three hundred exercises. Most chapters contain warmup exercises that

do not depend on the material in the chapter and can be assigned before the
chapter is read. Some exercises are central, in the sense that current or later
material depends on them. Those exercises are marked with ♣, and we give
detailed hints for them at the end of the book. Most chapters also contain
lists of open research problems.

It turns out that even a fifth grader can write an interesting paper on integer-
point enumeration [192], while the subject lends itself to deep investigations
that attract the current efforts of leading researchers. Thus, it is an area of
mathematics that attracts our innocent childhood questions as well as our
refined insight and deeper curiosity. The level of study is highly appropriate
for a junior/senior undergraduate course in mathematics. In fact, this book
is ideally suited to be used for a *capstone course*. Because the three topics
outlined above lend themselves to more sophisticated exploration, our book
has also been used effectively for an introductory graduate course.

To help the reader fully appreciate the scope of the connections between
continuous volume and discrete volume, we begin the discourse in two dimensions, where we can easily draw pictures and quickly experiment. We gently
introduce the functions we need in higher dimensions (Dedekind sums) by
looking at the coin-exchange problem geometrically as the discrete volume of
a generalized triangle, called a simplex.

The initial techniques are quite simple, essentially nothing more than
expanding rational functions into partial fractions. Thus, the book is easily
accessible to a student who has completed a standard college calculus and
linear algebra curriculum. It would be useful to have a basic understanding of
partial fraction expansions, infinite series, open and closed sets in $\mathbb{R}^d$, complex
numbers (in particular, roots of unity), and modular arithmetic.

An important computational tool that is harnessed throughout the text
is the *generating function* $f(x) = \sum_{m=0}^{\infty} a(m) x^m$, where the $a(m)$'s form a
sequence of numbers that we are interested in analyzing. When the infinite
sequence of numbers $a(m)$, $m = 0, 1, 2, \ldots$, is embedded into a single generating function $f(x)$, it is often true that for hitherto unforeseen reasons, we can
rewrite the whole sum $f(x)$ in a surprisingly compact form. It is the rewriting
of these generating functions that allows us to understand the combinatorics
of the relevant sequence $a(m)$. For us, the sequence of numbers might be the
number of ways to partition an integer into given coin denominations, or the
number of points in an increasingly large body, and so on. Here we find yet
another example of the interplay between the discrete and the continuous: we
are given a *discrete* set of numbers $a(m)$, and we then carry out analysis on
the generating function $f(x)$ in the *continuous* variable x.

What Is the Discrete Volume?

The physically intuitive description of the discrete volume given above rests
on a sound mathematical footing as soon as we introduce the notion of a

lattice. The grid is captured mathematically as the collection of all integer points in Euclidean space, namely $\mathbb{Z}^d = \{(x_1, \ldots, x_d) : \text{all } x_k \in \mathbb{Z}\}$. This discrete collection of equally spaced points is called a *lattice*. If we are given a geometric body $\mathcal{P} \subset \mathbb{R}^d$, its discrete volume is simply defined as the number of lattice points inside $\mathcal{P}$, that is, the number of elements in the set $\mathbb{Z}^d \cap \mathcal{P}$.

Intuitively, if we shrink the lattice by a factor k and count the number of newly shrunken lattice points inside $\mathcal{P}$, we obtain a better approximation for the volume of $\mathcal{P}$, relative to the volume of a single cell of the shrunken lattice. It turns out that after the lattice is shrunk by an integer factor k, the number $\#\left(\mathcal{P} \cap \frac{1}{k}\mathbb{Z}^d\right)$ of shrunken lattice points inside an *integral polytope* $\mathcal{P}$ is magically a polynomial in k. This counting function $\#\left(\mathcal{P} \cap \frac{1}{k}\mathbb{Z}^d\right)$ is known as the *Ehrhart polynomial* of $\mathcal{P}$. If we kept shrinking the lattice by taking a limit, we would of course end up with the continuous volume that is given by the usual Riemannian integral definition of calculus:

$$\mathrm{vol}\,\mathcal{P} = \lim_{k \to \infty} \#\left(\mathcal{P} \cap \frac{1}{k}\mathbb{Z}^d\right) \frac{1}{k^d}.$$

However, pausing at fixed dilations of the lattice gives surprising flexibility for the computation of the volume of $\mathcal{P}$ and for the number of lattice points that are contained in $\mathcal{P}$.

Thus, when the body $\mathcal{P}$ is an integral polytope, the error terms that measure the discrepancy between the discrete volume and the usual continuous volume are quite nice; they are given by Ehrhart polynomials, and these enumeration polynomials are the content of Chapter 3.

The Fourier–Dedekind Sums Are the Building Blocks: Number Theory

Every polytope has a discrete volume that is expressible in terms of certain finite sums that are known as *Dedekind sums*. Before giving their definition, we first motivate these sums with some examples that illustrate their building-block behavior for lattice-point enumeration. To be concrete, consider, for example, a 1-dimensional polytope given by an interval $\mathcal{P} = [0, a]$, where a is any positive real number. It is clear that we need the greatest integer function $\lfloor x \rfloor$ to help us enumerate the lattice points in $\mathcal{P}$, and indeed, the answer is $\lfloor a \rfloor + 1$.

Next, consider a 1-dimensional line segment that is sitting in the 2-dimensional plane. Let's choose our segment $\mathcal{P}$ so that it begins at the origin and ends at the lattice point (c, d). As becomes apparent after a moment's thought, the number of lattice points on this finite line segment involves an old friend, namely the greatest common divisor of c and d. The exact number of lattice points on the line segment is $\gcd(c, d) + 1$.

To unify both of these examples, consider a triangle $\mathcal{P}$ in the plane whose vertices have rational coordinates. It turns out that a certain finite sum is completely natural because it simultaneously extends both the greatest integer function and the greatest common divisor, although the latter is less obvious. An example of a Dedekind sum in two dimensions that arises naturally in the formula for the discrete volume of the rational triangle $\mathcal{P}$ is the following:

$$ s(a,b) = \sum_{m=1}^{b-1} \left(\frac{m}{b} - \frac{1}{2} \right) \left(\frac{ma}{b} - \left\lfloor \frac{ma}{b} \right\rfloor - \frac{1}{2} \right). $$

The definition makes use of the greatest integer function. Why do these sums also resemble the greatest common divisor? Luckily, the Dedekind sums satisfy a remarkable reciprocity law, quite similar to the Euclidean algorithm that computes the greatest common divisor. This reciprocity law allows the Dedekind sums to be computed in roughly $\log(b)$ steps rather than the b steps that are implied by the definition above. The reciprocity law for $s(a,b)$ lies at the heart of some amazing number theory that we treat in an elementary fashion, but that also comes from the deeper subject of modular forms and other modern tools.

We find ourselves in the fortunate position of viewing an important summit of an enormous mountain of ideas, submerged by the waters of geometry. As we delve more deeply into these waters, more and more hidden beauty unfolds for us, and the Dedekind sums are an indispensable tool that allow us to see farther as the waters get deeper.

The Relevant Solids Are Polytopes: Geometry

The examples we have used, namely line segments and polygons in the plane, are special cases of polytopes in all dimensions. One way to define a polytope is to consider the *convex hull* of a finite collection of points in Euclidean space $\mathbb{R}^d$. That is, suppose someone gives us a set of points $\mathbf{v}_1, \ldots, \mathbf{v}_n$ in $\mathbb{R}^d$. The polytope determined by the given points $\mathbf{v}_j$ is defined by all linear combinations $c_1\mathbf{v}_1 + c_2\mathbf{v}_2 + \cdots + c_n\mathbf{v}_n$, where the coefficients c_j are nonnegative real numbers that satisfy the relation $c_1 + c_2 + \cdots + c_n = 1$. This construction is called the *vertex description* of the polytope.

There is another equivalent definition, called the *hyperplane description* of the polytope. Namely, if someone hands us the linear inequalities that define a finite collection of half-spaces in $\mathbb{R}^d$, we can define the associated polytope as the simultaneous intersection of the half-spaces defined by the given inequalities.

There are some "obvious" facts about polytopes that are intuitively clear to most students but are, in fact, subtle and often nontrivial to prove from first principles. One of these facts, namely that every polytope has both a vertex and a hyperplane description, forms a crucial basis to the material we will develop in this book. We carefully prove this fact in the appendix. The statement is intuitively clear, so that novices can skip over its proof without any detriment to their ability to compute continuous and discrete volumes of polytopes. All theorems in the text (including those in the appendix) are proved from first principles, with the exception of Chapter 14, where we assume basic notions from complex analysis.

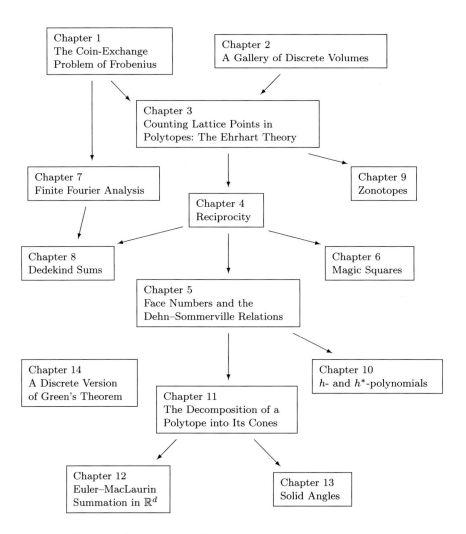

Fig. 0.2 The partially ordered set of chapter dependencies

The text naturally flows into two parts, which we now explicate.

Part I

We have taken great care in making the content of the chapters flow seamlessly from one to the next, over the span of the first six chapters.

- Chapters 1 and 2 introduce some basic notions of generating functions, in the visually compelling context of discrete geometry, with an abundance of detailed motivating examples.
- Chapters 3, 4, and 5 develop the full Ehrhart theory of discrete volumes of rational polytopes.
- Chapter 6 is a "dessert" chapter, in that it enables us to use the theory developed to treat the enumeration of *magic squares*, an ancient topic that enjoys active current research.

Part II

We now begin anew.

- Having attained experience with numerous examples and results about integral polytopes, we are ready to learn about the *Dedekind sums* of Chapter 8, which form the atomic pieces of the discrete-volume polynomials. On the other hand, to fully understand Dedekind sums, we need to understand *finite Fourier analysis*, which we therefore develop from first principles in Chapter 7, using only partial fractions.
- In Chapter 9, we study a concrete class of polytopes—projections of cubes, which go by the name *zonotopes*—whose discrete volume is tractable and has neat connections to number theory and graph theory.
- Chapter 10 develops inequalities among the coefficients of an Ehrhart polynomial, based on a *polynomial decomposition* formula that arises naturally from the arithmetic and combinatorial data of triangulations.
- Chapter 11 answers a simple yet tricky question: how does a finite geometric series in one dimension extend to higher-dimensional polytopes? *Brion's theorem* gives an elegant and decisive answer to this question.
- In Chapter 12, we extend the interplay between the continuous volume and the discrete volume of a polytope (already studied in detail in Part I) by introducing *Euler–Maclaurin summation* formulas in all dimensions. These formulas compare the continuous Fourier transform of a polytope to its discrete Fourier transform, yet the material is completely self-contained.
- Chapter 13 develops an exciting extension of Ehrhart theory that defines and studies the *solid angles* of a polytope; these are the natural extensions of 2-dimensional angles to higher dimensions.
- Finally, we end with another "dessert" chapter that uses complex-analytic methods to find an integral formula for the discrepancy between the discrete and continuous areas enclosed by a closed curve in the plane.

Because polytopes are both theoretically useful (in triangulated manifolds, for example) and practically essential (in computer graphics, for example) we use them to link results in number theory and combinatorics. Many research papers have been written on these interconnections, and it is impossible to capture them all here, especially since some are being written even as we are writing this sentence! However, we hope that these modest beginnings will give the reader who is unfamiliar with these fields a good sense of their beauty, inexorable connectedness, and utility. We have written a gentle invitation to what we consider a gorgeous world of counting and of links between the fields of combinatorics, number theory, and geometry for the general mathematical reader.

There are a number of excellent books that have a nontrivial intersection with ours and contain material that complements the topics discussed here. We heartily recommend the monographs of Barvinok [20, 21] (on general convexity topics and Ehrhart theory), Bruns–Gubeladze [73] (on commutative algebra and K-theory connected with polytopes), De Loera–Hemmecke–Köppe [95] (on connections to optimization), De Loera–Rambau–Santos [98] (on triangulations of point configurations), Ehrhart [111] (the historic introduction to Ehrhart theory), Ewald [112] (on connections to algebraic geometry), Hibi [134] (on the interplay of algebraic combinatorics with polytopes), Miller–Sturmfels [177] (on combinatorial commutative algebra), and Stanley [230] (on general enumerative problems in combinatorics).

Acknowledgments

We have had the good fortune of receiving help from many gracious people in the process of writing this book. First and foremost, we thank the students of the classes in which we could try out this material, at Binghamton University (SUNY), Nanyang Technological University, San Francisco State University, and Temple University. We are indebted to our MSRI/Banff 2005 graduate summer school students. We give special thanks to Kristin Camenga and Kevin Woods, who ran the problem sessions for this summer school, detected numerous typos, and provided us with many interesting suggestions for this book. We are grateful for the generous support for the summer school from the Mathematical Sciences Research Institute, the Pacific Institute of Mathematics, and the Banff International Research Station.

Many colleagues have supported this endeavor, and we are particularly grateful to everyone notifying us of mistakes and typos and to those who have made useful suggestions: Daniel Antonetti, Alexander Barvinok, Nathanael Berglund, Andrew Beyer, Tristram Bogart, Garry Bowlin, Benjamin Braun, Swee Hong Chan, Robin Chapman, Yitwah Cheung, Jessica Cuomo, Dimitros Dais, Jesus De Loera, David Desario, Mike Develin, Ricardo Diaz, Michael Dobbins, Jeff Doker, Han Minh Duong, Richard Ehrenborg, David Einstein, Joseph Gubeladze, Christian Haase, Mary Halloran, Friedrich Hirzebruch,

Brian Hopkins, Serkan Hoşten, Benjamin Howard, Piotr Maciak, Evgeny Materov (Chapter 5 rocks!), Asia Matthews, Peter McMullen, Martín Mereb, Ezra Miller, Mel Nathanson, Julian Pfeifle, Peter Pleasants, Jorge Ramírez Alfonsín, Bruce Reznick, Adrian Riskin, Junro Sato, Melissa Simmons, Richard Stanley, Bernd Sturmfels, Thorsten Theobald, Read Vanderbilt, Andrew Van Herick, Sven Verdoolaege, Michèle Vergne, Julie Von Bergen, Neil Weickel, Carl Woll, Zhiqiang Xu, Jon Yaggie, Ruriko Yoshida, Thomas Zaslavsky, Günter Ziegler, and two anonymous referees.

We are indebted to the Springer editorial staff, first and foremost to Mark Spencer, for facilitating the publication process in an always friendly and supportive spirit. We thank David Kramer for the impeccable copyediting, Frank Ganz for sharing his LaTeX expertise, and Felix Portnoy for the seamless production process. The Springer editors in Germany and Japan kindly suggested translations of our book, and we are particularly grateful to Kord Eickmeyer and Yoshio Okamoto for taking on the monumental task of translating this work.

Many colleagues have sent us suggestions and updates for this second edition; in addition to the ones listed above, we thank Ana Berrizbeitia, Steven Collazos, Aaron Dall, Maryam Farahmand, Katharina Jochemko, Michael Joswig, Victor Katsnelson, Jim Lawrence, Nhat Le Quang, Matthias Lenz, Pieter Moree, Bruce Sagan, Steven Sam, Raman Sanyal, Kim Seashore, and Ed Swartz. Special thanks go to Philippe Clauss for permission to include Eugène Ehrhart's self-portrait.

Perhaps the most visible change can be seen in the figures, whose new and much-improved looks owe their beauty to David Austin. We are huge fans of his work and thank him for his time and energy on this project.

We thank Eugene Ha for facilitating the publication process of this second edition. We will collect further corrections, updates, etc., at the Internet website math.sfsu.edu/beck/ccd.html.

Matthias Beck's work has been partially supported by the United States National Science Foundation (DMS-0810105 & DMS-1162638). Sinai Robins's work has been partially supported by the Singapore Ministry of Education (Tier 2 Research Grant MOE2011-T2-1-090), by the Institute for Constructive and Experimental Mathematics (ICERM), and by Brown University.

Matthias Beck would like to express his deepest gratitude to Tendai Chitewere for her patience, support, and unconditional love. He thanks his family for always being there for him. Sinai Robins would like to thank his mother, Michal Robins, and his brothers, Shani Robins and Gabriel Robins, for their relentless support and understanding during the completion of this project. We both thank all the cafés we have inhabited over the past 14 years for enabling us to turn their coffee into theorems.

San Francisco, CA   Matthias Beck
Providence, RI   Sinai Robins
September 2015

Contents

Part I
The Essentials of Discrete Volume Computations

Chapter 1
The Coin-Exchange Problem of Frobenius

The full beauty of the subject of generating functions emerges only from tuning in on both channels: the discrete and the continuous.

Herbert Wilf [252]

Suppose we are interested in an infinite sequence of numbers $(a_k)_{k=0}^{\infty}$ that arises geometrically or recursively. Is there a "good formula" for a_k as a function of k? Are there identities involving various a_k's? Embedding this sequence into the **generating function**

$$F(z) = \sum_{k \geq 0} a_k \, z^k$$

allows us to retrieve answers to the questions above in a surprisingly quick and elegant way. We can think of $F(z)$ as lifting our sequence a_k from its discrete setting into the continuous world of functions.

1.1 Why Use Generating Functions?

To illustrate these concepts, we warm up with the classic example of the **Fibonacci sequence** f_k, named after Leonardo Pisano Fibonacci (c. 1170– c. 1250)[1] and defined by the recurrence relation

$$f_0 = 0, \ f_1 = 1, \ \text{ and } \ f_{k+2} = f_{k+1} + f_k \ \text{ for } \ k \geq 0 \, .$$

This gives the sequence $(f_k)_{k=0}^{\infty} = (0, 1, 1, 2, 3, 5, 8, 13, 21, 34, \dots)$ (see also [1, Sequence A000045]). Now let's see what generating functions can do for us.

---

[1] For more information about Fibonacci, see
http://www-history.mcs.st-andrews.ac.uk/Biographies/Fibonacci.html.

Let

$$F(z) := \sum_{k \geq 0} f_k \, z^k.$$

We embed both sides of the recurrence identity into their generating functions:

$$\sum_{k \geq 0} f_{k+2} \, z^k = \sum_{k \geq 0} (f_{k+1} + f_k) \, z^k = \sum_{k \geq 0} f_{k+1} \, z^k + \sum_{k \geq 0} f_k \, z^k. \qquad (1.1)$$

The left–hand side of (1.1) is

$$\sum_{k \geq 0} f_{k+2} \, z^k = \frac{1}{z^2} \sum_{k \geq 0} f_{k+2} \, z^{k+2} = \frac{1}{z^2} \sum_{k \geq 2} f_k \, z^k = \frac{1}{z^2} \left(F(z) - z \right),$$

while the right–hand side of (1.1) is

$$\sum_{k \geq 0} f_{k+1} \, z^k + \sum_{k \geq 0} f_k \, z^k = \frac{1}{z} F(z) + F(z).$$

So (1.1) can be restated as

$$\frac{1}{z^2} \left(F(z) - z \right) = \frac{1}{z} F(z) + F(z),$$

or

$$F(z) = \frac{z}{1 - z - z^2}.$$

It is fun to check (e.g., with a computer) that when we expand the function F into a power series, we indeed obtain the Fibonacci numbers as coefficients:

$$\frac{z}{1 - z - z^2} = z + z^2 + 2 \, z^3 + 3 \, z^4 + 5 \, z^5 + 8 \, z^6 + 13 \, z^7 + 21 \, z^8 + 34 \, z^9 + \cdots.$$

Now we use our favorite method of handling rational functions: the partial fraction expansion. In our case, the denominator factors as $1 - z - z^2 = \left(1 - \frac{1+\sqrt{5}}{2} z \right) \left(1 - \frac{1-\sqrt{5}}{2} z \right)$, and the partial fraction expansion is (see Exercise 1.1)

$$F(z) = \frac{z}{1 - z - z^2} = \frac{1/\sqrt{5}}{1 - \frac{1+\sqrt{5}}{2} z} - \frac{1/\sqrt{5}}{1 - \frac{1-\sqrt{5}}{2} z}. \qquad (1.2)$$

The two terms suggest the use of the **geometric series**

$$\sum_{k \geq 0} x^k = \frac{1}{1 - x} \qquad (1.3)$$

(see Exercise 1.2) with $x = \frac{1+\sqrt{5}}{2} z$ and $x = \frac{1-\sqrt{5}}{2} z$, respectively:

$$F(z) = \frac{z}{1 - z - z^2} = \frac{1}{\sqrt{5}} \sum_{k \geq 0} \left(\frac{1 + \sqrt{5}}{2} z \right)^k - \frac{1}{\sqrt{5}} \sum_{k \geq 0} \left(\frac{1 - \sqrt{5}}{2} z \right)^k$$

$$= \sum_{k \geq 0} \frac{1}{\sqrt{5}} \left(\left(\frac{1 + \sqrt{5}}{2} \right)^k - \left(\frac{1 - \sqrt{5}}{2} \right)^k \right) z^k.$$

Comparing the coefficients of z^k in the definition of $F(z) = \sum_{k \geq 0} f_k z^k$ and the new expression above for $F(z)$, we discover the closed-form expression

$$f_k = \frac{1}{\sqrt{5}} \left(\frac{1 + \sqrt{5}}{2} \right)^k - \frac{1}{\sqrt{5}} \left(\frac{1 - \sqrt{5}}{2} \right)^k$$

for the Fibonacci sequence.

This method of decomposing a rational generating function into partial fractions is one of our key tools. Because we will use partial fractions time and again throughout this book, we record the result on which this method is based.

Theorem 1.1 (Partial fraction expansion). *Given a rational function*

$$F(z) := \frac{p(z)}{\prod_{k=1}^{m} (z - a_k)^{e_k}},$$

where p is a polynomial of degree less than $e_1 + e_2 + \cdots + e_m$ and the a_k are distinct complex numbers, there exists a decomposition

$$F(z) = \sum_{k=1}^{m} \left(\frac{c_{k,1}}{z - a_k} + \frac{c_{k,2}}{(z - a_k)^2} + \cdots + \frac{c_{k,e_k}}{(z - a_k)^{e_k}} \right),$$

where $c_{k,j} \in \mathbb{C}$ are unique.

One possible proof of this theorem is based on the fact that the polynomials form a Euclidean domain. For readers who are acquainted with this notion, we outline this proof in Exercise 1.37.

1.2 Two Coins

Let's imagine that we introduce a new coin system. Instead of using pennies, nickels, dimes, and quarters, let's say that we agree on using 4-cent, 7-cent, 9-cent, and 34-cent coins. The reader might point out the following flaw of this new system: certain amounts cannot be obtained (that is, created with the available coins), for example, 2 or 5 cents. On the other hand, this deficiency makes our new coin system more interesting than the old one, because we can ask the question, "which amounts can be obtained?" In fact, we will prove in Exercise 1.20 that there are only finitely many integer amounts that *cannot* be obtained using our new coin system. A natural question, first tackled by Ferdinand Georg Frobenius (1849–1917),[2] and James Joseph Sylvester (1814–1897)[3] is, what is the *largest* amount that cannot be changed? As mathematicians, we like to keep questions as general as possible, and so we ask, given coins of denominations $a_1, a_2, \ldots, a_d$ that are positive integers without any common factor, can you give a formula for the largest amount that cannot be obtained using the coins $a_1, a_2, \ldots, a_d$? This problem is known as the *Frobenius coin-exchange problem.*

To be precise, suppose we are given a set of positive integers

$$A = \{a_1, a_2, \ldots, a_d\}$$

with $\gcd(a_1, a_2, \ldots, a_d) = 1$. We call a positive integer n **representable** by the set A if there exist nonnegative integers $m_1, m_2, \ldots, m_d$ such that

$$n = m_1 a_1 + \cdots + m_d a_d.$$

In the language of coins, this means that we can obtain the amount n using the coins $a_1, a_2, \ldots, a_d$. The Frobenius problem (often called the *linear Diophantine problem of Frobenius*) asks us to find the largest integer that is not representable. We call this largest integer the **Frobenius number** and denote it by $g(a_1, \ldots, a_d)$. The following theorem gives us a pretty formula for $d = 2$.

Theorem 1.2. *If a_1 and a_2 are relatively prime positive integers, then*

$$g(a_1, a_2) = a_1 a_2 - a_1 - a_2.$$

This simple-looking formula for g inspired a great deal of research into formulas for $g(a_1, a_2, \ldots, a_d)$ with only limited success; see the notes at the end of this chapter. For $d = 2$, Sylvester gave the following result.

Theorem 1.3 (Sylvester's theorem). *Let a_1 and a_2 be relatively prime positive integers. Exactly half of the integers between 1 and $(a_1 - 1)(a_2 - 1)$ are representable by $\{a_1, a_2\}$.*

Our goal in this chapter is to prove these two theorems (and a little more) using the machinery of partial fractions. We approach the Frobenius problem through the study of the **restricted partition function**

$$p_A(n) := \#\left\{(m_1, \ldots, m_d) \in \mathbb{Z}^d : \text{ all } m_j \geq 0, \ m_1 a_1 + \cdots + m_d a_d = n\right\},$$

---

[2] For more information about Frobenius, see
http://www-history.mcs.st-andrews.ac.uk/Biographies/Frobenius.html.

[3] For more information about Sylvester, see
http://www-history.mcs.st-andrews.ac.uk/Biographies/Sylvester.html.

the number of partitions of n using only the elements of A as parts.[4] In view of this partition function, $g(a_1, \ldots, a_d)$ is the largest positive integer n for which $p_A(n) = 0$.

There is a beautiful geometric interpretation of the restricted partition function. The geometric description begins with the set

$$\mathcal{P} = \left\{(x_1, \ldots, x_d) \in \mathbb{R}^d : \text{ all } x_j \geq 0, \ x_1 a_1 + \cdots + x_d a_d = 1\right\}. \tag{1.4}$$

The nth **dilate** of a set $S \subseteq \mathbb{R}^d$ is

$$\{(nx_1, nx_2, \ldots, nx_d) : (x_1, \ldots, x_d) \in S\}.$$

The function $p_A(n)$ counts precisely those integer points in $\mathbb{Z}^d$ that lie in the nth integer dilate of the body $\mathcal{P}$. The set $\mathbb{Z}^d$ is an example of a *lattice*,[5] and so integer points are often called **lattice points**. The dilation process in this context is tantamount to replacing $x_1 a_1 + \cdots + x_d a_d = 1$ in the definition of $\mathcal{P}$ by $x_1 a_1 + \cdots + x_d a_d = n$. The set $\mathcal{P}$ turns out to be a *polytope*. We can easily picture $\mathcal{P}$ and its dilates for dimension $d \leq 3$; Figure 1.1 shows the 3-dimensional case.

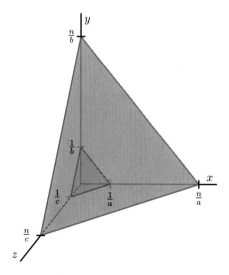

Fig. 1.1 The polytope $\mathcal{P}$ for $d = 3$.

---

[4] A **partition** of a positive integer n is a multiset (i.e., a set in which we allow repetition) $\{n_1, n_2, \ldots, n_k\}$ of positive integers such that $n = n_1 + n_2 + \cdots + n_k$. The numbers $n_1, n_2, \ldots, n_k$ are called the **parts** of the partition.

[5] A **lattice** is a discrete subgroup of $\mathbb{R}^d$, where d is a positive integer.

1.3 Partial Fractions and a Surprising Formula

We first concentrate on the case $d = 2$ and study

$$p_{\{a,b\}}(n) = \# \left\{ (k,l) \in \mathbb{Z}^2 : k,l \geq 0, \ ak + bl = n \right\}.$$

Recall that we require a and b to be relatively prime. To begin our discussion, we play around with generating functions. Consider the following product of two geometric series:

$$\left(\frac{1}{1 - z^a} \right) \left(\frac{1}{1 - z^b} \right) = \left(1 + z^a + z^{2a} + \cdots \right) \left(1 + z^b + z^{2b} + \cdots \right)$$

(see Exercise 1.2). If we multiply out all the terms, we obtain a power series all of whose exponents are linear combinations of a and b. In fact, the coefficient of z^n in this power series counts the number of ways that n can be written as a nonnegative linear combination of a and b. In other words, these coefficients are precisely evaluations of our counting function $p_{\{a,b\}}$:

$$\left(\frac{1}{1 - z^a} \right) \left(\frac{1}{1 - z^b} \right) = \sum_{k \geq 0} \sum_{l \geq 0} z^{ak} z^{bl} = \sum_{n \geq 0} p_{\{a,b\}}(n) \, z^n.$$

So this function is the generating function for the sequence of integers $\left(p_{\{a,b\}}(n) \right)_{n=0}^{\infty}$. The idea is now to study the compact function on the left.

We would like to uncover an interesting formula for $p_{\{a,b\}}(n)$ by looking at the generating function on the left more closely. To make our computational life easier, we study the *constant term* of a related series; namely, $p_{\{a,b\}}(n)$ is the constant term of

$$f(z) := \frac{1}{(1 - z^a)(1 - z^b) z^n} = \sum_{k \geq 0} p_{\{a,b\}}(k) \, z^{k-n}.$$

The latter series is not quite a power series, since it includes terms with negative exponents. These series are called *Laurent series*, after Pierre Alphonse Laurent (1813–1854). For a power series (centered at 0), we could simply evaluate the corresponding function at $z = 0$ to obtain the constant term; once we have negative exponents, such an evaluation is no longer possible. However, if we first subtract all terms with negative exponents, we obtain a power series whose constant term (which remains unchanged) can now be computed by evaluating this remaining function at $z = 0$.

To be able to compute this constant term, we will expand f into partial fractions. As a warmup to partial fraction decompositions, we first work out a 1-dimensional example. Let's denote the ath root of unity $e^{2\pi i/a}$ by ξ_a:

$$\xi_a := e^{2\pi i/a} = \cos \frac{2\pi}{a} + i \sin \frac{2\pi}{a} \, ;$$

then the collection of all a^{th} roots of unity comprises $1, \xi_a, \xi_a^2, \xi_a^3, \ldots, \xi_a^{a-1}$.

Example 1.4. Let's find the partial fraction expansion of $\frac{1}{1-z^a}$. The poles of this function are located at all a^{th} roots of unity ξ_a^k for $k = 0, 1, \ldots, a-1$. So we expand

$$\frac{1}{1-z^a} = \sum_{k=0}^{a-1} \frac{C_k}{z - \xi_a^k}.$$

To find the coefficients C_k, we proceed as follows:

$$C_k = \lim_{z \to \xi_a^k} \left(z - \xi_a^k\right)\left(\frac{1}{1-z^a}\right) = \lim_{z \to \xi_a^k} \frac{1}{-a\,z^{a-1}} = -\frac{\xi_a^k}{a},$$

where we have used L'Hôpital's rule in the penultimate equality. Therefore, we arrive at the expansion

$$\frac{1}{1-z^a} = -\frac{1}{a} \sum_{k=0}^{a-1} \frac{\xi_a^k}{z - \xi_a^k}. \qquad \square$$

Returning to restricted partitions, the poles of f are located at $z = 0$ with multiplicity n, at $z = 1$ with multiplicity 2, and at all the other a^{th} and b^{th} roots of unity with multiplicity 1, because a and b are relatively prime. Hence our partial fraction expansion looks like this:

$$f(z) = \frac{A_1}{z} + \frac{A_2}{z^2} + \cdots + \frac{A_n}{z^n} + \frac{B_1}{z-1} + \frac{B_2}{(z-1)^2} + \sum_{k=1}^{a-1} \frac{C_k}{z - \xi_a^k} + \sum_{j=1}^{b-1} \frac{D_j}{z - \xi_b^j}. \quad (1.5)$$

We invite the reader to compute the coefficients (Exercise 1.21)

$$C_k = -\frac{1}{a\left(1 - \xi_a^{kb}\right)\xi_a^{k(n-1)}}, \qquad (1.6)$$

$$D_j = -\frac{1}{b\left(1 - \xi_b^{ja}\right)\xi_b^{j(n-1)}}.$$

To compute B_2, we multiply both sides of (1.5) by $(z-1)^2$ and take the limit as $z \to 1$ to obtain

$$B_2 = \lim_{z \to 1} \frac{(z-1)^2}{(1-z^a)(1-z^b)\,z^n} = \frac{1}{ab},$$

by applying L'Hôpital's rule twice, for example. For the more interesting constant B_1, we compute

$$B_1 = \lim_{z \to 1} (z-1)\left(\frac{1}{(1-z^a)(1-z^b)\,z^n} - \frac{\frac{1}{ab}}{(z-1)^2}\right) = \frac{1}{ab} - \frac{1}{2a} - \frac{1}{2b} - \frac{n}{ab},$$

again by applying L'Hôpital's rule.

We do not need to compute the coefficients $A_1, \ldots, A_n$, since they contribute only to the terms with negative exponents, which we can safely neglect, because those terms are different from the constant term of f. Once we have the other coefficients, the constant term of the Laurent series of f is—as we said above—the following function evaluated at 0:

$$p_{\{a,b\}}(n) = \left(\frac{B_1}{z-1} + \frac{B_2}{(z-1)^2} + \sum_{k=1}^{a-1} \frac{C_k}{z - \xi_a^k} + \sum_{j=1}^{b-1} \frac{D_j}{z - \xi_b^j} \right) \Bigg|_{z=0}$$

$$= -B_1 + B_2 - \sum_{k=1}^{a-1} \frac{C_k}{\xi_a^k} - \sum_{j=1}^{b-1} \frac{D_j}{\xi_b^j} .$$

With (1.6) in hand, this simplifies to

$$p_{\{a,b\}}(n) = \frac{1}{2a} + \frac{1}{2b} + \frac{n}{ab} + \frac{1}{a} \sum_{k=1}^{a-1} \frac{1}{(1-\xi_a^{kb})\xi_a^{kn}} + \frac{1}{b} \sum_{j=1}^{b-1} \frac{1}{(1-\xi_b^{ja})\xi_b^{jn}} . \quad (1.7)$$

Encouraged by this initial success, we now proceed to analyze each sum in (1.7) with the hope of recognizing them as more familiar objects.

For the next step, we need to define the **greatest-integer function** $\lfloor x \rfloor$, which denotes the greatest integer less than or equal to x. A close sibling to this function is the **fractional-part function** $\{x\} = x - \lfloor x \rfloor$. To readers not familiar with the functions $\lfloor x \rfloor$ and $\{x\}$ we recommend working through Exercises 1.3–1.5.

What we will do next is to study a special case, namely $b = 1$. This is appealing, because $p_{\{a,1\}}(n)$ simply counts integer points in an interval:

$$p_{\{a,1\}}(n) = \#\left\{ (k,l) \in \mathbb{Z}^2 : k, l \geq 0, \ ak + l = n \right\}$$

$$= \#\left\{ k \in \mathbb{Z} : k \geq 0, \ ak \leq n \right\}$$

$$= \#\left\{ k \in \mathbb{Z} : 0 \leq k \leq \frac{n}{a} \right\}$$

$$= \left\lfloor \frac{n}{a} \right\rfloor + 1 .$$

(See Exercise 1.3.) On the other hand, in (1.7), we just computed a different expression for this function, so that

$$\frac{1}{2a} + \frac{1}{2} + \frac{n}{a} + \frac{1}{a} \sum_{k=1}^{a-1} \frac{1}{(1-\xi_a^k)\,\xi_a^{kn}} = p_{\{a,1\}}(n) = \left\lfloor \frac{n}{a} \right\rfloor + 1 .$$

With the help of the fractional-part function $\{x\} = x - \lfloor x \rfloor$, we have derived a formula for the following sum over the nontrivial a^{th} roots of unity:

$$\frac{1}{a}\sum_{k=1}^{a-1}\frac{1}{\left(1-\xi_a^k\right)\xi_a^{kn}} = -\left\{\frac{n}{a}\right\}+\frac{1}{2}-\frac{1}{2a}. \tag{1.8}$$

We are almost there: we invite the reader (Exercise 1.22) to show that

$$\frac{1}{a}\sum_{k=1}^{a-1}\frac{1}{\left(1-\xi_a^{bk}\right)\xi_a^{kn}} = \frac{1}{a}\sum_{k=1}^{a-1}\frac{1}{\left(1-\xi_a^{k}\right)\xi_a^{b^{-1}kn}}, \tag{1.9}$$

where b^{-1} is an integer such that $b^{-1}b \equiv 1 \bmod a$, and to conclude that

$$\frac{1}{a}\sum_{k=1}^{a-1}\frac{1}{\left(1-\xi_a^{bk}\right)\xi_a^{kn}} = -\left\{\frac{b^{-1}n}{a}\right\}+\frac{1}{2}-\frac{1}{2a}. \tag{1.10}$$

Now all that is left to do is to substitute this expression back into (1.7), which yields the following beautiful formula due to Peter Barlow (1776–1862)[6] and (in the form we state here) Tiberiu Popoviciu (1906–1975).

Theorem 1.5 (Barlow–Popoviciu formula). *If a and b are relatively prime, then*

$$p_{\{a,b\}}(n) = \frac{n}{ab} - \left\{\frac{b^{-1}n}{a}\right\} - \left\{\frac{a^{-1}n}{b}\right\} + 1,$$

where $b^{-1}b \equiv 1 \bmod a$ and $a^{-1}a \equiv 1 \bmod b$. $\qquad\square$

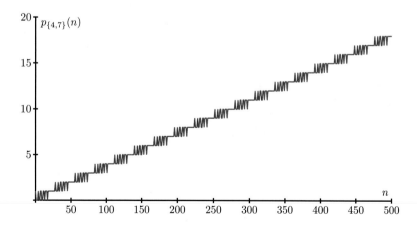

Fig. 1.2 The graph of $p_{\{4,7\}}(n)$.

We shall have more to say about the arithmetic nature of restricted partition functions in general; for now, let's note that the function $p_{\{a,b\}}(n)$ behaves roughly linearly, with a periodic part that stems from the fractional-part functions. Figure 1.2 shows the behavior of the function for $a = 4$, $b = 7$.

Theorem 1.5 can be neatly and easily rewritten using greatest-integer functions, as follows (Exercise 1.28):

Corollary 1.6. *Given relatively prime positive integers a and b, let p and q be positive integers such that $aq - bp = 1$. Then*

$$p_{\{a,b\}}(n) = \begin{cases} \left\lfloor \frac{qn}{b} \right\rfloor - \left\lfloor \frac{pn}{a} \right\rfloor & \text{if } a \nmid n, \\ \left\lfloor \frac{qn}{b} \right\rfloor - \frac{pn}{a} + 1 & \text{if } a \mid n. \end{cases}$$

1.4 Sylvester's Result

Before we apply Theorem 1.5 to obtain the classical Theorems 1.2 and 1.3, we return for a moment to the geometry behind the restricted partition function $p_{\{a,b\}}(n)$. In the 2-dimensional case (which is the setting of Theorem 1.5), we are counting integer points $(x, y) \in \mathbb{Z}^2$ on the line segments defined by the constraints

$$ax + by = n, \qquad x, y \geq 0.$$

Thus this line segment connects the points $(0, \frac{n}{b})$ and $(\frac{n}{a}, 0)$; we denote this line segment by $[(0, \frac{n}{b}), (\frac{n}{a}, 0)]$. (More generally, we define the closed line segment joining the points $\mathbf{x}$ and $\mathbf{y} \in \mathbb{R}^d$ as follows:

$$[\mathbf{x}, \mathbf{y}] := \{\lambda \mathbf{x} + (1 - \lambda) \mathbf{y} : 0 \leq \lambda \leq 1\},$$

and we will use the usual similar notation to denote an open line segment $(\mathbf{x}, \mathbf{y})$ and a half-open line segment $(\mathbf{x}, \mathbf{y}]$.)

As n increases, the line segment $[(0, \frac{n}{b}), (\frac{n}{a}, 0)]$ becomes dilated. It is not too far-fetched (although Exercise 1.13 teaches us to be careful with such statements) to expect that the likelihood for an integer point to lie on the line segment increases with n. In fact, one might even guess that the number of points on the line segment increases linearly with n, since the line segment is a 1-dimensional object. Theorem 1.5 quantifies the previous statement in a very precise form: $p_{\{a,b\}}(n)$ has the "leading term" $\frac{n}{ab}$, and the remaining terms are bounded as functions of n. Figure 1.3 shows the geometry behind the counting function $p_{\{4,7\}}(n)$ for the first few values of n. Note that the thick line segment for $n = 17 = 4 \cdot 7 - 4 - 7$ is the last one that does not contain any integer point.

Lemma 1.7. *If a and b are relatively prime positive integers and $n \in [1, ab-1]$ is not a multiple of a or b, then*

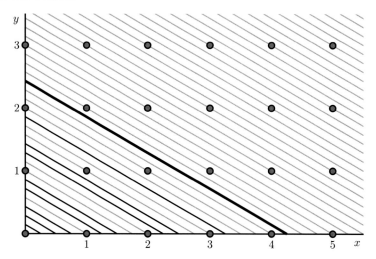

Fig. 1.3 The nonnegative integer solutions to $4x + 7y = n$ for $n = 0, 1, 2, \ldots$.

$$p_{\{a,b\}}(n) + p_{\{a,b\}}(ab - n) = 1.$$

In other words, for n between 1 and $ab - 1$ and not divisible by a or b, exactly one of the two integers n and $ab - n$ is representable in terms of a and b.

Proof. This identity follows directly from Theorem 1.5:

$$
\begin{aligned}
p_{\{a,b\}}(ab - n) &= \frac{ab - n}{ab} - \left\{ \frac{b^{-1}(ab - n)}{a} \right\} - \left\{ \frac{a^{-1}(ab - n)}{b} \right\} + 1 \\
&= 2 - \frac{n}{ab} - \left\{ \frac{-b^{-1}n}{a} \right\} - \left\{ \frac{-a^{-1}n}{b} \right\} \\
&\overset{(\star)}{=} -\frac{n}{ab} + \left\{ \frac{b^{-1}n}{a} \right\} + \left\{ \frac{a^{-1}n}{b} \right\} \\
&= 1 - p_{\{a,b\}}(n).
\end{aligned}
$$

Here, $(\star)$ follows from the fact that $\{-x\} = 1 - \{x\}$ if $x \notin \mathbb{Z}$ (see Exercise 1.5). $\qquad\square$

Proof of Theorem 1.2. We have to show that $p_{\{a,b\}}(ab - a - b) = 0$ and that $p_{\{a,b\}}(n) > 0$ for every $n > ab - a - b$. The first assertion follows from Exercise 1.24, which states that $p_{\{a,b\}}(a + b) = 1$, and Lemma 1.7. To prove the second assertion, we note that $\left\{ \frac{m}{a} \right\} \leq 1 - \frac{1}{a}$ for every integer m. Hence for every positive integer n,

$$p_{\{a,b\}}(ab - a - b + n) \geq \frac{ab - a - b + n}{ab} - \left(1 - \frac{1}{a}\right) - \left(1 - \frac{1}{b}\right) + 1 = \frac{n}{ab} > 0.$$

$\qquad\square$

Proof of Theorem 1.3. Recall that Lemma 1.7 states that for n between 1 and $ab - 1$ and not divisible by a or b, exactly one of n and $ab - n$ is representable. There are

$$ab - a - b + 1 = (a - 1)(b - 1)$$

integers between 1 and $ab - 1$ that are not divisible by a or b. Finally, we note that $p_{\{a,b\}}(n) > 0$ if n is a multiple of a or b, by the very definition of $p_{\{a,b\}}(n)$. Hence the number of nonrepresentable integers is $\frac{1}{2}(a - 1)(b - 1)$.   □

Note that we have proved even more. Essentially by Lemma 1.7, every positive integer less than ab has at most one representation. Hence, the representable integers less than ab are *uniquely* representable (see also Exercise 1.25).

1.5 Three and More Coins

What happens to the complexity of the Frobenius problem if we have more than two coins? Let's return to our restricted partition function

$$p_A(n) = \# \left\{ (m_1, \ldots, m_d) \in \mathbb{Z}^d : \text{ all } m_j \geq 0, \ m_1 a_1 + \cdots + m_d a_d = n \right\},$$

where $A = \{a_1, \ldots, a_d\}$. By the very same reasoning as in Section 1.3, we can easily write down the generating function for $p_A(n)$:

$$\sum_{n \geq 0} p_A(n) \, z^n = \left(\frac{1}{1 - z^{a_1}} \right) \left(\frac{1}{1 - z^{a_2}} \right) \cdots \left(\frac{1}{1 - z^{a_d}} \right).$$

We use the same methods that were exploited in Section 1.3 to recover our function $p_A(n)$ as the constant term of a useful generating function. Namely,

$$p_A(n) = \text{const} \left(\frac{1}{(1 - z^{a_1})(1 - z^{a_2}) \cdots (1 - z^{a_d}) \, z^n} \right).$$

We now expand the function on the right into partial fractions. For reasons of simplicity, we assume in the following that $a_1, \ldots, a_d$ are *pairwise* relatively prime; that is, no two of the integers $a_1, a_2, \ldots, a_d$ have a common factor. Then our partial fraction expansion looks like this:

$$
\begin{aligned}
f(z) &= \frac{1}{(1 - z^{a_1}) \cdots (1 - z^{a_d}) \, z^n} \\
&= \frac{A_1}{z} + \frac{A_2}{z^2} + \cdots + \frac{A_n}{z^n} + \frac{B_1}{z - 1} + \frac{B_2}{(z - 1)^2} + \cdots + \frac{B_d}{(z - 1)^d} \quad (1.11) \\
&\quad + \sum_{k=1}^{a_1 - 1} \frac{C_{1k}}{z - \xi_{a_1}^k} + \sum_{k=1}^{a_2 - 1} \frac{C_{2k}}{z - \xi_{a_2}^k} + \cdots + \sum_{k=1}^{a_d - 1} \frac{C_{dk}}{z - \xi_{a_d}^k}.
\end{aligned}
$$

By now, we are experienced in computing partial fraction coefficients, so that the reader will easily verify that (Exercise 1.31)

$$C_{1k} = -\frac{1}{a_1 \left(1 - \xi_{a_1}^{ka_2}\right) \left(1 - \xi_{a_1}^{ka_3}\right) \cdots \left(1 - \xi_{a_1}^{ka_d}\right) \xi_{a_1}^{k(n-1)}}. \tag{1.12}$$

As before, we do not have to compute the coefficients $A_1, \ldots, A_n$, because they do not contribute to the constant term of f. For the computation of $B_1, \ldots, B_d$, we may use a symbolic manipulation program such as Maple, Mathematica, or Sage. Again, once we have calculated these coefficients, we can compute the constant term of f by dropping all negative exponents and evaluating the remaining function at 0:

$$p_A(n) = \left(\frac{B_1}{z-1} + \cdots + \frac{B_d}{(z-1)^d} + \sum_{k=1}^{a_1-1} \frac{C_{1k}}{z - \xi_{a_1}^k} + \cdots + \sum_{k=1}^{a_d-1} \frac{C_{dk}}{z - \xi_{a_d}^k} \right) \Bigg|_{z=0}$$

$$= -B_1 + B_2 - \cdots + (-1)^d B_d - \sum_{k=1}^{a_1-1} \frac{C_{1k}}{\xi_{a_1}^k} - \cdots - \sum_{k=1}^{a_d-1} \frac{C_{dk}}{\xi_{a_d}^k}.$$

Substituting the expression we found for C_{1k} into the latter sum over the nontrivial a_1^{th} roots of unity, for example, gives rise to

$$\frac{1}{a_1} \sum_{k=1}^{a_1-1} \frac{1}{\left(1 - \xi_{a_1}^{ka_2}\right) \left(1 - \xi_{a_1}^{ka_3}\right) \cdots \left(1 - \xi_{a_1}^{ka_d}\right) \xi_{a_1}^{kn}}.$$

This motivates the definition of the **Fourier–Dedekind sum**

$$s_n(a_1, a_2, \ldots, a_m; b) := \frac{1}{b} \sum_{k=1}^{b-1} \frac{\xi_b^{kn}}{\left(1 - \xi_b^{ka_1}\right) \left(1 - \xi_b^{ka_2}\right) \cdots \left(1 - \xi_b^{ka_m}\right)}. \tag{1.13}$$

We will study these sums in detail in Chapter 8. With this definition, we have arrived at the following result.

Theorem 1.8. *The restricted partition function for* $A = \{a_1, a_2, \ldots, a_d\}$, *where the* a_k's *are pairwise relatively prime, can be computed as*

$$p_A(n) = -B_1 + B_2 - \cdots + (-1)^d B_d + s_{-n}(a_2, a_3, \ldots, a_d; a_1)$$
$$+ s_{-n}(a_1, a_3, a_4, \ldots, a_d; a_2) + \cdots + s_{-n}(a_1, a_2, \ldots, a_{d-1}; a_d).$$

Here $B_1, B_2, \ldots, B_d$ *are the partial fraction coefficients in the expansion* (1.11).
□

Example 1.9. We give the restricted partition functions for $d = 3$ and 4. These closed-form formulas have proven useful in the refined analysis of the periodicity that is inherent in the restricted partition function $p_A(n)$. For example, one can visualize the graph of $p_{\{a,b,c\}}(n)$ as a "wavy parabola," as its formula plainly shows (Figure 1.4 gives an example, the case $A = \{4, 7, 15\}$):

$$p_{\{a,b,c\}}(n) = \frac{n^2}{2abc} + \frac{n}{2}\left(\frac{1}{ab} + \frac{1}{ac} + \frac{1}{bc}\right) + \frac{1}{12}\left(\frac{3}{a} + \frac{3}{b} + \frac{3}{c} + \frac{a}{bc} + \frac{b}{ac} + \frac{c}{ab}\right)$$

$$+ \frac{1}{a}\sum_{k=1}^{a-1} \frac{1}{\left(1 - \xi_a^{kb}\right)\left(1 - \xi_a^{kc}\right)\xi_a^{kn}} + \frac{1}{b}\sum_{k=1}^{b-1} \frac{1}{\left(1 - \xi_b^{kc}\right)\left(1 - \xi_b^{ka}\right)\xi_b^{kn}}$$

$$+ \frac{1}{c}\sum_{k=1}^{c-1} \frac{1}{\left(1 - \xi_c^{ka}\right)\left(1 - \xi_c^{kb}\right)\xi_c^{kn}} \,,$$

$$p_{\{a,b,c,d\}}(n) = \frac{n^3}{6abcd} + \frac{n^2}{4}\left(\frac{1}{abc} + \frac{1}{abd} + \frac{1}{acd} + \frac{1}{bcd}\right)$$

$$+ \frac{n}{12}\left(\frac{3}{ab} + \frac{3}{ac} + \frac{3}{ad} + \frac{3}{bc} + \frac{3}{bd} + \frac{3}{cd} + \frac{a}{bcd} + \frac{b}{acd} + \frac{c}{abd} + \frac{d}{abc}\right)$$

$$+ \frac{1}{24}\left(\frac{a}{bc} + \frac{a}{bd} + \frac{a}{cd} + \frac{b}{ad} + \frac{b}{ac} + \frac{b}{cd} + \frac{c}{ab} + \frac{c}{ad} + \frac{c}{bd}\right.$$

$$\left. + \frac{d}{ab} + \frac{d}{ac} + \frac{d}{bc}\right) - \frac{1}{8}\left(\frac{1}{a} + \frac{1}{b} + \frac{1}{c} + \frac{1}{d}\right)$$

$$+ \frac{1}{a}\sum_{k=1}^{a-1} \frac{1}{\left(1 - \xi_a^{kb}\right)\left(1 - \xi_a^{kc}\right)\left(1 - \xi_a^{kd}\right)\xi_a^{kn}}$$

$$+ \frac{1}{b}\sum_{k=1}^{b-1} \frac{1}{\left(1 - \xi_b^{kc}\right)\left(1 - \xi_b^{kd}\right)\left(1 - \xi_b^{ka}\right)\xi_b^{kn}}$$

$$+ \frac{1}{c}\sum_{k=1}^{c-1} \frac{1}{\left(1 - \xi_c^{kd}\right)\left(1 - \xi_c^{ka}\right)\left(1 - \xi_c^{kb}\right)\xi_c^{kn}}$$

$$+ \frac{1}{d}\sum_{k=1}^{d-1} \frac{1}{\left(1 - \xi_d^{ka}\right)\left(1 - \xi_d^{kb}\right)\left(1 - \xi_d^{kc}\right)\xi_d^{kn}} \,. \qquad \square$$

Notes

1. The theory of generating functions has a long and powerful tradition. We only touch on its utility. For those readers who would like to dig a little deeper into the vast generating-function garden, we strongly recommend Herb Wilf's *generatingfunctionology* [252] and László Lovász's *Combinatorial Problems and Exercises* [165]. The reader might wonder why we do not stress convergence aspects of the generating functions we play with. Almost all of

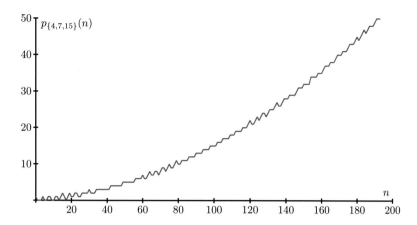

Fig. 1.4 The graph of $p_{\{4,7,15\}}(n)$.

our series are geometric series and have trivial convergence properties. In the spirit of not muddying the waters of lucid mathematical exposition, we omit such convergence details.

2. The Frobenius problem is named after Georg Frobenius, who apparently liked to raise this problem in his lectures [60]. Theorem 1.2 is one of the famous folklore results and might be one of the most misquoted theorems in all of mathematics. People usually cite James J. Sylvester's problem in [239], but his paper contains Theorem 1.3 rather than Theorem 1.2. In fact, Sylvester's problem had previously appeared as a theorem in [238]. It is not known who first discovered or proved Theorem 1.2. It is very conceivable that Sylvester knew about it when he came up with Theorem 1.3.

3. The linear Diophantine problem of Frobenius should not be confused with the *postage-stamp problem*. The latter problem asks for a similar determination, but adds an additional independent bound on the size of the integer solutions to the linear equation.

4. Theorem 1.5 has a fascinating history. The earliest appearance of this result (in the form of Corollary 1.6) that we are aware of is in a book on elementary number theory, from 1811, by Peter Barlow [18, p. 323–325]. The version we state in Theorem 1.5 (which is our first example of a *quasipolynomial*, a term we will see frequently in the coming chapters) seems to go back to a paper by Tiberiu Popoviciu [195]. The Barlow–Popoviciu formula has since been resurrected at least twice [215, 245].

Guglielmo Libri (1803–1869),[7] a mathematical prodigy, wrote down a version of the Barlow–Popoviciu formula in terms of trigonometric functions. His exceedingly colorful life is perhaps best summed up in the following quotataion from [205]:

> Admirable in the salons and incomparably friendly, flexible, with gentle epigrams of sweet humour, elegant flattery, a good writer in both French and Italian, a profound mathematician, geometer, physicist, knowing history through and through, a very analytic and comparative mind ... more expert than an auctioneer or a bookseller in the science of books, this man had only one misfortune: he was essentially a thief.

5. Fourier–Dedekind sums first surfaced implicitly in Sylvester's work (see, e.g., [237]) and explicitly in connection with restricted partition functions in [143]. They were rediscovered in [37], in connection with the Frobenius problem. The papers [113, 208] contain interesting connections to Bernoulli and Euler polynomials. We will resume the study of Fourier–Dedekind sums in Chapter 8.

6. As we mentioned above, the Frobenius problem for $d \geq 3$ is much harder than the case $d = 2$ that we have discussed. Certainly beyond $d = 3$, the Frobenius problem is wide open, though much effort has been put into its study. The literature on the Frobenius problem is vast, and there is still much room for improvement. The interested reader might consult the comprehensive monograph [201], which surveys the references to almost all articles dealing with the Frobenius problem and gives about 40 open problems and conjectures related to the Frobenius problem. To give a flavor, we mention two landmark results that go beyond $d = 2$.

The first one concerns the generating function $r(z) := \sum_{k \in R} z^k$, where R is the set of all integers representable by a given set of relatively prime positive integers $a_1, a_2, \ldots, a_d$. It is not hard to see (Exercise 1.36) that $r(z) = p(z)/(1 - z^{a_1})(1 - z^{a_2}) \cdots (1 - z^{a_d})$ for some polynomial p. This rational generating function contains all the information about the Frobenius problem; for example, the Frobenius number is the total degree of the rational function $\frac{1}{1-z} - r(z)$. Hence the Frobenius problem reduces to finding the polynomial p, the numerator of r. Marcel Morales [179, 180] and Graham Denham [102] discovered the remarkable fact that for $d = 3$, the polynomial p has either four or six terms. Moreover, they gave semiexplicit formulas for p. The Morales–Denham theorem implies that the Frobenius number in the case $d = 3$ is quickly computable, a result that is originally due, in various disguises, to Jürgen Herzog [132], Harold Greenberg [124], and J. Leslie Davison [93]. As much as there seems to be a well-defined border between the cases $d = 2$ and $d = 3$, there also seems to be such a border between the cases $d = 3$ and $d = 4$: Henrik Bresinsky [64] proved that for $d \geq 4$, there is no absolute

---

[7] For more information about Libri, see
http://www-history.mcs.st-andrews.ac.uk/Biographies/Libri.html.

bound on the number of terms in the numerator p, in sharp contrast to the Morales–Denham theorem.

On the other hand, Alexander Barvinok and Kevin Woods [23] proved that for fixed d, the rational generating function $r(z)$ can be written as a "short" sum of rational functions; in particular, r can be efficiently computed when d is fixed. A corollary of this fact is that the Frobenius number can be efficiently computed when d is fixed; this theorem is due to Ravi Kannan [144]. On the other hand, Jorge Ramírez-Alfonsín [200] proved that trying to compute the Frobenius number efficiently is hopeless if d is left as a variable.

While the above results settle the theoretical complexity of the computation of the Frobenius number, practical algorithms are a completely different matter. Both Kannan's and Barvinok–Woods's ideas seem complex enough that nobody has yet tried to implement them. The currently most competitive algorithms include [49, 59, 207].

Exercises

1.1. ♣ Check the partial fraction expansion (1.2):

$$\frac{z}{1 - z - z^2} = \frac{1/\sqrt{5}}{1 - \frac{1+\sqrt{5}}{2} z} - \frac{1/\sqrt{5}}{1 - \frac{1-\sqrt{5}}{2} z}.$$

1.2. ♣ Suppose z is a complex number, and n a positive integer. Show that

$$(1 - z)\left(1 + z + z^2 + \cdots + z^n\right) = 1 - z^{n+1},$$

and use this to prove that if $|z| < 1$, then

$$\sum_{k \geq 0} z^k = \frac{1}{1 - z}.$$

1.3. ♣ Find a formula for the number of lattice points in $[a, b]$ for arbitrary real numbers a and b.

1.4. Prove the following. Unless stated differently, $n \in \mathbb{Z}$ and $x, y \in \mathbb{R}$.

(a) $\lfloor x + n \rfloor = \lfloor x \rfloor + n$.
(b) $\lfloor x \rfloor + \lfloor y \rfloor \leq \lfloor x + y \rfloor \leq \lfloor x \rfloor + \lfloor y \rfloor + 1$.
(c) $\lfloor x \rfloor + \lfloor -x \rfloor = \begin{cases} 0 & \text{if } x \in \mathbb{Z}, \\ -1 & \text{otherwise.} \end{cases}$
(d) For $n \in \mathbb{Z}_{>0}$, $\left\lfloor \frac{\lfloor x \rfloor}{n} \right\rfloor = \left\lfloor \frac{x}{n} \right\rfloor$.
(e) $-\lfloor -x \rfloor$ is the least integer greater than or equal to x, denoted by $\lceil x \rceil$.
(f) $\lfloor x + \frac{1}{2} \rfloor$ is the nearest integer to x (and if two integers are equally close to x, it is the larger of the two).

(g) $\lfloor x \rfloor + \lfloor x + \frac{1}{2} \rfloor = \lfloor 2x \rfloor$.

(h) If m and n are positive integers, $\lfloor \frac{m}{n} \rfloor$ is the number of integers among $1, \ldots, m$ that are divisible by n.

(i) ♣ If $m \in \mathbb{Z}_{>0}$, $n \in \mathbb{Z}$, then $\lfloor \frac{n-1}{m} \rfloor = -\lfloor \frac{-n}{m} \rfloor - 1$.

(j) ♣ If $m \in \mathbb{Z}_{>0}$, $n \in \mathbb{Z}$, then $\lfloor \frac{n-1}{m} \rfloor + 1$ is the least integer greater than or equal to $\frac{n}{m}$.

1.5. Rewrite in terms of the fractional-part function as many of the above identities as you can make sense of.

1.6. Suppose m and n are relatively prime positive integers. Prove that

$$\sum_{k=0}^{m-1} \left\lfloor \frac{kn}{m} \right\rfloor = \sum_{j=0}^{n-1} \left\lfloor \frac{jm}{n} \right\rfloor = \frac{1}{2}(m-1)(n-1).$$

1.7. Prove the following identities. They will come in handy at least twice: when we study partial fractions, and when we discuss finite Fourier series. For $\phi, \psi \in \mathbb{R}$, $n \in \mathbb{Z}_{>0}$, $m \in \mathbb{Z}$,

(a) $e^{i0} = 1$

(b) $e^{i\phi} e^{i\psi} = e^{i(\phi+\psi)}$

(c) $\frac{1}{e^{i\phi}} = e^{-i\phi}$

(d) $e^{i(\phi+2\pi)} = e^{i\phi}$

(e) $e^{2\pi i} = 1$

(f) $\left| e^{i\phi} \right| = 1$

(g) $\frac{d}{d\phi} e^{i\phi} = i e^{i\phi}$

(h) $\displaystyle\sum_{k=0}^{n-1} e^{2\pi i k m/n} = \begin{cases} n & \text{if } n \mid m, \\ 0 & \text{otherwise} \end{cases}$

(i) If $n > 1$ then $\displaystyle\sum_{k=1}^{n-1} k\, e^{2\pi i k/n} = \frac{n}{e^{2\pi i/n} - 1}$.

1.8. Suppose $m, n \in \mathbb{Z}$ and $n > 0$. Find a closed form for

$$\sum_{k=0}^{n-1} \left\{ \frac{k}{n} \right\} e^{2\pi i k m/n} = \sum_{k=0}^{n-1} \frac{k}{n} e^{2\pi i k m/n}$$

(as a function of m and n).

1.9. ♣ Suppose m and n are relatively prime integers, and n is positive. Show that

$$\left\{ e^{2\pi i m k/n} : 0 \leq k < n \right\} = \left\{ e^{2\pi i j/n} : 0 \leq j < n \right\}$$

and

$$\left\{ e^{2\pi i m k/n} : 0 < k < n \right\} = \left\{ e^{2\pi i j/n} : 0 < j < n \right\}.$$

Conclude that if f is any complex-valued function, then

$$\sum_{k=0}^{n-1} f\left(e^{2\pi imk/n}\right) = \sum_{j=0}^{n-1} f\left(e^{2\pi ij/n}\right)$$

and

$$\sum_{k=1}^{n-1} f\left(e^{2\pi imk/n}\right) = \sum_{j=1}^{n-1} f\left(e^{2\pi ij/n}\right).$$

1.10. Suppose n is a positive integer. If you know what a *group* is, prove that the set $\left\{e^{2\pi ik/n} : 0 \le k < n\right\}$ forms a cyclic group of order n (under multiplication in $\mathbb{C}$).

1.11. Fix $n \in \mathbb{Z}_{>0}$. For an integer m, let $(m \bmod n)$ denote the least nonnegative integer in $G_1 := \mathbb{Z}_n$ to which m is congruent. Let's denote by $\star$ addition modulo n, and by $\circ$ the following composition:

$$\left\{\frac{m_1}{n}\right\} \circ \left\{\frac{m_2}{n}\right\} = \left\{\frac{m_1 + m_2}{n}\right\},$$

defined on the set $G_2 := \left\{\left\{\frac{m}{n}\right\} : m \in \mathbb{Z}\right\}$. Define the following functions:

$$\phi\left((m \bmod n)\right) = e^{2\pi im/n},$$
$$\psi\left(e^{2\pi im/n}\right) = \left\{\frac{m}{n}\right\},$$
$$\chi\left(\left\{\frac{m}{n}\right\}\right) = (m \bmod n).$$

Prove the following:

$$\phi\left((m_1 \bmod n) \star (m_2 \bmod n)\right) = \phi\left((m_1 \bmod n)\right)\phi\left((m_2 \bmod n)\right),$$
$$\psi\left(e^{2\pi im_1/n}e^{2\pi im_2/n}\right) = \psi\left(e^{2\pi im_1/n}\right) \circ \psi\left(e^{2\pi im_2/n}\right),$$
$$\chi\left(\left\{\frac{m_1}{n}\right\} \circ \left\{\frac{m_2}{n}\right\}\right) = \chi\left(\left\{\frac{m_1}{n}\right\}\right) \star \chi\left(\left\{\frac{m_2}{n}\right\}\right).$$

Prove that the three maps defined above, namely ϕ, ψ, and χ, are one-to-one. Again, for the reader who is familiar with the notion of a *group*, let G_3 be the group of nth roots of unity. What we have shown is that the three groups G_1, G_2, and G_3 are all isomorphic. It is very useful to cycle among these three isomorphic groups.

1.12. ♣ Given integers a, b, c, d, form the line segment $[(a, b), (c, d)] \subset \mathbb{R}^2$ joining the points (a, b) and (c, d). Show that the number of integer points on this line segment is $\gcd(a - c, b - d) + 1$.

1.13. Give an example of a line with

(a) no lattice points;
(b) one lattice point;
(c) an infinite number of lattice points.

In each case, state—if appropriate—necessary conditions about the (ir)rationality of the slope.

1.14. Suppose a line $y = mx + b$ passes through the lattice points (p_1, q_1) and (p_2, q_2). Prove that it also passes through the lattice points

$$\bigl(p_1 + k(p_2 - p_1), q_1 + k(q_2 - q_1)\bigr), \quad k \in \mathbb{Z}.$$

1.15. Given positive irrational numbers p and q with $\frac{1}{p} + \frac{1}{q} = 1$, show that $\mathbb{Z}_{>0}$ is the disjoint union of the two integer sequences $\{\lfloor pn \rfloor : n \in \mathbb{Z}_{>0}\}$ and $\{\lfloor qn \rfloor : n \in \mathbb{Z}_{>0}\}$. This theorem from 1894 is due to Lord Rayleigh and was rediscovered in 1926 by Sam Beatty. Sequences of the form $\{\lfloor pn \rfloor : n \in \mathbb{Z}_{>0}\}$ are often called **Beatty sequences**.

1.16. Let $a, b, c, d \in \mathbb{Z}$. We say that $\{(a, b), (c, d)\}$ is a **lattice basis** of $\mathbb{Z}^2$ if every lattice point $(m, n) \in \mathbb{Z}^2$ can be written as

$$(m, n) = p\,(a, b) + q\,(c, d)$$

for some $p, q \in \mathbb{Z}$. Prove that if $\{(a, b), (c, d)\}$ and $\{(e, f), (g, h)\}$ are lattice bases of $\mathbb{Z}^2$, then there exists an integer matrix M with determinant ± 1 such that

$$\begin{pmatrix} a & b \\ c & d \end{pmatrix} = M \begin{pmatrix} e & f \\ g & h \end{pmatrix}.$$

Conclude that the determinant of $\begin{pmatrix} a & b \\ c & d \end{pmatrix}$ is ± 1.

1.17. ♣ Prove that a triangle with vertices on the integer lattice has no other interior/boundary lattice points if and only if it has area $\frac{1}{2}$. (*Hint:* You may begin by "doubling" the triangle to form a parallelogram.)

1.18. Let's define a **northeast lattice path** as a path through lattice points that uses only the steps $(1, 0)$ and $(0, 1)$. Let L_n be the line defined by $x + 2y = n$. Prove that the number of northeast lattice paths from the origin to a lattice point on L_n is the $(n + 1)^{\text{th}}$ Fibonacci number f_{n+1}.

1.19. Compute the coefficients of the Taylor series of $\frac{1}{(1-z)^2}$ expanded at $z = 0$

(a) by a counting argument,
(b) by differentiating the geometric series.

Generalize.

1.20. ♣ Prove that if $a_1, a_2, \ldots, a_d \in \mathbb{Z}_{>0}$ do not have a common factor, then the Frobenius number $g(a_1, \ldots, a_d)$ is well defined.

1.21. ♣ Compute the partial fraction coefficients (1.6).

1.22. ♣ Prove (1.9): for relatively prime positive integers a and b,

$$\frac{1}{a} \sum_{k=1}^{a-1} \frac{1}{\left(1 - \zeta_a^{bk}\right) \zeta_a^{kn}} = \frac{1}{a} \sum_{k=1}^{a-1} \frac{1}{\left(1 - \zeta_a^{k}\right) \zeta_a^{b^{-1}kn}},$$

where $b^{-1}b \equiv 1 \bmod a$, and deduce from this (1.10), namely,

$$\frac{1}{a} \sum_{k=1}^{a-1} \frac{1}{\left(1 - \zeta_a^{bk}\right) \zeta_a^{kn}} = -\left\{\frac{b^{-1}n}{a}\right\} + \frac{1}{2} - \frac{1}{2a}.$$

(*Hint:* Use Exercise 1.9.)

1.23. Prove that for relatively prime positive integers a and b,

$$p_{\{a,b\}}(n + ab) = p_{\{a,b\}}(n) + 1.$$

1.24. ♣ Show that if a and b are relatively prime integers ≥ 2, then

$$p_{\{a,b\}}(a + b) = 1.$$

1.25. To extend the Frobenius problem, we say that an integer n is k-**representable** if $p_A(n) = k$; that is, n can be represented in exactly k ways using the integers in the set A. Define $g_k = g_k(a_1, \ldots, a_d)$ to be the largest k-representable integer. Prove:

(a) Let $d = 2$. For every $k \in \mathbb{Z}_{\geq 0}$, there is an N such that all integers larger than N have at least k representations (and hence $g_k(a, b)$ is well defined).
(b) $g_k(a, b) = (k + 1)ab - a - b$.
(c) Given $k \geq 2$, the smallest k-representable integer is $ab(k - 1)$.
(d) The smallest interval containing all uniquely representable integers is $[\min(a, b), g_1(a, b)]$.
(e) Given $k \geq 2$, the smallest interval containing all k-representable integers is $[g_{k-2}(a, b) + a + b, g_k(a, b)]$.
(f) There are exactly $ab - 1$ integers that are uniquely representable. Given $k \geq 2$, there are exactly ab k-representable integers.
(g) Extend all of this to $d \geq 3$ (see Open Problem 1.43).

1.26. Find a formula for $p_{\{a\}}(n)$.

1.27. Prove the following recurrence formula for $n \in \mathbb{Z}_{\geq 0}$:

$$p_{\{a_1,\ldots,a_d\}}(n) = \sum_{m=0}^{\lfloor \frac{n}{a_d} \rfloor} p_{\{a_1,\ldots,a_{d-1}\}}(n - m\, a_d).$$

Use it in the case $d = 2$ to give an alternative proof of Theorem 1.2.

1.28. Prove the equivalence of Theorem 1.5 and Corollary 1.6.

1.29. Give an alternative proof of Corollary 1.6 (i.e., one that does not use partial fraction expansions of generating functions).

1.30. Prove the following extension of Theorem 1.5: Suppose $\gcd(a, b) = d$. Then

$$p_{\{a,b\}}(n) = \begin{cases} \frac{nd}{ab} - \left\{\frac{\beta n}{a}\right\} - \left\{\frac{\alpha n}{b}\right\} + 1 & \text{if } d|n, \\ 0 & \text{otherwise,} \end{cases}$$

where $\beta \frac{b}{d} \equiv 1 \bmod \frac{a}{d}$ and $\alpha \frac{a}{d} \equiv 1 \bmod \frac{b}{d}$.

1.31. ♣ Compute the partial fraction coefficient (1.12).

1.32. Find a formula for $p_{\{a,b,c\}}(n)$ for the case $\gcd(a, b, c) \neq 1$.

1.33. ♣ With $A = \{a_1, a_2, \ldots, a_d\} \subset \mathbb{Z}_{>0}$, let

$$p_A^\circ(n) := \# \left\{ (m_1, \ldots, m_d) \in \mathbb{Z}^d : \text{ all } m_j > 0, \ m_1 a_1 + \cdots + m_d a_d = n \right\};$$

that is, $p_A^\circ(n)$ counts the number of partitions of the positive integer n using only the elements of A as parts, *where each part is used at least once.* Find formulas for p_A° for $A = \{a\}, A = \{a, b\}, A = \{a, b, c\}, A = \{a, b, c, d\}$, where a, b, c, d are pairwise relatively prime positive integers. Observe that in all examples, the counting functions p_A and p_A° satisfy the algebraic relation

$$p_A^\circ(-n) = (-1)^{d-1} p_A(n).$$

(What we mean here is the following: using the algebraic expression for $p_A^\circ(n)$, substitute n by $-n$, even though negative arguments do not make sense in terms of the combinatorial definition of $p_A^\circ(n)$. Then compare this formula with the algebraic expression for $(-1)^{d-1} p_A(n)$.)

1.34. Prove that $p_A^\circ(n) = p_A(n - a_1 - a_2 - \cdots - a_d)$ for $n \geq a_1 + a_2 + \cdots + a_d$. (Here, as usual, $A = \{a_1, a_2, \ldots, a_d\}$.) Conclude that in the examples of Exercise 1.33, the algebraic relation

$$p_A(-n) = (-1)^{d-1} p_A(n - a_1 - a_2 - \cdots - a_d)$$

holds.

1.35. For relatively prime positive integers a, b, let

$$R := \{am + bn : m, n \in \mathbb{Z}_{\geq 0}\},$$

the set of all integers representable by a and b. Prove that

$$\sum_{k \in R} z^k = \frac{1 - z^{ab}}{(1 - z^a)(1 - z^b)}.$$

Use this rational generating function to give alternative proofs of Theorems 1.2 and 1.3.

1.36. For relatively prime positive integers $a_1, a_2, \ldots, a_d$, let

$$R := \{m_1 a_1 + m_2 a_2 + \cdots + m_d a_d : m_1, m_2, \ldots, m_d \in \mathbb{Z}_{\geq 0}\},$$

the set of all integers representable by $a_1, a_2, \ldots, a_d$. Prove that

$$r(z) := \sum_{k \in R} z^k = \frac{p(z)}{(1 - z^{a_1})(1 - z^{a_2}) \cdots (1 - z^{a_d})}$$

for some polynomial p.

1.37. Prove Theorem 1.1: Given a rational function $\frac{p(z)}{\prod_{k=1}^m (z - a_k)^{e_k}}$, where p is a polynomial of degree less than $e_1 + e_2 + \cdots + e_m$ and the a_k are distinct, there exists a decomposition

$$\sum_{k=1}^m \left(\frac{c_{k,1}}{z - a_k} + \frac{c_{k,2}}{(z - a_k)^2} + \cdots + \frac{c_{k,e_k}}{(z - a_k)^{e_k}} \right),$$

where the $c_{k,j} \in \mathbb{C}$ are unique.

Here is an outline of one possible proof. Recall that the set of polynomials (over $\mathbb{R}$ or $\mathbb{C}$) forms a *Euclidean domain*, that is, given two polynomials $a(z), b(z)$, there exist polynomials $q(z), r(z)$ with $\deg(r) < \deg(b)$ such that

$$a(z) = b(z) q(z) + r(z).$$

Applying this procedure repeatedly (the *Euclidean algorithm*) gives the greatest common divisor of $a(z)$ and $b(z)$ as a linear combination of $a(z)$ and $b(z)$, that is, there exist polynomials $c(z)$ and $d(z)$ such that $a(z)c(z) + b(z)d(z) = \gcd(a(z), b(z))$.

Step 1: Apply the Euclidean algorithm to show that there exist polynomials u_1, u_2 such that

$$u_1(z)(z - a_1)^{e_1} + u_2(z)(z - a_2)^{e_2} = 1.$$

Step 2: Deduce that there exist polynomials v_1, v_2 with $\deg(v_k) < e_k$ such that

$$\frac{p(z)}{(z - a_1)^{e_1}(z - a_2)^{e_2}} = \frac{v_1(z)}{(z - a_1)^{e_1}} + \frac{v_2(z)}{(z - a_2)^{e_2}}.$$

(*Hint:* Long division.)

Step 3: Repeat this procedure to obtain a partial fraction decomposition for

$$\frac{p(z)}{(z - a_1)^{e_1}(z - a_2)^{e_2}(z - a_3)^{e_3}}.$$

Open Problems

1.38. Come up with a new approach or a new algorithm for the Frobenius problem in the $d = 4$ case.

1.39. There are a very good lower [93] and several upper bounds [201, Chapter 3] for the Frobenius number. Come up with improved upper bounds.

1.40. Solve Vladimir I. Arnold's Problems 1999–2008 through 1999–2011 [11]. To give a flavor, we mention two of the problems explicitly:

(a) Explore the statistics of $g(a_1, a_2, \ldots, a_d)$ for typical large $a_1, a_2, \ldots, a_d$. It is conjectured that $g(a_1, a_2, \ldots, a_d)$ grows asymptotically like a constant times $\sqrt[d-1]{a_1 a_2 \cdots a_d}$. (See [4, 5, 115].)
(b) Determine what fraction of the integers in the interval $[0, g(a_1, a_2, \ldots, a_d)]$ are representable, for typical large $a_1, a_2, \ldots, a_d$. It is conjectured that this fraction is asymptotically equal to $\frac{1}{d}$. (Theorem 1.3 implies that this conjecture is true in the case $d = 2$.)

1.41. Study vector generalizations of the Frobenius problem [203, 221].

1.42. There are several special cases of $A = \{a_1, a_2, \ldots, a_d\}$ for which the Frobenius problem is solved, for example arithmetic sequences [201, Chapter 3]; see also [216, 246, 247]. Study these special cases in light of the generating function $r(x)$, defined in the notes and in Exercise 1.36.

1.43. Study the generalized Frobenius number g_k (defined in Exercise 1.25; see also [41, 44, 217]), e.g., in light of the Morales–Denham theorem mentioned in the notes. Derive formulas for special cases, e.g., arithmetic sequences.

1.44. For which $0 \leq n \leq b - 1$ is $s_n(a_1, a_2, \ldots, a_d; b) = 0$? (See also Open Problem 8.24.)

Chapter 2
A Gallery of Discrete Volumes

Few things are harder to put up with than a good example.

Mark Twain (1835–1910)

A unifying theme of this book is the study of the number of integer points in polytopes, where the polytopes live in a real Euclidean space $\mathbb{R}^d$. The integer points $\mathbb{Z}^d$ form a lattice in $\mathbb{R}^d$, and we often call the integer points **lattice points**. This chapter carries us through concrete instances of lattice-point enumeration in various integral and rational polytopes. There is a tremendous amount of research taking place along these lines, even as the reader is looking at these pages.

2.1 The Language of Polytopes

A polytope in dimension 1 is a closed interval; the number of integer points in $\left[\frac{a}{b}, \frac{c}{d}\right]$ is easily seen to be $\left\lfloor \frac{c}{d} \right\rfloor - \left\lfloor \frac{a-1}{b} \right\rfloor$ (Exercise 2.1; here we assume that $a, b, c, d \in \mathbb{Z}$ with $\frac{a}{b} < \frac{c}{d}$). A 2-dimensional convex polytope is a **convex polygon**: a compact convex subset of $\mathbb{R}^2$ bounded by a simple closed curve that is made up of finitely many line segments.

In general dimension d, a **convex polytope** is the convex hull of finitely many points in $\mathbb{R}^d$. To be precise, for a finite point set $\{\mathbf{v}_1, \mathbf{v}_2, \ldots, \mathbf{v}_n\} \subset \mathbb{R}^d$, the polytope $\mathcal{P}$ is the smallest convex set containing those points; that is,

$$\mathcal{P} = \{\lambda_1 \mathbf{v}_1 + \lambda_2 \mathbf{v}_2 + \cdots + \lambda_n \mathbf{v}_n : \text{ all } \lambda_k \geq 0 \text{ and } \lambda_1 + \lambda_2 + \cdots + \lambda_n = 1\}.$$

This definition is called the **vertex description** of $\mathcal{P}$, and we use the notation

$$\mathcal{P} = \text{conv} \{\mathbf{v}_1, \mathbf{v}_2, \ldots, \mathbf{v}_n\},$$

the convex hull of $\mathbf{v}_1, \mathbf{v}_2, \ldots, \mathbf{v}_n$. In particular, a polytope is a *closed* subset of $\mathbb{R}^d$. Many polytopes that we will study, however, are not defined in this way, but rather as bounded intersections of finitely many half-spaces and hyperplanes. One example is the polytope $\mathcal{P}$ defined by (1.4) in Chapter 1. (A set, bounded or not, that can be described as the intersection of finitely many half-spaces and hyperplanes is a **polyhedron**.) This **hyperplane description** of a polytope is, in fact, equivalent to the vertex description. The fact that every polytope has both a vertex and a hyperplane description is highly nontrivial, both algorithmically and conceptually. We carefully work out a proof in Appendix A.

The **dimension** of a polytope $\mathcal{P}$ is the dimension of the affine space

$$\operatorname{span} \mathcal{P} := \{ \mathbf{x} + \lambda (\mathbf{y} - \mathbf{x}) : \mathbf{x}, \mathbf{y} \in \mathcal{P}, \, \lambda \in \mathbb{R} \}$$

spanned by $\mathcal{P}$. If $\mathcal{P}$ has dimension d, we use the notation $\dim \mathcal{P} = d$ and call $\mathcal{P}$ a d-polytope. Note that $\mathcal{P} \subset \mathbb{R}^d$ does not necessarily have dimension d. For example, the polytope $\mathcal{P}$ defined by (1.4) has dimension $d - 1$.

For a convex polytope $\mathcal{P} \subset \mathbb{R}^d$, we say that the hyperplane $H = \{ \mathbf{x} \in \mathbb{R}^d : \mathbf{a} \cdot \mathbf{x} = b \}$ is a **supporting hyperplane** of $\mathcal{P}$ if $\mathcal{P}$ lies entirely on one side of H, that is,

$$\mathcal{P} \subset \{ \mathbf{x} \in \mathbb{R}^d : \mathbf{a} \cdot \mathbf{x} \leq b \} \qquad \text{or} \qquad \mathcal{P} \subset \{ \mathbf{x} \in \mathbb{R}^d : \mathbf{a} \cdot \mathbf{x} \geq b \} .$$

A **face** of $\mathcal{P}$ is a set of the form $\mathcal{P} \cap H$, where H is a supporting hyperplane of $\mathcal{P}$. Note that $\mathcal{P}$ itself is a face of $\mathcal{P}$, corresponding to the **degenerate hyperplane** $\mathbb{R}^d$,[1] and the empty set $\varnothing$ is a face of $\mathcal{P}$, corresponding to a hyperplane that does not meet $\mathcal{P}$. The $(d-1)$-dimensional faces are called **facets**, the 1-dimensional faces **edges**, and the 0-dimensional faces **vertices** of $\mathcal{P}$. Vertices are the "extreme points" of a polytope.

A convex d-polytope has at least $d + 1$ vertices. A convex d-polytope with exactly $d + 1$ vertices is called a d-**simplex**. Every 1-dimensional convex polytope is a 1-simplex, namely, a line segment. The 2-dimensional simplices are the triangles, the 3-dimensional simplices the tetrahedra.

A convex polytope $\mathcal{P}$ is called **integral** if all of its vertices have integer coordinates,[2] and $\mathcal{P}$ is called **rational** if all of its vertices have rational coordinates.

---

[1] In the remainder of the book, we will reserve the term *hyperplane* for nondegenerate hyperplanes, i.e., sets of the form $\{ \mathbf{x} \in \mathbb{R}^d : \mathbf{a} \cdot \mathbf{x} = b \}$, where not all of the entries of $\mathbf{a}$ are zero.

[2] Integral polytopes are also called **lattice polytopes**.

2.2 The Unit Cube

As a warmup example, we begin with the **unit d-cube** $\square := [0,1]^d$, which simultaneously offers simple geometry and an endless fountain of research questions. The vertex description of $\square$ is given by the set of 2^d vertices $\{(x_1, x_2, \ldots, x_d) : \text{all } x_k = 0 \text{ or } 1\}$. The hyperplane description is

$$\square = \left\{ (x_1, x_2, \ldots, x_d) \in \mathbb{R}^d : 0 \leq x_k \leq 1 \text{ for all } k = 1, 2, \ldots, d \right\}.$$

Thus, there are the $2d$ bounding hyperplanes $x_1 = 0$, $x_1 = 1$, $x_2 = 0$, $x_2 = 1$, $\ldots$, $x_d = 0$, $x_d = 1$.

We now compute the discrete volume of an integer dilate of $\square$. That is, we seek the number of integer points $t\square \cap \mathbb{Z}^d$ for all $t \in \mathbb{Z}_{>0}$. Here $t\mathcal{P}$ denotes the dilated polytope

$$\{(tx_1, tx_2, \ldots, tx_d) : (x_1, x_2, \ldots, x_d) \in \mathcal{P}\},$$

for a polytope $\mathcal{P}$. What is the discrete volume of $\square$? We dilate by the positive integer t, as depicted in Figure 2.1, and count:

$$\# \left(t\square \cap \mathbb{Z}^d \right) = \# \left([0,t]^d \cap \mathbb{Z}^d \right) = (t+1)^d.$$

Fig. 2.1 The 6th dilate of $\square$ in dimension 2.

We generally denote the **lattice-point enumerator** for the tth dilate of $\mathcal{P} \subset \mathbb{R}^d$ by

$$L_{\mathcal{P}}(t) := \# \left(t\mathcal{P} \cap \mathbb{Z}^d \right),$$

a useful object that we also call the **discrete volume** of $\mathcal{P}$. We may also think of leaving $\mathcal{P}$ fixed and shrinking the integer lattice:

$$L_{\mathcal{P}}(t) = \# \left(\mathcal{P} \cap \frac{1}{t} \mathbb{Z}^d \right).$$

With this convention, $L_\square(t) = (t+1)^d$, a polynomial in the integer variable t. Notice that the coefficients of this polynomial are the **binomial coefficients** $\binom{d}{k}$, defined through

$$\binom{m}{n} := \frac{m(m-1)(m-2)\cdots(m-n+1)}{n!} \tag{2.1}$$

for $m \in \mathbb{C}, n \in \mathbb{Z}_{>0}$.

What about the *interior* $\square^\circ$ of the cube? The number of interior integer points in $t\,\square^\circ$ is

$$L_{\square^\circ}(t) = \#\left(t\,\square^\circ \cap \mathbb{Z}^d\right) = \#\left((0,t)^d \cap \mathbb{Z}^d\right) = (t-1)^d.$$

Notice that this polynomial equals $(-1)^d L_\square(-t)$, the evaluation of the polynomial $L_\square(t)$ at negative integers, up to a sign.

We now introduce another important tool for analyzing a polytope $\mathcal{P}$, namely the generating function of $L_\mathcal{P}$:

$$\mathrm{Ehr}_\mathcal{P}(z) := 1 + \sum_{t \geq 1} L_\mathcal{P}(t)\, z^t.$$

This generating function is also called the **Ehrhart series** of $\mathcal{P}$.

In our case, the Ehrhart series of $\mathcal{P} = \square$ takes on a special form. To illustrate, we define the **Eulerian number** $A(d,k)$ through[3]

$$\sum_{j \geq 0} j^d\, z^j = \frac{\sum_{k=0}^d A(d,k)\, z^k}{(1-z)^{d+1}}. \tag{2.2}$$

It is not hard to see that the polynomial $\sum_{k=1}^d A(d,k)\, z^k$ is the numerator of the rational function

$$\left(z\frac{d}{dz}\right)^d \left(\frac{1}{1-z}\right) = \underbrace{z\frac{d}{dz}\cdots z\frac{d}{dz}}_{d \text{ times}}\left(\frac{1}{1-z}\right).$$

The Eulerian numbers have many fascinating properties, including

$$A(d,k) = A(d,d+1-k),$$
$$A(d,k) = (d-k+1)\,A(d-1,k-1) + k\,A(d-1,k),$$
$$\sum_{k=0}^d A(d,k) = d!, \tag{2.3}$$

---

[3] There are two slightly conflicting definitions of *Eulerian numbers* in the literature: sometimes, they are defined through $\sum_{j \geq 0}(j+1)^d\, z^j = \frac{\sum_{k=0}^d A(d,k)z^k}{(1-z)^{d+1}}$ instead of (2.2).

$$A\left(d,k\right)=\sum_{j=0}^{k}(-1)^{j}\binom{d+1}{j}(k-j)^{d}.$$

The first few Eulerian numbers $A\left(d,k\right)$ for $0\le k\le d$ are

$d = 0$: 1
$d = 1$: 0 1
$d = 2$: 0 1 1
$d = 3$: 0 1 4 1
$d = 4$: 0 1 11 11 1
$d = 5$: 0 1 26 66 26 1
$d = 6$: 0 1 57 302 302 57 1

(see also [1, Sequence A008292]).

With this definition, we can now express the Ehrhart series of $\square$ in terms of Eulerian numbers:

$$\mathrm{Ehr}_{\square}(z)=1+\sum_{t\ge1}(t+1)^{d}\,z^{t}=\sum_{t\ge0}(t+1)^{d}\,z^{t}=\frac{1}{z}\sum_{t\ge1}t^{d}\,z^{t}$$
$$=\frac{\sum_{k=1}^{d}A\left(d,k\right)z^{k-1}}{(1-z)^{d+1}}.$$

To summarize, we have proved the following theorem.

Theorem 2.1. *Let $\square$ be the unit d-cube.*

(a) *The lattice-point enumerator of $\square$ is the polynomial*

$$L_{\square}(t)=(t+1)^{d}=\sum_{k=0}^{d}\binom{d}{k}t^{k}.$$

(b) *Its evaluation at negative integers yields the relation*

$$(-1)^{d}L_{\square}(-t)=L_{\square^{\circ}}(t)\,.$$

(c) *The Ehrhart series of $\square$ is* $\mathrm{Ehr}_{\square}(z)=\frac{\sum_{k=1}^{d}A(d,k)z^{k-1}}{(1-z)^{d+1}}.$ $\square$

2.3 The Standard Simplex

The **standard simplex** Δ in dimension d is the convex hull of the $d+1$ points $\mathbf{e}_{1},\mathbf{e}_{2},\ldots,\mathbf{e}_{d}$ and the origin; here $\mathbf{e}_{j}$ is the unit vector $(0,\ldots,1,\ldots,0)$, with a 1 in the jth position. Figure 2.2 shows Δ for $d=3$. On the other hand, Δ can also be realized by its hyperplane description, namely

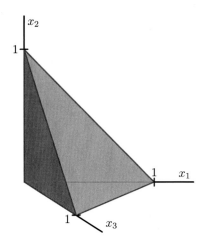

Fig. 2.2 The standard
simplex Δ in dimension 3.

$$\Delta = \left\{(x_1, x_2 \ldots, x_d) \in \mathbb{R}^d : x_1 + x_2 + \cdots + x_d \leq 1 \text{ and all } x_k \geq 0\right\}.$$

In the case of the standard simplex, the dilate $t\Delta$ is now given by

$$t\Delta = \left\{(x_1, x_2, \ldots, x_d) \in \mathbb{R}^d : x_1 + x_2 + \cdots + x_d \leq t \text{ and all } x_k \geq 0\right\}.$$

To compute the discrete volume of Δ, we would like to use the methods developed in Chapter 1, but there is an extra twist. The counting functions in Chapter 1 were defined by equalities, whereas the standard simplex is defined by an *inequality*. We are trying to count all integer solutions $(m_1, m_2, \ldots, m_d) \in \mathbb{Z}_{\geq 0}^d$ to

$$m_1 + m_2 + \cdots + m_d \leq t. \tag{2.4}$$

To translate this inequality in d variables into an equality in $d + 1$ variables, we introduce a *slack variable* $m_{d+1} \in \mathbb{Z}_{\geq 0}$, which picks up the difference between the right-hand and left-hand sides of (2.4). So the number of solutions $(m_1, m_2, \ldots, m_d) \in \mathbb{Z}_{\geq 0}^d$ to (2.4) equals the number of solutions $(m_1, m_2, \ldots, m_{d+1}) \in \mathbb{Z}_{\geq 0}^{d+1}$ to

$$m_1 + m_2 + \cdots + m_{d+1} = t.$$

Now the methods of Chapter 1 apply:

$$\# \left(t\Delta \cap \mathbb{Z}^d \right) = \text{const} \left(\left(\sum_{m_1 \geq 0} z^{m_1} \right) \left(\sum_{m_2 \geq 0} z^{m_2} \right) \cdots \left(\sum_{m_{d+1} \geq 0} z^{m_{d+1}} \right) z^{-t} \right)$$

$$= \text{const} \left(\frac{1}{(1-z)^{d+1} z^t} \right). \tag{2.5}$$

In contrast with Chapter 1, we do not require a partial fraction expansion but simply use the **binomial series**

$$\frac{1}{(1-z)^{d+1}} = \sum_{k \geq 0} \binom{d+k}{d} z^k \tag{2.6}$$

for $d \geq 0$. The constant-term identity (2.5) requires us to find the coefficient of z^t in the binomial series (2.6), which is $\binom{d+t}{d}$. Hence the discrete volume of Δ is given by $L_\Delta(t) = \binom{d+t}{d}$, a polynomial in the integer variable t of degree d. Incidentally, the coefficients of this polynomial function in t have an alternative life in traditional combinatorics:

$$L_\Delta(t) = \frac{1}{d!} \sum_{k=0}^{d} (-1)^{d-k} \, \text{stirl}(d+1, k+1) \, t^k,$$

where $\text{stirl}(n, j)$ is the *Stirling number of the first kind* (see Exercise 2.11). We also notice that (2.6) is, by definition, the Ehrhart series of Δ.

Let's repeat this computation for the *interior* Δ° of the standard d-simplex. Now we introduce a slack variable $m_{d+1} > 0$, so that strict inequality is forced:

$$L_{\Delta^\circ}(t) = \# \left\{ (m_1, m_2, \ldots, m_d) \in \mathbb{Z}_{>0}^d : m_1 + m_2 + \cdots + m_d < t \right\}$$
$$= \# \left\{ (m_1, m_2, \ldots, m_{d+1}) \in \mathbb{Z}_{>0}^{d+1} : m_1 + m_2 + \cdots + m_{d+1} = t \right\}.$$

Now

$$L_{\Delta^\circ}(t) = \text{const} \left(\left(\sum_{m_1 > 0} z^{m_1} \right) \left(\sum_{m_2 > 0} z^{m_2} \right) \cdots \left(\sum_{m_{d+1} > 0} z^{m_{d+1}} \right) z^{-t} \right)$$

$$= \text{const} \left(\left(\frac{z}{1-z} \right)^{d+1} z^{-t} \right)$$

$$= \text{const} \left(z^{d+1-t} \sum_{k \geq 0} \binom{d+k}{d} z^k \right) = \binom{t-1}{d}.$$

It is a fun exercise to prove that

$$(-1)^d \binom{d-t}{d} = \binom{t-1}{d} \tag{2.7}$$

(see Exercise 2.10). We have arrived at our destination:

Theorem 2.2. *Let Δ be the standard d-simplex.*

(a) *The lattice-point enumerator of Δ is the polynomial $L_\Delta(t) = \binom{d+t}{d}$.*
(b) *Its evaluation at negative integers yields $(-1)^d L_\Delta(-t) = L_{\Delta^\circ}(t)$.*
(c) *The Ehrhart series of Δ is $\mathrm{Ehr}_\Delta(z) = \frac{1}{(1-z)^{d+1}}$.*   □

2.4 The Bernoulli Polynomials as Lattice-Point Enumerators of Pyramids

There is a fascinating connection between the Bernoulli polynomials and certain pyramids over unit cubes. The **Bernoulli polynomials** $B_k(x)$ are defined through the generating function

$$\frac{z\,e^{xz}}{e^z - 1} = \sum_{k \geq 0} \frac{B_k(x)}{k!} z^k \tag{2.8}$$

and are ubiquitous in the study of the Riemann zeta function, among other objects; they are named after Jacob Bernoulli (1654–1705).[4] The Bernoulli polynomials will play a prominent role in Chapter 12 in the context of Euler–Maclaurin summation. The first few Bernoulli polynomials are

$$B_0(x) = 1\,,$$
$$B_1(x) = x - \frac{1}{2}\,,$$
$$B_2(x) = x^2 - x + \frac{1}{6}\,,$$
$$B_3(x) = x^3 - \frac{3}{2}x^2 + \frac{1}{2}x\,,$$
$$B_4(x) = x^4 - 2x^3 + x^2 - \frac{1}{30}\,,$$
$$B_5(x) = x^5 - \frac{5}{2}x^4 + \frac{5}{3}x^3 - \frac{1}{6}x\,,$$
$$B_6(x) = x^6 - 3x^5 + \frac{5}{2}x^4 - \frac{1}{2}x^2 + \frac{1}{42}\,.$$

The **Bernoulli numbers** are $B_k := B_k(0)$ (see also [1, Sequences A000367 & A002445]) and have the generating function

$$\frac{z}{e^z - 1} = \sum_{k \geq 0} \frac{B_k}{k!} z^k\,.$$

---

[4] For more information about Bernoulli, see
http://www-history.mcs.st-and.ac.uk/Mathematicians/Bernoulli_Jacob.html.

Lemma 2.3. *For integers $d \geq 1$ and $n \geq 2$,*

$$\sum_{k=0}^{n-1} k^{d-1} = \frac{1}{d} \left(B_d(n) - B_d \right).$$

Proof. We play with the generating function of $\frac{B_d(n) - B_d}{d!}$:

$$\sum_{d \geq 0} \frac{B_d(n) - B_d}{d!} z^d = z \frac{e^{nz} - 1}{e^z - 1} = z \sum_{k=0}^{n-1} e^{kz} = z \sum_{k=0}^{n-1} \sum_{j \geq 0} \frac{(kz)^j}{j!}$$

$$= \sum_{j \geq 0} \left(\sum_{k=0}^{n-1} k^j \right) \frac{z^{j+1}}{j!} = \sum_{j \geq 1} \left(\sum_{k=0}^{n-1} k^{j-1} \right) \frac{z^j}{(j-1)!}.$$

Now compare coefficients on both sides. □

Consider a $(d-1)$-dimensional unit cube embedded into $\mathbb{R}^d$ and form a d-dimensional pyramid by adjoining one more vertex at $(0, 0, \ldots, 0, 1)$, as depicted in Figure 2.3. More precisely, this geometric object has the following hyperplane description:

$$\mathcal{P} = \left\{ (x_1, x_2, \ldots, x_d) \in \mathbb{R}^d : 0 \leq x_1, x_2, \ldots, x_{d-1} \leq 1 - x_d \leq 1 \right\}. \quad (2.9)$$

By definition, $\mathcal{P}$ is contained in the unit d-cube; in fact, its vertices are a subset of the vertices of the d-cube.

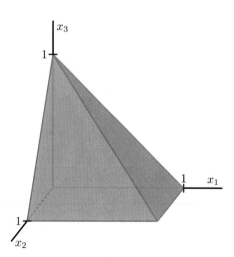

Fig. 2.3 The pyramid $\mathcal{P}$ in dimension 3.

We now count lattice points in integer dilates of $\mathcal{P}$. This number equals

$$\# \left\{ (m_1, m_2, \ldots, m_d) \in \mathbb{Z}^d : 0 \leq m_k \leq t - m_d \leq t \text{ for } k = 1, 2, \ldots, d-1 \right\}.$$

In this case, we just count the solutions to $0 \le m_k \le t - m_d \le t$ directly: once we choose the integer m_d (between 0 and t), we have $t - m_d + 1$ independent choices for each of the integers $m_1, m_2, \ldots, m_{d-1}$. Hence

$$L_{\mathcal{P}}(t) = \sum_{m_d=0}^{t} (t - m_d + 1)^{d-1} = \sum_{k=1}^{t+1} k^{d-1} = \frac{1}{d} (B_d(t+2) - B_d), \quad (2.10)$$

by Lemma 2.3. (Here we need to require $d \ge 2$.) This is, naturally, a polynomial in t.

We now turn our attention to the number of *interior* lattice points in $\mathcal{P}$:

$$L_{\mathcal{P}^\circ}(t) = \# \left\{ (m_1, m_2, \ldots, m_d) \in \mathbb{Z}^d : \begin{array}{l} 0 < m_k < t - m_d < t \\ \text{for all } k = 1, 2, \ldots, d-1 \end{array} \right\}.$$

By a similar counting argument,

$$L_{\mathcal{P}^\circ}(t) = \sum_{m_d=1}^{t-1} (t - m_d - 1)^{d-1} = \sum_{k=0}^{t-2} k^{d-1} = \frac{1}{d} (B_d(t-1) - B_d).$$

Incidentally, the Bernoulli polynomials are known (Exercise 2.15) to have the symmetry

$$B_d(1 - x) = (-1)^d B_d(x). \quad (2.11)$$

This identity coupled with the fact (Exercise 2.16) that

$$B_d = 0 \text{ for all odd } d \ge 3 \quad (2.12)$$

gives the relation

$$L_{\mathcal{P}}(-t) = \frac{1}{d} (B_d(-t + 2) - B_d) = \frac{1}{d} (B_d (1 - (t - 1)) - B_d)$$

$$= (-1)^d \frac{1}{d} (B_d(t - 1) - B_d) = (-1)^d L_{\mathcal{P}^\circ}(t).$$

Next we compute the Ehrhart series of $\mathcal{P}$. We can actually do this in somewhat greater generality. Namely, for a polytope $\mathcal{Q} \subset \mathbb{R}^{d-1}$ with vertices $\mathbf{v}_1, \mathbf{v}_2, \ldots, \mathbf{v}_m$, define $\mathrm{Pyr}(\mathcal{Q})$, the **pyramid over** $\mathcal{Q}$, as the convex hull of $(\mathbf{v}_1, 0), (\mathbf{v}_2, 0), \ldots, (\mathbf{v}_m, 0), (0, \ldots, 0, 1)$ in $\mathbb{R}^d$. In our example above, the d-polytope $\mathcal{P}$ is equal to $\mathrm{Pyr}(\square)$ for the unit $(d-1)$-cube $\square$. The number of integer points in $t \, \mathrm{Pyr}(\mathcal{Q})$ is, by construction,

$$L_{\mathrm{Pyr}(\mathcal{Q})}(t) = 1 + L_{\mathcal{Q}}(1) + L_{\mathcal{Q}}(2) + \cdots + L_{\mathcal{Q}}(t) = 1 + \sum_{j=1}^{t} L_{\mathcal{Q}}(j),$$

because in $t \, \mathrm{Pyr}(\mathcal{Q})$, there is one lattice point with x_d-coordinate t, we have $L_{\mathcal{Q}}(1)$ lattice points with x_d-coordinate $t - 1$, there are $L_{\mathcal{Q}}(2)$ lattice points with x_d-coordinate $t-2$, etc., up to $L_Q(t)$ lattice points with $x_d = 0$. Figure 2.4 shows an instance for a pyramid over a square.

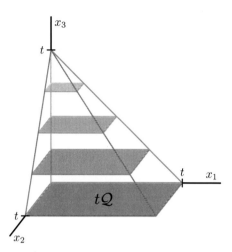

Fig. 2.4 Counting the lattice points in $t\operatorname{Pyr}(Q)$.

This identity for $L_{\operatorname{Pyr}(Q)}(t)$ allows us to compute the Ehrhart series of $\operatorname{Pyr}(Q)$ from the Ehrhart series of Q:

Theorem 2.4. $\operatorname{Ehr}_{\operatorname{Pyr}(Q)}(z) = \dfrac{\operatorname{Ehr}_Q(z)}{1 - z}$.

Proof.

$$\operatorname{Ehr}_{\operatorname{Pyr}(Q)}(z) = 1 + \sum_{t \geq 1} L_{\operatorname{Pyr}(Q)}(t)\, z^t = 1 + \sum_{t \geq 1}\left(1 + \sum_{j=1}^{t} L_Q(j)\right) z^t$$

$$= \sum_{t \geq 0} z^t + \sum_{t \geq 1}\sum_{j=1}^{t} L_Q(j)\, z^t = \frac{1}{1 - z} + \sum_{j \geq 1} L_Q(j) \sum_{t \geq j} z^t$$

$$= \frac{1}{1 - z} + \sum_{j \geq 1} L_Q(j)\frac{z^j}{1 - z} = \frac{1 + \sum_{j \geq 1} L_Q(j)\, z^j}{1 - z} . \qquad \square$$

Our pyramid $\mathcal{P}$ that began this section is a pyramid over the unit $(d-1)$-cube, and so

$$\operatorname{Ehr}_{\mathcal{P}}(z) = \frac{1}{1 - z}\frac{\sum_{k=1}^{d-1} A\,(d-1,k)\, z^{k-1}}{(1 - z)^d} = \frac{\sum_{k=1}^{d-1} A\,(d-1,k)\, z^{k-1}}{(1 - z)^{d+1}} . \quad (2.13)$$

Let's summarize what we have proved for the pyramid over the unit cube.

Theorem 2.5. *Given $d \geq 2$, let $\mathcal{P}$ be the d-pyramid*

$$\mathcal{P} = \left\{ (x_1, x_2, \ldots, x_d) \in \mathbb{R}^d : 0 \leq x_1, x_2, \ldots, x_{d-1} \leq 1 - x_d \leq 1 \right\}.$$

(a) *The lattice-point enumerator of $\mathcal{P}$ is the polynomial*

$$L_{\mathcal{P}}(t) = \frac{1}{d} \left(B_d(t+2) - B_d \right).$$

(b) *Its evaluation at negative integers yields $(-1)^d L_{\mathcal{P}}(-t) = L_{\mathcal{P}^\circ}(t)$.*

(c) *The Ehrhart series of $\mathcal{P}$ is* $\operatorname{Ehr}_{\mathcal{P}}(z) = \frac{\sum_{k=1}^{d-1} A(d-1,k) z^{k-1}}{(1-z)^{d+1}}.$ $\square$

Patterns are emerging...

2.5 The Lattice-Point Enumerators of the Cross-Polytopes

Consider the **cross-polytope** $\Diamond$ in $\mathbb{R}^d$ given by the hyperplane description

$$\Diamond := \left\{ (x_1, x_2, \ldots, x_d) \in \mathbb{R}^d : |x_1| + |x_2| + \cdots + |x_d| \leq 1 \right\}. \qquad (2.14)$$

Figure 2.5 shows the 3-dimensional instance of $\Diamond$, an octahedron. The vertices of $\Diamond$ are $(\pm 1, 0, \ldots, 0), (0, \pm 1, 0, \ldots, 0), \ldots, (0, \ldots, 0, \pm 1)$.

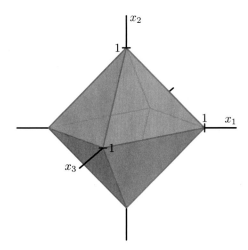

Fig. 2.5 The cross-polytope $\Diamond$ in dimension 3.

To compute the discrete volume of $\Diamond$, we use a process similar to that of Section 2.4. Namely, for a $(d-1)$-polytope $\mathcal{Q}$ with vertices $\mathbf{v}_1, \mathbf{v}_2, \ldots, \mathbf{v}_m$

such that the origin is in $\mathcal{Q}$, we define $\mathrm{BiPyr}(\mathcal{Q})$, the **bipyramid over** $\mathcal{Q}$, as the convex hull of

$$(\mathbf{v}_1, 0), (\mathbf{v}_2, 0), \ldots, (\mathbf{v}_m, 0), (0, \ldots, 0, 1), \quad \text{and} \quad (0, \ldots, 0, -1).$$

In our example above, the d-dimensional cross-polytope is the bipyramid over the $(d-1)$-dimensional cross-polytope. The number of integer points in $t\,\mathrm{BiPyr}(\mathcal{Q})$ is, by construction,

$$L_{\mathrm{BiPyr}(\mathcal{Q})}(t) = 2 + 2L_{\mathcal{Q}}(1) + 2L_{\mathcal{Q}}(2) + \cdots + 2L_{\mathcal{Q}}(t-1) + L_{\mathcal{Q}}(t)$$

$$= 2 + 2\sum_{j=1}^{t-1} L_{\mathcal{Q}}(j) + L_{\mathcal{Q}}(t).$$

This identity allows us to compute the Ehrhart series of $\mathrm{BiPyr}(\mathcal{Q})$ from the Ehrhart series of $\mathcal{Q}$, in a manner similar to the proof of Theorem 2.4. We leave the proof of the following result as Exercise 2.23.

Theorem 2.6. *If $\mathcal{Q}$ contains the origin, then* $\mathrm{Ehr}_{\mathrm{BiPyr}(\mathcal{Q})}(z) = \frac{1+z}{1-z}\,\mathrm{Ehr}_{\mathcal{Q}}(z)$.

$\square$

This theorem allows us to compute the Ehrhart series of $\Diamond$ effortlessly: The cross-polytope $\Diamond$ in dimension 0 is the origin, with Ehrhart series $\frac{1}{1-z}$. The higher-dimensional cross-polytopes can be computed recursively through Theorem 2.6 as

$$\mathrm{Ehr}_{\Diamond}(z) = \frac{(1+z)^d}{(1-z)^{d+1}}.$$

Since $\mathrm{Ehr}_{\Diamond}(z) = 1 + \sum_{t \geq 1} L_{\Diamond}(t)\,z^t$, we can retrieve $L_{\Diamond}(t)$ by expanding $\mathrm{Ehr}_{\Diamond}(z)$ into its power series at $z = 0$:

$$\mathrm{Ehr}_{\Diamond}(z) = \frac{(1+z)^d}{(1-z)^{d+1}} = \frac{\sum_{k=0}^{d}\binom{d}{k}z^k}{(1-z)^{d+1}}$$

$$= \sum_{k=0}^{d}\binom{d}{k}z^k \sum_{t \geq 0}\binom{t+d}{d}z^t = \sum_{k=0}^{d}\binom{d}{k}\sum_{t \geq k}\binom{t-k+d}{d}z^t$$

$$= \sum_{k=0}^{d}\binom{d}{k}\sum_{t \geq 0}\binom{t-k+d}{d}z^t.$$

In the last step, we used the fact that $\binom{t-k+d}{d} = 0$ for $0 \leq t < k$. But then

$$1 + \sum_{t \geq 1} L_{\Diamond}(t)\,z^t = \sum_{t \geq 0}\sum_{k=0}^{d}\binom{d}{k}\binom{t-k+d}{d}z^t,$$

and hence $L_{\Diamond}(t) = \sum_{k=0}^{d}\binom{d}{k}\binom{t-k+d}{d}$ for all $t \geq 1$.

We finish this section by counting the *interior* lattice points in $t\diamond$. We begin by noticing, since t is an integer, that

$$
\begin{aligned}
L_{\diamond^\circ}(t) &= \# \left\{ (m_1, m_2, \ldots, m_d) \in \mathbb{Z}^d : |m_1| + |m_2| + \cdots + |m_d| < t \right\} \\
&= \# \left\{ (m_1, m_2, \ldots, m_d) \in \mathbb{Z}^d : |m_1| + |m_2| + \cdots + |m_d| \le t - 1 \right\} \\
&= L_\diamond(t - 1) \,.
\end{aligned}
$$

On the other hand, we can use (2.7):

$$
\begin{aligned}
L_\diamond(-t) &= \sum_{k=0}^{d} \binom{d}{k} \binom{-t - k + d}{d} \\
&= \sum_{k=0}^{d} \binom{d}{k} (-1)^d \binom{t - 1 + k}{d} \\
&= (-1)^d \sum_{k=0}^{d} \binom{d}{d - k} \binom{t - 1 + d - k}{d} \\
&= (-1)^d L_\diamond(t - 1) \,.
\end{aligned}
$$

Comparing the last two computations, we see that $(-1)^d L_\diamond(-t) = L_{\diamond^\circ}(t)$. Let's summarize:

Theorem 2.7. *Let $\diamond$ be the cross-polytope in $\mathbb{R}^d$.*

(a) *The lattice-point enumerator of $\diamond$ is the polynomial*

$$
L_\diamond(t) = \sum_{k=0}^{d} \binom{d}{k} \binom{t - k + d}{d} \,.
$$

(b) *Its evaluation at negative integers yields $(-1)^d L_\diamond(-t) = L_{\diamond^\circ}(t)$.*

(c) *The Ehrhart series of $\mathcal{P}$ is $\mathrm{Ehr}_\diamond(z) = \frac{(1+z)^d}{(1-z)^{d+1}}$.*   □

2.6 Pick's Theorem

Returning to basic concepts, we now give a complete account of $L_\mathcal{P}$ for all integral convex polygons $\mathcal{P}$ in $\mathbb{R}^2$. Denote the number of integer points inside the polygon $\mathcal{P}$ by I, and the number of integer points on the boundary of $\mathcal{P}$ by B. The following result, called *Pick's theorem* in honor of its discoverer Georg Alexander Pick (1859–1942), presents the astonishing fact that the area A of $\mathcal{P}$ can be computed simply by counting lattice points:

Theorem 2.8 (Pick's theorem). *For an integral convex polygon,*

$$A = I + \frac{1}{2}B - 1.$$

Proof. We begin by proving that Pick's identity has an additive character: we can decompose $\mathcal{P}$ into the union of two integral polygons $\mathcal{P}_1$ and $\mathcal{P}_2$ by joining two vertices of $\mathcal{P}$ with a line segment, as shown in Figure 2.6.

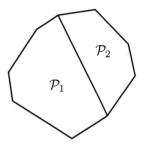

Fig. 2.6 Decomposition of
a polygon into two.

We claim that the validity of Pick's identity for $\mathcal{P}$ follows from the validity of Pick's identity for $\mathcal{P}_1$ and $\mathcal{P}_2$. Denote the area, number of interior lattice points, and number of boundary lattice points of $\mathcal{P}_k$ by A_k, I_k, and B_k, respectively, for $k = 1, 2$. Clearly,

$$A = A_1 + A_2.$$

Furthermore, if we denote the number of lattice points on the edge common to $\mathcal{P}_1$ and $\mathcal{P}_2$ by L, then

$$I = I_1 + I_2 + L - 2 \qquad \text{and} \qquad B = B_1 + B_2 - 2L + 2.$$

Thus

$$I + \frac{1}{2}B - 1 = I_1 + I_2 + L - 2 + \frac{1}{2}B_1 + \frac{1}{2}B_2 - L + 1 - 1$$
$$= I_1 + \frac{1}{2}B_1 - 1 + I_2 + \frac{1}{2}B_2 - 1.$$

This proves the claim. Note that our proof also shows that the validity of Pick's identity for $\mathcal{P}_1$ follows from the validity of Pick's identity for $\mathcal{P}$ and $\mathcal{P}_2$.

Now, every convex polygon can be decomposed into triangles that share a common vertex, as illustrated in Figure 2.7. Hence it suffices to prove Pick's theorem for triangles. Further simplifying the picture, we can embed every integral triangle into an integral rectangle, as suggested by Figure 2.8.

This reduces the proof of Pick's theorem to proving the theorem for integral rectangles whose edges are parallel to the coordinate axes, and for rectangular

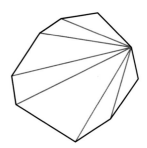

Fig. 2.7 Triangulation of
a polygon.

triangles two of whose edges are parallel to the coordinate axes. These two
cases are left to the reader as Exercise 2.25.   □

 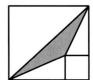

Fig. 2.8 Embedding a
triangle in a rectangle.

Pick's theorem allows us to count not only the lattice points strictly inside
the polygon $\mathcal{P}$, but also the total number of lattice points contained in $\mathcal{P}$,
because this number is

$$I + B = A - \frac{1}{2}B + 1 + B = A + \frac{1}{2}B + 1. \tag{2.15}$$

From this identity, it is now easy to describe the lattice-point enumerator $L_\mathcal{P}$:

Theorem 2.9. *Suppose $\mathcal{P}$ is an integral convex polygon with area A and B
lattice points on its boundary.*

(a) *The lattice-point enumerator of $\mathcal{P}$ is the polynomial*

$$L_\mathcal{P}(t) = A\,t^2 + \frac{1}{2}B\,t + 1.$$

(b) *Its evaluation at negative integers yields the relation*

$$L_\mathcal{P}(-t) = L_{\mathcal{P}^\circ}(t).$$

(c) *The Ehrhart series of $\mathcal{P}$ is*

$$\mathrm{Ehr}_\mathcal{P}(z) = \frac{\left(A - \frac{B}{2} + 1\right)z^2 + \left(A + \frac{B}{2} - 2\right)z + 1}{(1 - z)^3}.$$

Note that in the numerator of the Ehrhart series, the coefficient of z^2 is $L_{\mathcal{P}^\circ}(1)$, and the coefficient of z is $L_{\mathcal{P}}(1) - 3$.

Proof. Statement (a) will follow from (2.15) if we can prove that the area of $t\mathcal{P}$ is At^2 and that the number of boundary points on $t\mathcal{P}$ is Bt, which is the content of Exercise 2.26. Statement (b) follows with $L_{\mathcal{P}^\circ}(t) = L_{\mathcal{P}}(t) - Bt$. Finally, the Ehrhart series is

$$\mathrm{Ehr}_{\mathcal{P}}(z) = 1 + \sum_{t \geq 1} L_{\mathcal{P}}(t)\, z^t$$

$$= \sum_{t \geq 0} \left(At^2 + \frac{B}{2}t + 1 \right) z^t$$

$$= A \frac{z^2 + z}{(1-z)^3} + \frac{B}{2}\frac{z}{(1-z)^2} + \frac{1}{1-z}$$

$$= \frac{\left(A - \frac{B}{2} + 1\right)z^2 + \left(A + \frac{B}{2} - 2\right)z + 1}{(1-z)^3}. \qquad \square$$

2.7 Polygons with Rational Vertices

In this section we will establish formulas for the number of integer points in a *rational* convex polygon and its integral dilates.

A natural first step is to fix a triangulation of the polygon $\mathcal{P}$, which reduces our problem to that of counting integer points in rational *triangles*. However, this procedure merits some remarks. After counting lattice points in the triangles, we need to reassemble the triangles to form the polygon. But then we need to take care of the overcounting on line segments (where the triangles meet). Computing the number of lattice points on rational line segments is considerably easier than enumerating lattice points in 2-dimensional regions; however, it is still nontrivial (see Theorem 1.5).

After triangulating $\mathcal{P}$, we can further simplify the picture by embedding an arbitrary rational triangle in a rational rectangle, as in Figure 2.8. To compute lattice points in a triangle, we can first count the points in a rectangle with edges parallel to the coordinate axes, and then subtract the number of points in three right triangles, each with two edges are parallel to the axes, and possibly another rectangle, as shown in Figure 2.8. Since rectangles are easy to deal with (see Exercise 2.2), the problem reduces to finding a formula for a right triangle two of whose edges are parallel to the coordinate axes.

We now adjust and expand our generating-function machinery to these right triangles. Such a triangle $\mathcal{T}$ is a subset of $\mathbb{R}^2$ consisting of all points (x, y) satisfying

$$x \geq \frac{a}{d}, \ y \geq \frac{b}{d}, \ ex + fy \leq r$$

for some integers a, b, d, e, f, r (with $ea + fb \leq rd$; otherwise, the triangle would be empty). Because the lattice-point count is invariant under horizontal and vertical integer translations and under flipping about the x- or y-axis, we may assume that $a, b, d, e, f, r \geq 0$ and $a, b < d$. (One should meditate about this fact for a minute.) Thus we arrive at the triangle $\mathcal{T}$ depicted in Figure 2.9.

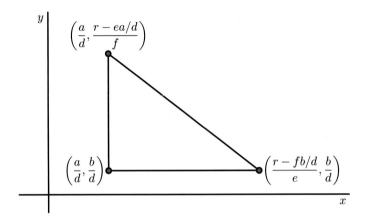

Fig. 2.9 A right rational triangle.

To make our life a little easier, let's assume for the moment that e and f are relatively prime; we will deal with the general case in the exercises. So let

$$\mathcal{T} = \left\{ (x, y) \in \mathbb{R}^2 : x \geq \frac{a}{d}, \ y \geq \frac{b}{d}, \ ex + fy \leq r \right\}. \qquad (2.16)$$

To derive a formula for

$$L_{\mathcal{T}}(t) = \#\left\{ (m, n) \in \mathbb{Z}^2 : m \geq \frac{ta}{d}, \ n \geq \frac{tb}{d}, \ em + fn \leq tr \right\},$$

we want to use methods similar to those in Chapter 1. As in Section 2.3, we introduce a slack variable s:

$$L_{\mathcal{T}}(t) = \#\left\{ (m, n) \in \mathbb{Z}^2 : m \geq \frac{ta}{d}, \ n \geq \frac{tb}{d}, \ em + fn \leq tr \right\}$$

$$= \#\left\{ (m, n, s) \in \mathbb{Z}^3 : m \geq \frac{ta}{d}, \ n \geq \frac{tb}{d}, \ s \geq 0, \ em + fn + s = tr \right\}.$$

This counting function can now, as earlier, be interpreted as the coefficient of z^{tr} in the function

$$\left(\sum_{m \geq \frac{ta}{d}} z^{em} \right) \left(\sum_{n \geq \frac{tb}{d}} z^{fn} \right) \left(\sum_{s \geq 0} z^{s} \right).$$

Here the subscript (e.g., $m \geq \frac{ta}{d}$) under a summation sign means *sum over all integers satisfying this condition*. For example, in the first sum, we begin with the least integer greater than or equal to $\frac{ta}{d}$, which is denoted by $\lceil \frac{ta}{d} \rceil$ (and is equal to $\lfloor \frac{ta-1}{d} \rfloor + 1$ by Exercise 1.4(j)). Hence the above generating function can be rewritten as

$$\left(\sum_{m \geq \lceil \frac{ta}{d} \rceil} z^{em} \right) \left(\sum_{n \geq \lceil \frac{tb}{d} \rceil} z^{fn} \right) \left(\sum_{s \geq 0} z^{s} \right) = \frac{z^{\lceil \frac{ta}{d} \rceil e}}{1 - z^e} \frac{z^{\lceil \frac{tb}{d} \rceil f}}{1 - z^f} \frac{1}{1 - z}$$

$$= \frac{z^{u+v}}{(1 - z^e)(1 - z^f)(1 - z)},$$
(2.17)

where we have introduced, for ease of notation,

$$u := \left\lceil \frac{ta}{d} \right\rceil e \qquad \text{and} \qquad v := \left\lceil \frac{tb}{d} \right\rceil f.$$
(2.18)

To extract the coefficient of z^{tr} of our generating function (2.17), we use familiar methods. As usual, we shift this coefficient to a constant term:

$$L_{\mathcal{T}}(t) = \text{const} \left(\frac{z^{u+v-tr}}{(1 - z^e)(1 - z^f)(1 - z)} \right)$$

$$= \text{const} \left(\frac{1}{(1 - z^e)(1 - z^f)(1 - z)z^{tr-u-v}} \right).$$

Before we apply the partial fraction machinery to this function, we should make sure that it is indeed a proper rational function, that is, that the total degree satisfies

$$u + v - tr - e - f - 1 < 0$$
(2.19)

(see Exercise 2.33). Then we expand into partial fractions (here we are using our assumption that e and f do not have any common factors):

$$\frac{1}{(1 - z^e)(1 - z^f)(1 - z)z^{tr-u-v}}$$

$$= \sum_{j=1}^{e-1} \frac{A_j}{z - \xi_e^j} + \sum_{j=1}^{f-1} \frac{B_j}{z - \xi_f^j} + \sum_{k=1}^{3} \frac{C_k}{(z-1)^k} + \sum_{k=1}^{tr-u-v} \frac{D_k}{z^k}.$$
(2.20)

As we have seen numerous times before, the coefficients D_k do not contribute to the constant term, so that we obtain

$$L_T(t) = -\sum_{j=1}^{e-1} \frac{A_j}{\xi_e^j} - \sum_{l=1}^{f-1} \frac{B_l}{\xi_f^l} - C_1 + C_2 - C_3. \tag{2.21}$$

We invite the reader to compute the coefficients appearing in this formula (Exercise 2.34):

$$A_j = -\frac{\xi_e^{j(v-tr+1)}}{e\left(1 - \xi_e^{jf}\right)\left(1 - \xi_e^j\right)},$$

$$B_l = -\frac{\xi_f^{l(u-tr+1)}}{f\left(1 - \xi_f^{le}\right)\left(1 - \xi_f^l\right)},$$

$$C_1 = -\frac{(u+v-tr)^2}{2ef} + \frac{u+v-tr}{2}\left(-\frac{1}{ef} + \frac{1}{e} + \frac{1}{f}\right) + \frac{1}{4}\left(\frac{1}{e} + \frac{1}{f} - 1\right)$$
$$- \frac{1}{12}\left(\frac{e}{f} + \frac{1}{ef} + \frac{f}{e}\right), \tag{2.22}$$

$$C_2 = -\frac{u+v-tr+1}{ef} + \frac{1}{2e} + \frac{1}{2f},$$

$$C_3 = -\frac{1}{ef}.$$

Putting these ingredients into (2.21) yields the following formula for our lattice-point count.

Theorem 2.10. *For the rectangular rational triangle T given by (2.16), where e and f are relatively prime,*

$$L_T(t) = \frac{1}{2ef}(tr - u - v)^2 + \frac{1}{2}(tr - u - v)\left(\frac{1}{e} + \frac{1}{f} + \frac{1}{ef}\right)$$
$$+ \frac{1}{4}\left(1 + \frac{1}{e} + \frac{1}{f}\right) + \frac{1}{12}\left(\frac{e}{f} + \frac{f}{e} + \frac{1}{ef}\right)$$
$$+ \frac{1}{e}\sum_{j=1}^{e-1} \frac{\xi_e^{j(v-tr)}}{\left(1 - \xi_e^{jf}\right)\left(1 - \xi_e^j\right)} + \frac{1}{f}\sum_{l=1}^{f-1} \frac{\xi_f^{l(u-tr)}}{\left(1 - \xi_f^{le}\right)\left(1 - \xi_f^l\right)}. \qquad \square$$

This identity can be rephrased in terms of the Fourier–Dedekind sum that we introduced in (1.13):

$$L_T(t) = \frac{1}{2ef}(tr - u - v)^2 + \frac{1}{2}(tr - u - v)\left(\frac{1}{e} + \frac{1}{f} + \frac{1}{ef}\right)$$
$$+ \frac{1}{4}\left(1 + \frac{1}{e} + \frac{1}{f}\right) + \frac{1}{12}\left(\frac{e}{f} + \frac{f}{e} + \frac{1}{ef}\right)$$
$$+ s_{v-tr}(f, 1; e) + s_{u-tr}(e, 1; f).$$

The general formula for $L_\mathcal{T}$—not assuming that e and f are relatively prime—is the content of Exercise 2.36.

Let's pause for a moment and study the nature of $L_\mathcal{T}$ as a function of t. Aside from the last two finite sums (which will be put in the spotlight in Chapter 8) and the appearance of u and v, the function $L_\mathcal{T}$ is a quadratic polynomial in t. And in those two sums, t appears only in the exponent of roots of unity, namely as the exponent of ξ_e and ξ_f. As a function of t, ξ_e^t is *periodic* with period e, and similarly, ξ_f^t is periodic with period f. We should also remember that u and v are functions of t; but they can be easily written in terms of the fractional-part function, which again gives rise to periodic functions in t. So $L_\mathcal{T}(t)$ is a (quadratic) "polynomial" in t, whose coefficients are periodic functions in t. This is reminiscent of the counting functions of Chapter 1, which showed a similar periodic-polynomial behavior. Inspired by both examples, we define a **quasipolynomial** Q as an expression of the form

$$Q(t) = c_n(t)\, t^n + \cdots + c_1(t)\, t + c_0(t)\,,$$

where $c_0, \ldots, c_n$ are periodic functions in t. The **degree** of Q is n,[5] and the least common period of $c_0, \ldots, c_n$ is the **period** of Q. Alternatively, for a quasipolynomial Q, there exist a positive integer k and polynomials $p_0, p_1, \ldots, p_{k-1}$ such that

$$Q(t) = \begin{cases} p_0(t) & \text{if } t \equiv 0 \bmod k, \\ p_1(t) & \text{if } t \equiv 1 \bmod k, \\ \vdots & \\ p_{k-1}(t) & \text{if } t \equiv k-1 \bmod k. \end{cases}$$

The minimal such k is the period of Q, and for this minimal k, the polynomials $p_0, p_1, \ldots, p_{k-1}$ are the **constituents** of Q.

By the triangulation and embedding-in-a-box arguments that began this section, we can now state a general structural result for rational polygons.

Theorem 2.11. *Let $\mathcal{P}$ be any rational polygon. Then $L_\mathcal{P}(t)$ is a quasipolynomial of degree 2. Its leading coefficient is the area of $\mathcal{P}$ (in particular, it is a constant).*

We have the technology at this point to study the period of L_P; we let the reader enjoy the ensuing details (see Exercise 2.37).

Proof. By Exercises 2.2 and 2.36 (the general form of Theorem 2.10), the theorem holds for rational rectangles and right triangles whose edges are parallel to the axes. Now use the additivity of both degree-2 quasipolynomials and areas along with Theorem 1.5.   □

---

[5] Here we tacitly assume that c_n is not the zero function.

2.8 Euler's Generating Function for General Rational Polytopes

By now, we have computed several instances of counting functions by setting up a generating function that fits the particular problem in which we are interested. In this section, we set up such a generating function for the lattice-point enumerator of an arbitrary rational polytope. Such a polytope is given by its hyperplane description as an intersection of half-spaces and hyperplanes. The half-spaces are algebraically given by linear inequalities, the hyperplanes by linear equations. If the polytope is rational, we can choose the coefficients of these inequalities and equations to be integers (Exercise 2.7). To unify both descriptions, we can introduce slack variables to turn the half-space inequalities into equalities. Furthermore, by translating our polytope into the nonnegative orthant (we can always shift a polytope by an integer vector without changing the lattice-point count), we may assume that all points in the polytope have nonnegative coordinates. In summary, after a harmless integer translation, we can describe every rational polytope $\mathcal{P}$ as

$$\mathcal{P} = \left\{ \mathbf{x} \in \mathbb{R}^d_{\geq 0} : \mathbf{A}\mathbf{x} = \mathbf{b} \right\} \tag{2.23}$$

for some integral matrix $\mathbf{A} \in \mathbb{Z}^{m \times d}$ and some integer vector $\mathbf{b} \in \mathbb{Z}^m$. (Note that d is not necessarily the dimension of $\mathcal{P}$.) To describe the tth dilate of $\mathcal{P}$, we simply scale a point $\mathbf{x} \in \mathcal{P}$ by $\frac{1}{t}$, or alternatively, multiply $\mathbf{b}$ by t:

$$t\mathcal{P} = \left\{ \mathbf{x} \in \mathbb{R}^d_{\geq 0} : \mathbf{A}\frac{\mathbf{x}}{t} = \mathbf{b} \right\} = \left\{ \mathbf{x} \in \mathbb{R}^d_{\geq 0} : \mathbf{A}\mathbf{x} = t\mathbf{b} \right\}.$$

Hence the lattice-point enumerator of $\mathcal{P}$ is the counting function

$$L_{\mathcal{P}}(t) = \# \left\{ \mathbf{x} \in \mathbb{Z}^d_{\geq 0} : \mathbf{A}\mathbf{x} = t\mathbf{b} \right\}. \tag{2.24}$$

Example 2.12. Suppose $\mathcal{P}$ is the quadrilateral with vertices $(0,0)$, $(2,0)$, $(1,1)$, and $\left(0, \frac{3}{2}\right)$ pictured in Figure 2.10. The half-space-inequality description of $\mathcal{P}$ is

$$\mathcal{P} = \left\{ (x_1, x_2) \in \mathbb{R}^2 : x_1, x_2 \geq 0, \begin{array}{l} x_1 + 2x_2 \leq 3, \\ x_1 + x_2 \leq 2 \end{array} \right\}.$$

Thus,

$$L_{\mathcal{P}}(t) = \# \left\{ (x_1, x_2) \in \mathbb{Z}^2 : x_1, x_2 \geq 0, \begin{array}{l} x_1 + 2x_2 \leq 3t, \\ x_1 + x_2 \leq 2t \end{array} \right\}$$

$$= \# \left\{ (x_1, x_2, x_3, x_4) \in \mathbb{Z}^4 : x_1, x_2, x_3, x_4 \geq 0, \begin{array}{l} x_1 + 2x_2 + x_3 = 3t, \\ x_1 + x_2 + x_4 = 2t \end{array} \right\}$$

$$= \# \left\{ \mathbf{x} \in \mathbb{Z}^4_{\geq 0} : \begin{pmatrix} 1 & 2 & 1 & 0 \\ 1 & 1 & 0 & 1 \end{pmatrix} \mathbf{x} = \begin{pmatrix} 3t \\ 2t \end{pmatrix} \right\}.$$

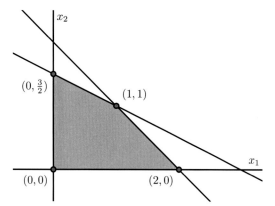

Fig. 2.10 The quadrilateral $\mathcal{P}$ from Example 2.12.

Using the ideas from Sections 1.3, 1.5, 2.3, and 2.7, we now construct a generating function for this counting function. In those previous sections, the lattice-point enumerator could be described with only *one* nontrivial linear equation, whereas now we have a system of such linear constraints. However, we can use the same approach of encoding the linear equation into geometric series; we just need more than one variable. When we expand the function

$$f(z_1, z_2) := \frac{1}{(1 - z_1 z_2)(1 - z_1^2 z_2)(1 - z_1)(1 - z_2) z_1^{3t} z_2^{2t}}$$

into geometric series,

$$f(z_1, z_2) =$$

$$= \left(\sum_{n_1 \geq 0} (z_1 z_2)^{n_1} \right) \left(\sum_{n_2 \geq 0} (z_1^2 z_2)^{n_2} \right) \left(\sum_{n_3 \geq 0} z_1^{n_3} \right) \left(\sum_{n_4 \geq 0} z_2^{n_4} \right) \frac{1}{z_1^{3t} z_2^{2t}}$$

$$= \sum_{n_1, \ldots, n_4 \geq 0} z_1^{n_1 + 2n_2 + n_3 - 3t} z_2^{n_1 + n_2 + n_4 - 2t}.$$

When we compute the constant term (in both z_1 and z_2), we are counting solutions $(n_1, n_2, n_3, n_4) \in \mathbb{Z}_{\geq 0}^4$ to

$$\begin{pmatrix} 1 & 2 & 1 & 0 \\ 1 & 1 & 0 & 1 \end{pmatrix} \begin{pmatrix} n_1 \\ n_2 \\ n_3 \\ n_4 \end{pmatrix} = \begin{pmatrix} 3t \\ 2t \end{pmatrix},$$

that is, the constant term of $f(z_1, z_2)$ counts the integer points in $\mathcal{P}$:

$$L_{\mathcal{P}}(t) = \text{const} \ \frac{1}{(1 - z_1 z_2)(1 - z_1^2 z_2)(1 - z_1)(1 - z_2) z_1^{3t} z_2^{2t}}.$$

We invite the reader to actually compute this constant term (Exercise 2.38). It turns out to be

$$\frac{7}{4}t^2 + \frac{5}{2}t + \frac{7 + (-1)^t}{8}.$$ □

Returning to the general case of a polytope $\mathcal{P}$ given by (2.23), we denote the columns of $\mathbf{A}$ by $\mathbf{c}_1, \mathbf{c}_2, \ldots, \mathbf{c}_d$. Let $\mathbf{z} = (z_1, z_2, \ldots, z_m)$ and expand the function

$$\frac{1}{(1 - \mathbf{z}^{\mathbf{c}_1})(1 - \mathbf{z}^{\mathbf{c}_2}) \cdots (1 - \mathbf{z}^{\mathbf{c}_d}) \mathbf{z}^{t\mathbf{b}}}$$ (2.25)

in terms of geometric series:

$$\left(\sum_{n_1 \geq 0} \mathbf{z}^{n_1 \mathbf{c}_1} \right) \left(\sum_{n_2 \geq 0} \mathbf{z}^{n_2 \mathbf{c}_2} \right) \cdots \left(\sum_{n_d \geq 0} \mathbf{z}^{n_d \mathbf{c}_d} \right) \frac{1}{\mathbf{z}^{t\mathbf{b}}}.$$

Here we use the abbreviating notation $\mathbf{z}^{\mathbf{a}} := z_1^{a_1} z_2^{a_2} \cdots z_m^{a_m}$ for the vectors $\mathbf{z} = (z_1, z_2, \ldots, z_m) \in \mathbb{C}^m$ and $\mathbf{a} = (a_1, a_2, \ldots, a_m) \in \mathbb{Z}^m$. In multiplying out everything, a typical exponent will look like

$$n_1 \mathbf{c}_1 + n_2 \mathbf{c}_2 + \cdots + n_d \mathbf{c}_d - t\mathbf{b} = \mathbf{A}\mathbf{n} - t\mathbf{b},$$

where $\mathbf{n} = (n_1, n_2, \ldots, n_d) \in \mathbb{Z}_{\geq 0}^d$. That is, if we take the constant term of our generating function (2.25), we are counting integer vectors $\mathbf{n} \in \mathbb{Z}_{\geq 0}^d$ satisfying

$$\mathbf{A}\mathbf{n} - t\mathbf{b} = \mathbf{0}, \qquad \text{that is,} \qquad \mathbf{A}\mathbf{n} = t\mathbf{b}.$$

So this constant term will pick up exactly the number of lattice points $\mathbf{n} \in \mathbb{Z}_{\geq 0}^d$ in $t\mathcal{P}$:

Theorem 2.13 (Euler's generating function). *Suppose the rational polytope $\mathcal{P}$ is given by (2.23). Then the lattice-point enumerator of $\mathcal{P}$ can be computed as follows:*

$$L_{\mathcal{P}}(t) = \text{const} \left(\frac{1}{(1 - \mathbf{z}^{\mathbf{c}_1})(1 - \mathbf{z}^{\mathbf{c}_2}) \cdots (1 - \mathbf{z}^{\mathbf{c}_d}) \mathbf{z}^{t\mathbf{b}}} \right).$$ □

We finish this section with rephrasing this constant-term identity in terms of Ehrhart series.

Corollary 2.14. *Suppose the rational polytope $\mathcal{P}$ is given by (2.23). Then the Ehrhart series of $\mathcal{P}$ can be computed as*

$$\text{Ehr}_{\mathcal{P}}(x) = \text{const} \left(\frac{1}{(1 - \mathbf{z}^{\mathbf{c}_1})(1 - \mathbf{z}^{\mathbf{c}_2}) \cdots (1 - \mathbf{z}^{\mathbf{c}_d}) \left(1 - \frac{x}{\mathbf{z}^{\mathbf{b}}}\right)} \right).$$

Proof. By Theorem 2.13,

$$\mathrm{Ehr}_{\mathcal{P}}(x) = \sum_{t \geq 0} \mathrm{const} \left(\frac{1}{\left(1 - \mathbf{z}^{\mathbf{c}_1}\right)\left(1 - \mathbf{z}^{\mathbf{c}_2}\right) \cdots \left(1 - \mathbf{z}^{\mathbf{c}_d}\right) \mathbf{z}^{t\mathbf{b}}} \right) x^t$$

$$= \mathrm{const} \left(\frac{1}{\left(1 - \mathbf{z}^{\mathbf{c}_1}\right)\left(1 - \mathbf{z}^{\mathbf{c}_2}\right) \cdots \left(1 - \mathbf{z}^{\mathbf{c}_d}\right)} \sum_{t \geq 0} \frac{x^t}{\mathbf{z}^{t\mathbf{b}}} \right)$$

$$= \mathrm{const} \left(\frac{1}{\left(1 - \mathbf{z}^{\mathbf{c}_1}\right)\left(1 - \mathbf{z}^{\mathbf{c}_2}\right) \cdots \left(1 - \mathbf{z}^{\mathbf{c}_d}\right)} \frac{1}{1 - \frac{x}{\mathbf{z}^{\mathbf{b}}}} \right). \qquad \square$$

Notes

1. Convex polytopes are beautiful objects with a rich history and interesting theory, which we have only glimpsed here. For good introductions to polytopes, we recommend [68, 126, 258]. Polytopes appear in a vast range of current research areas, including Gröbner bases and commutative algebra [236], combinatorial optimization [95, 213], integral geometry [149], K-theory [73], and geometry of numbers [220].

2. The distinction between the vertex and hyperplane description of a convex polytope leads to an interesting algorithmic question; namely, how quickly can we retrieve the first piece of data from the second and vice versa [213, 258]?

3. Ehrhart series are named after Eugène Ehrhart (1906–2000),[6] who proved the main structure theorems which we will see in Chapter 3. The Ehrhart series of a polytope is an example of a *Hilbert–Poincaré series*. These series appear in the study of *graded algebras* (see, for example, [134, 229]); in the Ehrhart case, this algebra lives in $\mathbb{C}[z_1^{\pm 1}, z_2^{\pm 1}, \ldots, z_d^{\pm 1}, z_{d+1}]$ and is generated by the monomials $\mathbf{z}^{\mathbf{m}}$, where $\mathbf{m}$ ranges over all integer points in cone($\mathcal{P}$), the cone over $\mathcal{P}$, which we will define in Chapter 3. Ehrhart series also appear in the context of *toric varieties*, a vast and fruitful subject [91, 116].

4. The Eulerian numbers $A(d, k)$ are named after Leonhard Euler (1707–1783)[7] and arise naturally in the statistics of permutations: $A(d, k)$ counts permutations of $\{1, 2, \ldots, d\}$ with $k - 1$ ascents. For more on $A(d, k)$, see [87, Section 6.5] and [139]; for more connections between $A(d, k)$ and Ehrhart theory, see [32].

5. The pyramids of Section 2.4 have an interpretation as *order polytopes* [230]. A curious fact about the lattice-point enumerators of these pyramids is that they have arbitrarily large real roots as the dimension grows [36].

---

[6] For more information about Ehrhart, see http://icps.u-strasbg.fr/~clauss/Ehrhart.html.

[7] For more information about Euler, see
http://www-history.mcs.st-andrews.ac.uk/Biographies/Euler.html.

6. The counting function $L_\diamond$ for the cross-polytope can, incidentally, also be written as
$$\sum_{k=0}^{\min(d,t)} 2^k \binom{d}{k}\binom{t}{k}.$$

In particular, $L_\diamond$ is symmetric in d and t. The cross-polytope counting functions bear a connection to Laguerre polynomials, the d-dimensional harmonic oscillator, and the Riemann hypothesis. This connection appeared in [75], where Daniel Bump, Kwok–Kwong Choi, Pär Kurlberg, and Jeffrey Vaaler also found a curious fact about the roots of the polynomials $L_\diamond$: they all have real part $-\frac{1}{2}$ (an instance of a *local Riemann hypothesis*). This fact was proved independently by Peter Kirschenhofer, Attila Pethő, and Robert Tichy [148]; see also the notes in Chapter 4.

7. Theorem 2.8 marks the beginning of the general study of lattice-point enumeration in polytopes. Its amazingly simple statement was discovered by Georg Alexander Pick (1859–1942)[8] in 1899 [190]. Pick's theorem holds also for a nonconvex polygon, provided its boundary forms a simple curve. In Chapter 14, we prove a generalization of Pick's theorem that includes nonconvex curves.

8. The results of Section 2.7 appeared in [43]. We will see in Chapter 8 that the finite sums over roots of unity can be rephrased in terms of *Dedekind–Rademacher sums*, which—as we will also see in Chapter 8—can be computed very quickly. The theorems of Section 2.7 will then imply that the discrete volume of every rational polygon can be computed efficiently.

9. If we replace $t\mathbf{b}$ in (2.24) by a variable integer vector $\mathbf{v}$, the counting function
$$f(\mathbf{v}) = \#\left\{\mathbf{x} \in \mathbb{Z}_{\geq 0}^d : \mathbf{A}\mathbf{x} = \mathbf{v}\right\}$$

is called a *vector partition function*: it counts partitions of the vector $\mathbf{v}$ in terms of the columns of $\mathbf{A}$. Vector partition functions are the multivariate analogues of our lattice-point enumerators $L_{\mathcal{P}}(t)$. They have many interesting properties and give rise to intriguing open questions [30, 58, 90, 235, 240].

10. While Leonhard Euler most likely did not think of lattice-point enumeration in the sense of Ehrhart, we attribute Theorem 2.13 to him, since he certainly worked with generating functions of this type, probably thinking of them as vector partition functions. Percy MacMahon (1854–1929)[9] developed powerful machinery for manipulating multivariate generating functions [167]; his viewpoint and motivation came from integer partitions, but his work

---

[8] For more information about Pick, see
http://www-history.mcs.st-andrews.ac.uk/Biographies/Pick.html.

[9] For more information about MacMahon, see
http://www-history.mcs.st-andrews.ac.uk/Biographies/MacMahon.html.

can be applied to more general linear-constraint settings, such as vector partition functions. The potential of Euler's generating function for Ehrhart polynomials was already realized by Ehrhart [109, 111]. Several modern approaches to computing Ehrhart polynomials are based on Theorem 2.13 (see, for example, [29, 67, 159]).

Exercises

2.1. ♣ Fix positive integers a, b, c, d such that $\gcd(a, b) = \gcd(c, d) = 1$ and $\frac{a}{b} < \frac{c}{d}$, and let $\mathcal{P}$ be the interval $\left[\frac{a}{b}, \frac{c}{d}\right]$ (so $\mathcal{P}$ is a 1-dimensional rational convex polytope). Compute $L_{\mathcal{P}}(t) = \#\,(t\mathcal{P} \cap \mathbb{Z})$ and $L_{\mathcal{P}^{\circ}}(t)$ and show directly that $L_{\mathcal{P}}(t)$ and $L_{\mathcal{P}^{\circ}}(t)$ are quasipolynomials with period $\mathrm{lcm}(b, d)$ that satisfy

$$L_{\mathcal{P}^{\circ}}(-t) = -L_{\mathcal{P}}(t).$$

(*Hint:* Exercise 1.4(i).)

2.2. ♣ Fix positive rational numbers a_1, b_1, a_2, b_2 and let $\mathcal{R}$ be the rectangle with vertices (a_1, b_1), (a_2, b_1), (a_2, b_2), and (a_1, b_2). Compute $L_{\mathcal{R}}(t)$ and $\mathrm{Ehr}_{\mathcal{R}}(z)$.

2.3. Fix positive integers a and b, and let $\mathcal{T}$ be a triangle with vertices $(0, 0)$, $(a, 0)$, and $(0, b)$.

(a) Compute $L_{\mathcal{T}}(t)$ and $\mathrm{Ehr}_{\mathcal{T}}(z)$.

(b) Use (a) to derive the following formula for the greatest common divisor of a and b:

$$\gcd(a, b) = 2 \sum_{k=1}^{b-1} \left\lfloor \frac{ka}{b} \right\rfloor + a + b - ab.$$

(*Hint:* Exercise 1.12.)

2.4. Prove that for two polytopes $\mathcal{P} \subset \mathbb{R}^m$ and $\mathcal{Q} \subset \mathbb{R}^n$,

$$\#\,\big((\mathcal{P} \times \mathcal{Q}) \cap \mathbb{Z}^{m+n}\big) = \#\,(\mathcal{P} \cap \mathbb{Z}^m) \cdot \#\,(\mathcal{Q} \cap \mathbb{Z}^n).$$

Hence, $L_{\mathcal{P} \times \mathcal{Q}}(t) = L_{\mathcal{P}}(t)\, L_{\mathcal{Q}}(t)$.

2.5. Prove that if $\mathcal{F}$ is a face of $\mathcal{P}$ and $\mathcal{G}$ is a face of $\mathcal{F}$, then $\mathcal{G}$ is also a face of $\mathcal{P}$. (That is, the face relation is transitive.)

2.6. ♣ Suppose Δ is a d-simplex with vertices $V = \{\mathbf{v}_1, \mathbf{v}_2, \ldots, \mathbf{v}_{d+1}\}$. Prove that for every nonempty subset $W \subseteq V$, conv W is a face of Δ, and conversely, that every face of Δ is of the form conv W for some $W \subseteq V$. Conclude the following corollaries from this characterization of the faces of a simplex:

(a) A face of a simplex is again a simplex.

(b) The intersection of two faces of a simplex Δ is again a face of Δ.

2.7. ♣ Prove that a rational convex polytope can be described by a system of linear inequalities and equations with *integral* coefficients.

2.8. ♣ Prove the properties (2.3) of the Eulerian numbers for all integers $1 \leq k \leq d$, namely:

(a) $A(d, k) = A(d, d + 1 - k)$
(b) $A(d, k) = (d - k + 1) A(d - 1, k - 1) + k A(d - 1, k)$

(c) $\displaystyle\sum_{k=0}^{d} A(d, k) = d!$

(d) $A(d, k) = \displaystyle\sum_{j=0}^{k} (-1)^j \binom{d+1}{j} (k - j)^d$.

2.9. ♣ Prove (2.6); namely, for $d \geq 0$, $\frac{1}{(1-z)^{d+1}} = \sum_{k \geq 0} \binom{d+k}{d} z^k$.

2.10. ♣ Prove (2.7): For $t, k \in \mathbb{Z}$ and $d \in \mathbb{Z}_{>0}$,

$$(-1)^d \binom{-t + k}{d} = \binom{t + d - 1 - k}{d}.$$

2.11. The **Stirling numbers of the first kind**, $\mathrm{stirl}(n, m)$, are defined through the finite generating function

$$x(x - 1) \cdots (x - n + 1) = \sum_{m=0}^{n} \mathrm{stirl}(n, m) \, x^m.$$

(See also [1, Sequence A008275].) Prove that

$$\frac{1}{d!} \sum_{k=0}^{d} (-1)^{d-k} \, \mathrm{stirl}(d + 1, k + 1) \, t^k = \binom{d + t}{d},$$

the lattice-point enumerator for the standard d-simplex.

2.12. Give a direct proof that the number of solutions $(m_1, m_2, \ldots, m_{d+1}) \in \mathbb{Z}_{\geq 0}^{d+1}$ to $m_1 + m_2 + \cdots + m_{d+1} = t$ equals $\binom{d+t}{d}$. (*Hint:* think of t objects lined up and separated by d walls.)

2.13. Compute $L_\mathcal{P}(t)$, where $\mathcal{P}$ is the regular tetrahedron with vertices $(0, 0, 0), (1, 1, 0), (1, 0, 1), (0, 1, 1)$.

2.14. ♣ Prove that the power series

$$\sum_{k \geq 0} \frac{B_k}{k!} z^k$$

that defines the Bernoulli numbers has radius of convergence 2π.

2.15. ♣ Prove (2.11); namely, $B_d(1 - x) = (-1)^d B_d(x)$.

2.16. ♣ Prove (2.12); namely, $B_d = 0$ for all odd $d \geq 3$.

2.17. Show that for each positive integer n,

$$n\, x^{n-1} = \sum_{k=1}^{n} \binom{n}{k} B_{n-k}(x)\,.$$

This gives us a change of basis for the polynomials of degree $\leq n$, allowing us to represent every polynomial as a sum of Bernoulli polynomials.

2.18. As a complement to the previous exercise, show that we also have a change of basis in the other direction. Namely, we can represent a single Bernoulli polynomial in terms of the monomials as follows:

$$B_n(x) = \sum_{k=0}^{n} \binom{n}{k} B_k\, x^{n-k}\,.$$

2.19. Show that for all positive integers m, n and for all $x \in \mathbb{R}$,

$$\frac{1}{m} \sum_{k=0}^{m-1} B_n\left(x + \frac{k}{m}\right) = m^{-n} B_n(mx)\,.$$

(This is a *Hecke-operator*-type identity, originally found by Joseph Ludwig Raabe in 1851.)

2.20. Show that $B_n(x+1) - B_n(x) = n\, x^{n-1}$.

2.21. An alternative way to define the Bernoulli polynomials is to give elementary properties that uniquely characterize them. Show that the following three properties uniquely determine the Bernoulli polynomials, as defined in the text by (2.8):

(a) $B_0(x) = 1$.
(b) $\frac{dB_n(x)}{dx} = n\, B_{n-1}(x)$, for all $n \geq 1$.
(c) $\int_0^1 B_n(x)\, dx = 0$, for all $n \geq 1$.

2.22. Use (2.13) to derive the following identity, which expresses a Bernoulli polynomial in terms of Eulerian numbers and binomial coefficients:

$$\frac{1}{d}\left(B_d(t+2) - B_d\right) = A\,(d-1, d-1) \binom{t+d-2}{d}$$

$$+ A\,(d-1, d-2) \binom{t+d-3}{d} + \cdots + A\,(d-1, 1) \binom{t}{d}\,.$$

2.23. ♣ Prove Theorem 2.6: $\mathrm{Ehr}_{\mathrm{BiPyr}(\mathcal{Q})}(z) = \frac{1+z}{1-z}\,\mathrm{Ehr}_{\mathcal{Q}}(z)$.

2.24. A **Delannoy path** is a path through lattice points in the plane with steps $(1,0)$, $(0,1)$, and $(1,1)$ (i.e., "east," "north," and "northeast"). Find a recurrence for the number $D(m,n)$ of Delannoy paths from the origin to the point (m,n), and use it to compute the two-variable generating function

$$\sum_{m \geq 0} \sum_{n \geq 0} D(m,n) \, z^m w^n = \frac{1}{1 - z - w - zw} .$$

Conclude from this generating function that the Ehrhart polynomial $L_\Diamond(t)$ of the d-dimensional cross-polytope equals $D(t,d)$. (*Hint:* start with the dth derivative of the generating function of $D(t,d)$ with respect to w.)

2.25. ♣ Let $\mathcal{R}$ be an integral rectangle whose edges are parallel to the coordinate axes, and let $\mathcal{T}$ be a rectangular triangle two of whose edges are parallel to the coordinate axes. Show that Pick's theorem holds for $\mathcal{R}$ and $\mathcal{T}$.

2.26. ♣ Suppose $\mathcal{P}$ is an integral polygon with area A and with B lattice points on its boundary. Show that the area of $t\mathcal{P}$ is At^2, and the number of boundary points on $t\mathcal{P}$ is Bt. (*Hint:* Exercise 1.12.)

2.27. Let $\mathcal{P}$ be the self-intersecting polygon defined by the line segments $[(0,0),(4,2)]$, $[(4,2),(4,0)]$, $[(4,0),(0,2)]$, and $[(0,2),(0,0)]$. Show that Pick's theorem does not hold for $\mathcal{P}$.

2.28. Suppose that $\mathcal{P}$ and $\mathcal{Q}$ are integral polygons, and that $\mathcal{Q}$ lies entirely inside $\mathcal{P}$. Then the area bounded by the boundaries of $\mathcal{P}$ and $\mathcal{Q}$, denoted by $\mathcal{P} - \mathcal{Q}$, is a "doubly connected polygon." Find and prove the analogue of Pick's theorem for $\mathcal{P} - \mathcal{Q}$. Generalize your formula to a polygon with n "holes" (instead of one).

2.29. Show that every convex integral polygon with more than four vertices must have an interior lattice point.

2.30. Consider the rhombus

$$\mathcal{R} = \{(x,y) : a|x| + b|y| \leq ab\} ,$$

where a and b are fixed positive integers. Find a formula for $L_\mathcal{R}(t)$.

2.31. We define the nth **Farey sequence** to be the sequence, in order from smallest to largest, of all the rational numbers $\frac{a}{b}$ in the interval $[0,1]$ such that a and b are coprime and $b \leq n$. For instance, the sixth Farey sequence is $\frac{0}{1}, \frac{1}{6}, \frac{1}{5}, \frac{1}{4}, \frac{1}{3}, \frac{2}{5}, \frac{1}{2}, \frac{3}{5}, \frac{2}{3}, \frac{3}{4}, \frac{4}{5}, \frac{5}{6}, \frac{1}{1}$.

(a) For two consecutive fractions $\frac{a}{b}$ and $\frac{c}{d}$ in a Farey sequence, prove that $bc - ad = 1$.

(b) For three consecutive fractions $\frac{a}{b}$, $\frac{c}{d}$, and $\frac{e}{f}$ in a Farey sequence, show that $\frac{c}{d} = \frac{a+e}{b+f}$.

2.32. Let $\lceil x \rceil$ denote the smallest integer greater than or equal to x. Prove that for all positive integers a and b,

$$a + (-1)^b \sum_{m=0}^{a} (-1)^{\lceil \frac{bm}{a} \rceil} \equiv b + (-1)^a \sum_{n=0}^{b} (-1)^{\lceil \frac{an}{b} \rceil} \bmod 4 \,.$$

(*Hint:* This is a variation of Exercise 1.6. One way to obtain this identity is by counting lattice points in a certain triangle, keeping track only of the parity.)

2.33. ♣ Verify (2.19).

2.34. ♣ Compute the partial fraction coefficients (2.22).

2.35. ♣ Let a, b be positive integers. Show that

$$\frac{1}{1 - z^{ab}} = -\frac{\xi_a^k}{ab} \left(z - \xi_a^k \right)^{-1} + \frac{ab - 1}{2ab} + \text{terms with positive powers of } \left(z - \xi_a^k \right).$$

2.36. ♣ Let $\mathcal{T}$ be given by (2.16), and let $c = \gcd(e, f)$. Prove that

$$
\begin{aligned}
L_{\mathcal{T}}(t) = {} & \frac{1}{2ef} \left(tr - u - v \right)^2 + \frac{1}{2} \left(tr - u - v \right) \left(\frac{1}{e} + \frac{1}{f} + \frac{1}{ef} \right) \\
& + \frac{1}{4} \left(1 + \frac{1}{e} + \frac{1}{f} \right) + \frac{1}{12} \left(\frac{e}{f} + \frac{f}{e} + \frac{1}{ef} \right) \\
& + \left(\frac{1}{2e} + \frac{1}{2f} - \frac{u + v - tr}{ef} \right) \sum_{k=1}^{c-1} \frac{\xi_c^{-ktr}}{1 - \xi_c^k} - \frac{1}{ef} \sum_{k=1}^{c-1} \frac{\xi_c^{k(-tr+1)}}{(1 - \xi_c^k)^2} \\
& + \frac{1}{e} \sum_{\substack{j=1 \\ \frac{e}{c} \nmid j}}^{e-1} \frac{\xi_e^{j(v-tr)}}{\left(1 - \xi_e^{jf} \right) \left(1 - \xi_e^j \right)} + \frac{1}{f} \sum_{\substack{l=1 \\ \frac{f}{c} \nmid l}}^{f-1} \frac{\xi_f^{l(u-tr)}}{\left(1 - \xi_f^{le} \right) \left(1 - \xi_f^l \right)} \,.
\end{aligned}
$$

2.37. Let $\mathcal{P}$ be a rational polygon, and let d be the least common multiple of the denominators of the vertices of $\mathcal{P}$. Prove directly (using Exercise 2.36) that the period of $L_{\mathcal{P}}$ divides d.

2.38. ♣ Finish the calculation in Example 2.12, that is, compute

$$\text{const } \frac{1}{(1 - z_1 z_2) \left(1 - z_1^2 z_2 \right) (1 - z_1) (1 - z_2) \, z_1^{3t} z_2^{2t}} \,.$$

2.39. Compute the vector partition function of the quadrilateral given in Example 2.12, that is, compute the counting function

$$f(v_1, v_2) := \# \left\{ \mathbf{x} \in \mathbb{Z}_{\geq 0}^4 : \begin{pmatrix} 1 & 2 & 1 & 0 \\ 1 & 1 & 0 & 1 \end{pmatrix} \mathbf{x} = \begin{pmatrix} v_1 \\ v_2 \end{pmatrix} \right\}$$

for $v_1, v_2 \in \mathbb{Z}$. (This function depends on the relationship between v_1 and v_2.)

2.40. Search on the Internet for the program `polymake` [118]. You can download it for free. Experiment.

Open Problems

2.41. Choose $d + 1$ of the 2^d vertices of the unit d-cube $\square$, and let Δ be the simplex defined by their convex hull.

(a) Which choice of vertices maximizes $\operatorname{vol} \Delta$?
(b) What is the maximum volume of such a Δ?

2.42. Find classes of integral d-polytopes $(\mathcal{P}_d)_{d \geq 1}$ for which $L_{\mathcal{P}_d}(t)$ is symmetric in d and t. (The standard simplices Δ and the cross-polytopes $\diamond$ form two such classes.)

2.43. We mentioned already in the notes that all the roots of the polynomials $L_\diamond$ have real part $-\frac{1}{2}$; see [75, 148]. Find other classes of polytopes whose lattice-point enumerator exhibits such special behavior.

Chapter 3
Counting Lattice Points in Polytopes: The Ehrhart Theory

Ubi materia, ibi geometria.

Johannes Kepler (1571–1630)

Fig. 3.1 Self-portrait of
Eugène Ehrhart.

Given the profusion of examples that gave rise to the polynomial behavior of the integer-point counting function $L_{\mathcal{P}}(t)$ for special polytopes $\mathcal{P}$, we now ask whether there is a general structure theorem. As the ideas unfold, the reader is invited to look back at Chapters 1 and 2 as appetizers and indeed as special cases of the theorems developed below.

3.1 Triangulations

Because most of the proofs that follow work like a charm for a simplex, we first dissect a polytope into simplices. This dissection is captured by the following definition.

A **triangulation** of a convex d-polytope $\mathcal{P}$ is a finite collection T of d-simplices with the following properties:

- $\mathcal{P} = \bigcup_{\Delta \in T} \Delta$.
- For every $\Delta_1, \Delta_2 \in T$, $\Delta_1 \cap \Delta_2$ is a face common to both Δ_1 and Δ_2.

Figure 3.2 exhibits two triangulations of the 3-cube. We say that $\mathcal{P}$ can be **triangulated using no new vertices** if there exists a triangulation T such that the vertices of every $\Delta \in T$ are vertices of $\mathcal{P}$.

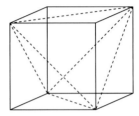

Fig. 3.2 Two (very different) triangulations of the 3-cube.

Theorem 3.1 (Existence of triangulations). *Every convex polytope can be triangulated using no new vertices.*

This theorem seems intuitively obvious, but it is not entirely trivial to prove. We will introduce a specific type of triangulation for a given polytope $\mathcal{P}$, in the hope of achieving a maximum level of concreteness. Namely, given the polytope $\mathcal{P} \subseteq \mathbb{R}^d$ with vertices $\mathbf{v}_1, \mathbf{v}_2, \ldots, \mathbf{v}_n$, randomly choose $h_1, h_2, \ldots, h_n \in \mathbb{R}$, and define a new polytope

$$\mathcal{Q} := \text{conv}\left\{(\mathbf{v}_1, h_1), (\mathbf{v}_2, h_2), \ldots, (\mathbf{v}_n, h_n)\right\} \subseteq \mathbb{R}^{d+1}. \tag{3.1}$$

The **lower hull** of $\mathcal{Q}$ consists of all points that are "visible from below," that is, all points $(x_1, x_2, \ldots, x_{d+1}) \in \mathcal{Q}$ for which there is no $\epsilon > 0$ such that $(x_1, x_2, \ldots, x_{d+1} - \epsilon) \in \mathcal{Q}$. A **lower face** of $\mathcal{Q}$ is a face of $\mathcal{Q}$ that is in the lower hull. Let $\pi : \mathbb{R}^{d+1} \to \mathbb{R}^d$ be defined through

$$\pi(x_1, x_2, \ldots, x_d, x_{d+1}) := (x_1, x_2, \ldots, x_d),$$

the projection that forgets the last coordinate. Exercise 3.3(a) implies that a lower face $\mathcal{F}$ projects to $\pi(\mathcal{F})$ bijectively; in particular, $\mathcal{F}$ and $\pi(\mathcal{F})$ have the same face structure (we say that they are **combinatorially equivalent**). In a moment, we will show that all lower faces of $\mathcal{Q}$ are simplices, and that their projections form a triangulation of $\mathcal{P}$. A triangulation that can be constructed from "lifting" a polytope in the above fashion is called **regular**. Figure 3.3 shows an example.

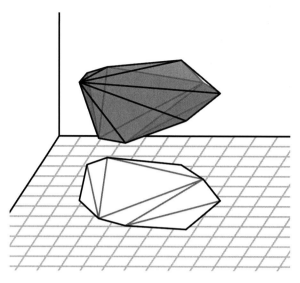

Fig. 3.3 Constructing a regular triangulation of a 9-gon.

Proof of Theorem 3.1. We may assume that $\mathcal{P} = \operatorname{conv}\{\mathbf{v}_1, \mathbf{v}_2, \ldots, \mathbf{v}_n\} \subseteq \mathbb{R}^d$ is full dimensional. Construct a lifted polytope $\mathcal{Q}$ as in (3.1); our first goal is to show that every lower face of $\mathcal{Q}$ is a simplex. To achieve this goal, it suffices to prove that every lower *facet* is a simplex. This, in turn, is equivalent to the statement that every affinely independent set of $d+1$ vertices of $\mathcal{P}$ gets lifted to a set of $d+1$ points in $\mathbb{R}^{d+1}$ that determine a hyperplane $H \subseteq \mathbb{R}^{d+1}$ that does not contain any other lifted vertex of $\mathcal{P}$. After relabeling, we may assume that the chosen $d+1$ affinely independent vertices are $\mathbf{v}_1, \mathbf{v}_2, \ldots, \mathbf{v}_{d+1}$, and so H is given by the equation

$$\det \begin{pmatrix} 1 & 1 & & 1 & 1 \\ \mathbf{v}_1 & \mathbf{v}_2 & \cdots & \mathbf{v}_{d+1} & \mathbf{x} \\ h_1 & h_2 & & h_{d+1} & x_{d+1} \end{pmatrix} = 0 \tag{3.2}$$

in the variables $x_1, x_2, \ldots, x_{d+1}$. If we now specialize $(\mathbf{x}, x_{d+1}) = (\mathbf{v}_j, h_j)$ for some $j > d+1$, the determinant on the left-hand side of (3.2) cannot be zero, since $h_1, h_2, \ldots, h_{d+1}, h_j$ were chosen randomly; in other words, the lifted vertex $(\mathbf{v}_j, h_j)$ is not on H for $j > d+1$. We have thus proved that every lower face of $\mathcal{Q}$ is a simplex. By our earlier bijective argument,

$$T := \{\pi(\mathcal{F}) : \mathcal{F} \text{ is a lower face of } \mathcal{Q}\}$$

consists of simplices contained in $\mathcal{P}$; we claim that T is a triangulation of $\mathcal{P}$.

To prove that $\mathcal{P} = \bigcup_{\Delta \in T} \Delta$, we have "only" to show that $\mathcal{P}^\circ \subseteq \bigcup_{\Delta \in T} \Delta$. So let $\mathbf{x} \in \mathcal{P}^\circ$; our goal is to construct a lower face $\mathcal{F}$ of $\mathcal{Q}$ such that $\mathbf{x} \in \pi(\mathcal{F})$.

Consider the line

$$\mathcal{L} := \{\mathbf{x} + \lambda\, \mathbf{e}_{d+1} : \lambda \in \mathbb{R}\}\,,$$

where $\mathbf{e}_{d+1}$ denotes the $(d+1)^{\text{st}}$ unit vector; that is, $\mathcal{L}$ is a vertical line through $\mathbf{x}$. Note that

$$\mathcal{L} \cap \mathcal{Q}^\circ \neq \varnothing\,, \qquad (3.3)$$

since $\mathbf{x} \in \mathcal{P}^\circ$. So $\mathcal{L} \cap \mathcal{Q}$ is a line segment with endpoints $(\mathbf{x}, y)$ and $(\mathbf{x}, z)$ for some $y < z$. Since $(\mathbf{x}, y)$ is on the boundary of $\mathcal{Q}$, it is contained in some (proper) face $\mathcal{F}$ of $\mathcal{Q}$. Thus we can find a supporting hyperplane

$$H = \{\mathbf{p} \in \mathbb{R}^{d+1} : \mathbf{a} \cdot \mathbf{p} = b\}$$

that defines $\mathcal{F}$ such that $\mathcal{Q} \subseteq \{\mathbf{p} \in \mathbb{R}^{d+1} : \mathbf{a} \cdot \mathbf{p} \geq b\}$. Note that $(\mathbf{x}, z)$ cannot lie on H, since otherwise, the entire line segment from $(\mathbf{x}, y)$ to $(\mathbf{x}, z)$ would be in H, contradicting (3.3). Thus

$$\mathbf{a} \cdot (\mathbf{x}, y) = b \qquad \text{and} \qquad \mathbf{a} \cdot (\mathbf{x}, z) > b\,,$$

which implies $a_{d+1}(z - y) > 0$ and so $a_{d+1} > 0$. But then we can use Exercise 3.3(b) to conclude that $\mathcal{F}$ is a *lower* face of $\mathcal{Q}$. By construction, $\mathbf{x} \in \pi(\mathcal{F})$, which proves $\mathcal{P} \subseteq \bigcup_{\Delta \in T} \Delta$.

That the second defining property for a triangulation holds for T, namely that for every $\Delta_1, \Delta_2 \in T$, $\Delta_1 \cap \Delta_2$ is a face common to both Δ_1 and Δ_2, is left to show in Exercise 3.3(c). □

3.2 Cones

A **pointed cone** $\mathcal{K} \subseteq \mathbb{R}^d$ is a set of the form

$$\mathcal{K} = \{\mathbf{v} + \lambda_1 \mathbf{w}_1 + \lambda_2 \mathbf{w}_2 + \cdots + \lambda_m \mathbf{w}_m : \lambda_1, \lambda_2, \ldots, \lambda_m \geq 0\}\,,$$

where $\mathbf{v}, \mathbf{w}_1, \mathbf{w}_2, \ldots, \mathbf{w}_m \in \mathbb{R}^d$ are such that there exists a hyperplane H for which $H \cap \mathcal{K} = \{\mathbf{v}\}$; that is, $\mathcal{K} \setminus \{\mathbf{v}\}$ lies strictly on one side of H. The vector $\mathbf{v}$ is called the **apex** of $\mathcal{K}$, and the $\mathbf{w}_k$ are the **generators** of $\mathcal{K}$. The cone is **rational** if $\mathbf{v}, \mathbf{w}_1, \mathbf{w}_2, \ldots, \mathbf{w}_m \in \mathbb{Q}^d$, in which case we may choose $\mathbf{w}_1, \mathbf{w}_2, \ldots, \mathbf{w}_m \in \mathbb{Z}^d$ by clearing denominators. The **dimension** of $\mathcal{K}$ is the dimension of the affine space spanned by $\mathcal{K}$; if $\mathcal{K}$ is of dimension d, we call it a d-**cone**. The d-cone $\mathcal{K}$ is **simplicial** if $\mathcal{K}$ has precisely d linearly independent generators.

Just as polytopes have a description as an intersection of half-spaces, so do pointed cones: a rational pointed d-cone is the d-dimensional intersection of finitely many half-spaces of the form

$$\{\mathbf{x} \in \mathbb{R}^d : a_1 x_1 + a_2 x_2 + \cdots + a_d x_d \leq b\}$$

with integral parameters $a_1, a_2, \ldots, a_d, b \in \mathbb{Z}$ such that the corresponding hyperplanes of the form

$$\left\{ \mathbf{x} \in \mathbb{R}^d \, : \, a_1 x_1 + a_2 x_2 + \cdots + a_d x_d = b \right\}$$

meet in exactly one point.

Cones are important for many reasons. The most practical for us is a process called *coning over a polytope*. Given a convex polytope $\mathcal{P} \subset \mathbb{R}^d$ with vertices $\mathbf{v}_1, \mathbf{v}_2, \ldots, \mathbf{v}_n$, we lift these vertices into $\mathbb{R}^{d+1}$ by adding a 1 as their last coordinate. So, let

$$\mathbf{w}_1 = (\mathbf{v}_1, 1), \ \mathbf{w}_2 = (\mathbf{v}_2, 1), \ \ldots, \ \mathbf{w}_n = (\mathbf{v}_n, 1).$$

Now we define the **cone over** $\mathcal{P}$ as

$$\text{cone}(\mathcal{P}) = \{\lambda_1 \mathbf{w}_1 + \lambda_2 \mathbf{w}_2 + \cdots + \lambda_n \mathbf{w}_n \, : \, \lambda_1, \lambda_2, \ldots, \lambda_n \geq 0\} \subset \mathbb{R}^{d+1}.$$

This pointed cone has the origin as apex, and we can recover our original polytope $\mathcal{P}$ (strictly speaking, the translated set $\{(\mathbf{x}, 1) : \mathbf{x} \in \mathcal{P}\}$) by cutting $\text{cone}(\mathcal{P})$ with the hyperplane $x_{d+1} = 1$, as shown in Figure 3.4.

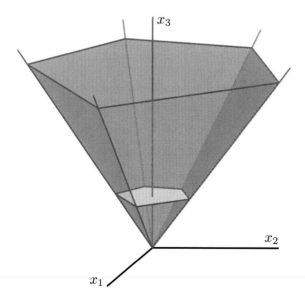

Fig. 3.4 Coning over a polytope.

By analogy with the language of polytopes, we say that the hyperplane $H = \{\mathbf{x} \in \mathbb{R}^d : \mathbf{a} \cdot \mathbf{x} = b\}$ is a **supporting hyperplane** of the pointed d-cone $\mathcal{K}$ if $\mathcal{K}$ lies entirely on one side of H, that is,

$$\mathcal{K} \subseteq \left\{ \mathbf{x} \in \mathbb{R}^d : \mathbf{a} \cdot \mathbf{x} \le b \right\} \qquad \text{or} \qquad \mathcal{K} \subseteq \left\{ \mathbf{x} \in \mathbb{R}^d : \mathbf{a} \cdot \mathbf{x} \ge b \right\}.$$

A **face** of $\mathcal{K}$ is a set of the form $\mathcal{K} \cap H$, where H is a supporting hyperplane of $\mathcal{K}$. The $(d-1)$-dimensional faces are called **facets**, and the 1-dimensional faces **edges**, of $\mathcal{K}$. The apex of $\mathcal{K}$ is its unique 0-dimensional face.

Just as polytopes can be triangulated into simplices, pointed cones can be triangulated into simplicial cones. So, a collection T of simplicial d-cones is a **triangulation** of the d-cone $\mathcal{K}$ if it satisfies the following:

- $\mathcal{K} = \bigcup_{\mathcal{S} \in T} \mathcal{S}$.

- For every $\mathcal{S}_1, \mathcal{S}_2 \in T$, $\mathcal{S}_1 \cap \mathcal{S}_2$ is a face common to both $\mathcal{S}_1$ and $\mathcal{S}_2$.

We say that $\mathcal{K}$ can be **triangulated using no new generators** if there exists a triangulation T such that the generators of every $\mathcal{S} \in T$ are generators of $\mathcal{K}$.

Theorem 3.2. *Every pointed cone can be triangulated into simplicial cones using no new generators.*

Proof. This theorem is really a corollary to Theorem 3.1. Given a pointed d-cone $\mathcal{K}$ with apex $\mathbf{v}$, there exists a hyperplane H that intersects $\mathcal{K}$ only at $\mathbf{v}$. Choose $\mathbf{w} \in \mathcal{K}^\circ$; then

$$\mathcal{P} := (\mathbf{w} - \mathbf{v} + H) \cap \mathcal{K}$$

is a $(d-1)$-polytope whose vertices are determined by the generators of $\mathcal{K}$. (This construction yields a variant of Figure 3.4.) Now triangulate $\mathcal{P}$ using no new vertices. Each simplex Δ_j in this triangulation gives naturally rise to a simplicial cone

$$\mathcal{S}_j := \left\{ \mathbf{v} + \lambda \mathbf{x} : \lambda \ge 0, \ \mathbf{x} \in \Delta_j \right\},$$

and these simplicial cones, by construction, triangulate $\mathcal{K}$. □

3.3 Integer-Point Transforms for Rational Cones

We want to encode the information contained by the lattice points in a set $S \subset \mathbb{R}^d$. It turns out that the following multivariate generating function allows us to do this in an efficient way if S is a rational cone or polytope:

$$\sigma_S(\mathbf{z}) = \sigma_S(z_1, z_2, \ldots, z_d) := \sum_{\mathbf{m} \in S \cap \mathbb{Z}^d} \mathbf{z}^{\mathbf{m}}.$$

The generating function σ_S simply lists all integer points in S in a special form: not as a list of vectors, but as a formal sum of Laurent monomials. We call σ_S the **integer-point transform** of S; the function σ_S also goes by the name *moment generating function* or simply *generating function* of S. The integer-point transform σ_S opens the door to both algebraic and analytic techniques.

Example 3.3. As a warmup example, consider the 1-dimensional cone $\mathcal{K} = [0, \infty)$. Its integer-point transform is our old friend

$$\sigma_{\mathcal{K}}(z) = \sum_{m \in [0,\infty) \cap \mathbb{Z}} z^m = \sum_{m \geq 0} z^m = \frac{1}{1-z}. \qquad \square$$

Example 3.4. Now we consider the 2-dimensional cone

$$\mathcal{K} := \{\lambda_1(1,1) + \lambda_2(-2,3) : \lambda_1, \lambda_2 \geq 0\} \subset \mathbb{R}^2$$

depicted in Figure 3.5. To obtain the integer-point transform $\sigma_{\mathcal{K}}$, we tile $\mathcal{K}$

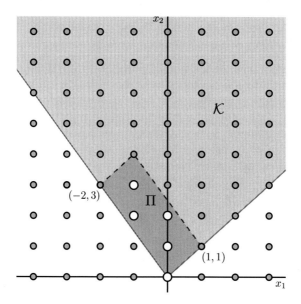

Fig. 3.5 The cone $\mathcal{K}$ from Example 3.4 and its fundamental parallelogram.

by copies of the *fundamental parallelogram*

$$\Pi := \{\lambda_1(1,1) + \lambda_2(-2,3) : 0 \leq \lambda_1, \lambda_2 < 1\} \subset \mathbb{R}^2.$$

More precisely, we translate Π by nonnegative integer linear combinations of the generators $(1,1)$ and $(-2,3)$, and these translates will exactly cover $\mathcal{K}$. How can we list the integer points in $\mathcal{K}$ as Laurent monomials? Let's first list all vertices of the translates of Π. These are nonnegative integer combinations of the generators $(1,1)$ and $(-2,3)$, so we can list them using geometric series:

$$\sum_{\substack{\mathbf{m}=j(1,1)+k(-2,3) \\ j,k \geq 0}} \mathbf{z}^{\mathbf{m}} = \sum_{j \geq 0} \sum_{k \geq 0} \mathbf{z}^{j(1,1)+k(-2,3)} = \frac{1}{\left(1 - z_1 z_2\right)\left(1 - z_1^{-2} z_2^3\right)}.$$

We now use the integer points $(m,n) \in \Pi$ to generate a subset of $\mathbb{Z}^2$ by adding to (m,n) nonnegative linear integer combinations of the generators $(1,1)$ and $(-2,3)$. Namely, we let

$$\mathcal{L}_{(m,n)} := \{(m,n) + j(1,1) + k(-2,3) : j,k \in \mathbb{Z}_{\geq 0}\}.$$

It is immediate that $\mathcal{K} \cap \mathbb{Z}^2$ is the disjoint union of the subsets $\mathcal{L}_{(m,n)}$ as (m,n) ranges over $\Pi \cap \mathbb{Z}^2 = \{(0,0),(0,1),(0,2),(\ 1,2),(\ 1,3)\}$. Hence

$$\sigma_{\mathcal{K}}(\mathbf{z}) = \left(1 + z_2 + z_2^2 + z_1^{-1} z_2^2 + z_1^{-1} z_2^3\right) \sum_{\substack{\mathbf{m}=j(1,1)+k(-2,3) \\ j,k \geq 0}} \mathbf{z}^{\mathbf{m}}$$

$$= \frac{1 + z_2 + z_2^2 + z_1^{-1} z_2^2 + z_1^{-1} z_2^3}{\left(1 - z_1 z_2\right)\left(1 - z_1^{-2} z_2^3\right)}. \qquad \square$$

Similar geometric series suffice to describe integer-point transforms for rational simplicial d-cones. The following result utilizes the geometric series in several directions simultaneously.

Theorem 3.5. *Suppose*

$$\mathcal{K} := \{\lambda_1 \mathbf{w}_1 + \lambda_2 \mathbf{w}_2 + \cdots + \lambda_d \mathbf{w}_d : \lambda_1, \lambda_2, \ldots, \lambda_d \geq 0\}$$

is a simplicial d-cone, where $\mathbf{w}_1, \mathbf{w}_2, \ldots, \mathbf{w}_d \in \mathbb{Z}^d$. *Then for* $\mathbf{v} \in \mathbb{R}^d$, *the integer-point transform* $\sigma_{\mathbf{v}+\mathcal{K}}$ *of the shifted cone* $\mathbf{v} + \mathcal{K}$ *is the rational function*

$$\sigma_{\mathbf{v}+\mathcal{K}}(\mathbf{z}) = \frac{\sigma_{\mathbf{v}+\Pi}(\mathbf{z})}{\left(1 - \mathbf{z}^{\mathbf{w}_1}\right)\left(1 - \mathbf{z}^{\mathbf{w}_2}\right) \cdots \left(1 - \mathbf{z}^{\mathbf{w}_d}\right)},$$

where Π *is the half-open parallelepiped*

$$\Pi := \{\lambda_1 \mathbf{w}_1 + \lambda_2 \mathbf{w}_2 + \cdots + \lambda_d \mathbf{w}_d : 0 \leq \lambda_1, \lambda_2, \ldots, \lambda_d < 1\}.$$

The half-open parallelepiped Π is called the **fundamental parallelepiped** of $\mathcal{K}$.

Proof. In $\sigma_{\mathbf{v}+\mathcal{K}}(\mathbf{z}) = \sum_{\mathbf{m} \in (\mathbf{v}+\mathcal{K}) \cap \mathbb{Z}^d} \mathbf{z}^{\mathbf{m}}$, we list each integer point $\mathbf{m}$ in $\mathbf{v} + \mathcal{K}$ as the Laurent monomial $\mathbf{z}^{\mathbf{m}}$. Such a lattice point can, by definition, be written as

$$\mathbf{m} = \mathbf{v} + \lambda_1 \mathbf{w}_1 + \lambda_2 \mathbf{w}_2 + \cdots + \lambda_d \mathbf{w}_d$$

for some numbers $\lambda_1, \lambda_2, \ldots, \lambda_d \geq 0$. Because the $\mathbf{w}_k$ form a basis of $\mathbb{R}^d$, this representation is unique. Let's write each λ_k in terms of its integer and fractional parts: $\lambda_k = \lfloor \lambda_k \rfloor + \{\lambda_k\}$. So

$$\mathbf{m} = \mathbf{v} + \left(\{\lambda_1\}\,\mathbf{w}_1 + \{\lambda_2\}\,\mathbf{w}_2 + \cdots + \{\lambda_d\}\,\mathbf{w}_d\right) + \lfloor\lambda_1\rfloor\,\mathbf{w}_1 + \lfloor\lambda_2\rfloor\,\mathbf{w}_2 + \cdots + \lfloor\lambda_d\rfloor\,\mathbf{w}_d\,,$$

and we should note that since $0 \leq \{\lambda_k\} < 1$, the vector

$$\mathbf{p} := \mathbf{v} + \{\lambda_1\}\,\mathbf{w}_1 + \{\lambda_2\}\,\mathbf{w}_2 + \cdots + \{\lambda_d\}\,\mathbf{w}_d$$

is in $\mathbf{v} + \Pi$. In fact, $\mathbf{p} \in \mathbb{Z}^d$, since $\mathbf{m}$ and $\lfloor\lambda_k\rfloor\,\mathbf{w}_k$ are all integer vectors. Again, the representation of $\mathbf{p}$ in terms of the $\mathbf{w}_k$ is unique. In summary, we have proved that every $\mathbf{m} \in \mathbf{v} + \mathcal{K} \cap \mathbb{Z}^d$ can be uniquely written as

$$\mathbf{m} = \mathbf{p} + k_1\mathbf{w}_1 + k_2\mathbf{w}_2 + \cdots + k_d\mathbf{w}_d \tag{3.4}$$

for some $\mathbf{p} \in (\mathbf{v} + \Pi) \cap \mathbb{Z}^d$ and some integers $k_1, k_2, \ldots, k_d \geq 0$. On the other hand, let's write the rational function on the right-hand side of the theorem as a product of series:

$$\frac{\sigma_{\mathbf{v}+\Pi}(\mathbf{z})}{(1 - \mathbf{z}^{\mathbf{w}_1}) \cdots (1 - \mathbf{z}^{\mathbf{w}_d})} = \left(\sum_{\mathbf{p} \in (\mathbf{v}+\Pi) \cap \mathbb{Z}^d} \mathbf{z}^{\mathbf{p}}\right)\left(\sum_{k_1 \geq 0} \mathbf{z}^{k_1\mathbf{w}_1}\right) \cdots \left(\sum_{k_d \geq 0} \mathbf{z}^{k_d\mathbf{w}_d}\right).$$

If we multiply everything out, a typical exponent will look exactly like (3.4). $\square$

Our proof contains a crucial geometric idea. Namely, we *tile* the cone $\mathbf{v} + \mathcal{K}$ with translates of $\mathbf{v} + \Pi$ by nonnegative integral combinations of the $\mathbf{w}_k$. It is this tiling that gives rise to the nice integer-point transform in Theorem 3.5. Computationally, we therefore favor cones over polytopes due to our ability to tile a simplicial cone with copies of the fundamental domain, as above. More reasons for favoring cones over polytopes appear in Chapters 10 and 11.

Theorem 3.5 shows that the real complexity of computing the integer-point transform $\sigma_{\mathbf{v}+\mathcal{K}}$ is embedded in the location of the lattice points in the parallelepiped $\mathbf{v} + \Pi$.

By mildly strengthening the hypothesis of Theorem 3.5, we obtain a slightly easier generating function, a result we shall need in Section 3.5 and Chapter 4.

Corollary 3.6. *Suppose*

$$\mathcal{K} := \{\lambda_1\mathbf{w}_1 + \lambda_2\mathbf{w}_2 + \cdots + \lambda_d\mathbf{w}_d : \lambda_1, \lambda_2, \ldots, \lambda_d \geq 0\}$$

is a simplicial d-cone, where $\mathbf{w}_1, \mathbf{w}_2, \ldots, \mathbf{w}_d \in \mathbb{Z}^d$, *and* $\mathbf{v} \in \mathbb{R}^d$, *such that the boundary of* $\mathbf{v} + \mathcal{K}$ *contains no integer point. Then*

$$\sigma_{\mathbf{v}+\mathcal{K}}(\mathbf{z}) = \frac{\sigma_{\mathbf{v}+\Pi}(\mathbf{z})}{(1 - \mathbf{z}^{\mathbf{w}_1})(1 - \mathbf{z}^{\mathbf{w}_2}) \cdots (1 - \mathbf{z}^{\mathbf{w}_d})}\,,$$

where Π *is the open parallelepiped*

$$\Pi := \{\lambda_1 \mathbf{w}_1 + \lambda_2 \mathbf{w}_2 + \cdots + \lambda_d \mathbf{w}_d : 0 < \lambda_1, \lambda_2, \ldots, \lambda_d < 1\}.$$

Proof. The proof of Theorem 3.5 goes through almost verbatim, except that $\mathbf{v} + \Pi$ now has no boundary lattice points, so that there is no harm in choosing Π to be open. $\qquad\square$

Since a general pointed cone can always be triangulated into simplicial cones, the integer-point transforms add up in an inclusion–exclusion manner (note that the intersection of simplicial cones in a triangulation is again a simplicial cone, by Exercise 3.5). Hence we have the following corollary.

Corollary 3.7. *For a pointed cone*

$$\mathcal{K} = \{\mathbf{v} + \lambda_1 \mathbf{w}_1 + \lambda_2 \mathbf{w}_2 + \cdots + \lambda_m \mathbf{w}_m : \lambda_1, \lambda_2, \ldots, \lambda_m > 0\}$$

with $\mathbf{v} \in \mathbb{R}^d$, $\mathbf{w}_1, \mathbf{w}_2, \ldots, \mathbf{w}_m \in \mathbb{Z}^d$, *the integer-point transform* $\sigma_{\mathcal{K}}(\mathbf{z})$ *evaluates to a rational function in the coordinates of* $\mathbf{z}$. $\qquad\square$

Philosophizing some more, one can show that the original infinite sum $\sigma_{\mathcal{K}}(\mathbf{z})$ converges only for $\mathbf{z}$ in a subset of $\mathbb{C}^d$, whereas the rational function that represents $\sigma_{\mathcal{K}}$ gives us its meromorphic continuation. In Chapters 4 and 11, we will make use of this continuation.

3.4 Expanding and Counting Using Ehrhart's Original Approach

Here is *the* fundamental theorem concerning the lattice-point count in an integral convex polytope.

Theorem 3.8 (Ehrhart's theorem). *If* $\mathcal{P}$ *is an integral convex d-polytope, then* $L_{\mathcal{P}}(t)$ *is a polynomial in t of degree d.*

This result is due to Eugène Ehrhart, in whose honor we call $L_{\mathcal{P}}$ the **Ehrhart polynomial** of $\mathcal{P}$. Naturally, there is an extension of Ehrhart's theorem to rational polytopes, which we will discuss in Section 3.8.

Our proof of Ehrhart's theorem uses generating functions of the form $\sum_{t \geq 0} f(t) z^t$, similar in spirit to those discussed at the beginning of Chapter 1. If f is a polynomial, this power series takes on a special form, which we invite the reader to prove (Exercise 3.13):

Lemma 3.9. *If*

$$\sum_{t \geq 0} f(t) z^t = \frac{g(z)}{(1 - z)^{d+1}},$$

then f is a polynomial of degree d if and only if g is a polynomial of degree at most d and $g(1) \neq 0$. $\qquad\square$

The reason we introduced generating functions of the form $\sigma_S(\mathbf{z}) = \sum_{\mathbf{m} \in S \cap \mathbb{Z}^d} \mathbf{z}^{\mathbf{m}}$ in Section 3.3 is that they are extremely handy for lattice-point problems. The connection to lattice points is evident, since we are summing over them. If we are interested in the lattice-point *count*, we simply evaluate σ_S at $\mathbf{z} = (1, 1, \ldots, 1)$:

$$\sigma_S(1, 1, \ldots, 1) = \sum_{\mathbf{m} \in S \cap \mathbb{Z}^d} \mathbf{1}^{\mathbf{m}} = \sum_{\mathbf{m} \in S \cap \mathbb{Z}^d} 1 = \# \left(S \cap \mathbb{Z}^d \right).$$

(Here we denote by $\mathbf{1}$ a vector all of whose components are 1.) Naturally, we should make this evaluation only if S is bounded; Theorem 3.5 already tells us that it is no fun evaluating $\sigma_{\mathcal{K}}(\mathbf{1})$ if $\mathcal{K}$ is a cone.

But the magic of the generating function σ_S does not stop there. To literally take it to the next level, we cone over a convex polytope $\mathcal{P}$. If $\mathcal{P} \subset \mathbb{R}^d$ has the vertices $\mathbf{v}_1, \mathbf{v}_2, \ldots, \mathbf{v}_n \in \mathbb{Z}^d$, recall that we lift these vertices into $\mathbb{R}^{d+1}$ by adding a 1 as their last coordinate. So let

$$\mathbf{w}_1 = (\mathbf{v}_1, 1), \quad \mathbf{w}_2 = (\mathbf{v}_2, 1), \quad \ldots, \quad \mathbf{w}_n = (\mathbf{v}_n, 1).$$

Then

$$\text{cone}(\mathcal{P}) = \{\lambda_1 \mathbf{w}_1 + \lambda_2 \mathbf{w}_2 + \cdots + \lambda_n \mathbf{w}_n : \lambda_1, \lambda_2, \ldots, \lambda_n \geq 0\} \subset \mathbb{R}^{d+1}.$$

Recall that we can recover our original polytope $\mathcal{P}$ by cutting $\text{cone}(\mathcal{P})$ with the hyperplane $x_{d+1} = 1$. We can recover more than just the original polytope in $\text{cone}(\mathcal{P})$: by cutting the cone with the hyperplane $x_{d+1} = 2$, we obtain a copy of $\mathcal{P}$ dilated by a factor of 2. (The reader should meditate on why this cut is a 2-dilate of $\mathcal{P}$.) More generally, we can cut the cone with the hyperplane $x_{d+1} = t$ and obtain $t\mathcal{P}$, as suggested by Figure 3.6.

Now let's form the integer-point transform $\sigma_{\text{cone}(\mathcal{P})}$ of $\text{cone}(\mathcal{P})$. By what we just said, we should look at different powers of z_{d+1}: there is one term (namely, 1), with z_{d+1}^0, corresponding to the origin; the terms with z_{d+1}^1 correspond to lattice points in $\mathcal{P}$ (listed as Laurent monomials in $z_1, z_2, \ldots, z_d$), the terms with z_{d+1}^2 correspond to points in $2\mathcal{P}$, etc. In short,

$$\sigma_{\text{cone}(\mathcal{P})}(z_1, z_2, \ldots, z_{d+1})$$
$$= 1 + \sigma_{\mathcal{P}}(z_1, \ldots, z_d) z_{d+1} + \sigma_{2\mathcal{P}}(z_1, \ldots, z_d) z_{d+1}^2 + \cdots$$
$$= 1 + \sum_{t \geq 1} \sigma_{t\mathcal{P}}(z_1, \ldots, z_d) z_{d+1}^t.$$

Specializing further for enumeration purposes, we recall that $\sigma_{\mathcal{P}}(1, 1, \ldots, 1) = \# \left(\mathcal{P} \cap \mathbb{Z}^d \right)$, and so

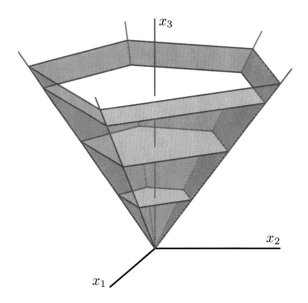

Fig. 3.6 Recovering dilates of $\mathcal{P}$ in $\operatorname{cone}(\mathcal{P})$.

$$\sigma_{\operatorname{cone}(\mathcal{P})} \left(1, 1, \ldots, 1, z_{d+1}\right) = 1 + \sum_{t \geq 1} \sigma_{t\mathcal{P}} \left(1, 1, \ldots, 1\right) z_{d+1}^t$$

$$= 1 + \sum_{t \geq 1} \# \left(t\mathcal{P} \cap \mathbb{Z}^d\right) z_{d+1}^t \, .$$

But by definition, the enumerators on the right-hand side are just evaluations of Ehrhart's counting function, that is, $\sigma_{\operatorname{cone}(\mathcal{P})} \left(1, 1, \ldots, 1, z_{d+1}\right)$ is nothing but the Ehrhart series of $\mathcal{P}$:

Lemma 3.10. $\sigma_{\operatorname{cone}(\mathcal{P})} \left(1, 1, \ldots, 1, z\right) = 1 + \sum_{t \geq 1} L_{\mathcal{P}}(t) \, z^t = \operatorname{Ehr}_{\mathcal{P}}(z) \, .$ $\qquad\square$

With this machinery at hand, we can prove Ehrhart's theorem.

Proof of Theorem 3.8. It suffices to prove the theorem for *simplices*, because we can triangulate any integral polytope into integral simplices, using no new vertices. Note that these simplices will intersect in lower-dimensional integral simplices.

By Lemma 3.9, it suffices to prove that for an integral d-simplex Δ,

$$\operatorname{Ehr}_{\Delta}(z) = 1 + \sum_{t \geq 1} L_{\Delta}(t) \, z^t = \frac{g(z)}{(1 - z)^{d+1}}$$

for some polynomial g of degree at most d with $g(1) \neq 0$. In Lemma 3.10, we showed that the Ehrhart series of Δ equals $\sigma_{\mathrm{cone}(\Delta)}(1, 1, \ldots, 1, z)$, so let's study the integer-point transform attached to $\mathrm{cone}(\Delta)$.

The simplex Δ has $d + 1$ vertices $\mathbf{v}_1, \mathbf{v}_2, \ldots, \mathbf{v}_{d+1}$, and so $\mathrm{cone}(\Delta) \subset \mathbb{R}^{d+1}$ is simplicial, with apex the origin and generators

$$\mathbf{w}_1 = (\mathbf{v}_1, 1), \ \mathbf{w}_2 = (\mathbf{v}_2, 1), \ \ldots, \ \mathbf{w}_{d+1} = (\mathbf{v}_{d+1}, 1) \in \mathbb{Z}^{d+1}.$$

Now we use Theorem 3.5:

$$\sigma_{\mathrm{cone}(\Delta)}(z_1, z_2, \ldots, z_{d+1}) = \frac{\sigma_\Pi(z_1, z_2, \ldots, z_{d+1})}{(1 - \mathbf{z}^{\mathbf{w}_1})(1 - \mathbf{z}^{\mathbf{w}_2}) \cdots (1 - \mathbf{z}^{\mathbf{w}_{d+1}})},$$

where $\Pi = \{\lambda_1 \mathbf{w}_1 + \lambda_2 \mathbf{w}_2 + \cdots + \lambda_{d+1} \mathbf{w}_{d+1} : 0 \leq \lambda_1, \lambda_2, \ldots, \lambda_{d+1} < 1\}$. This parallelepiped is bounded, so the attached generating function σ_Π is a Laurent polynomial in $z_1, z_2, \ldots, z_{d+1}$.

We claim that the z_{d+1}-degree of σ_Π is at most d. In fact, since the x_{d+1}-coordinate of each $\mathbf{w}_k$ is 1, the x_{d+1}-coordinate of a point in Π is $\lambda_1 + \lambda_2 + \cdots + \lambda_{d+1}$ for some $0 \leq \lambda_1, \lambda_2, \ldots, \lambda_{d+1} < 1$. But then $\lambda_1 + \lambda_2 + \cdots + \lambda_{d+1} < d + 1$, so if this sum is an integer, it is at most d, which implies that the z_{d+1}-degree of $\sigma_\Pi(z_1, z_2, \ldots, z_{d+1})$ is at most d. Consequently, $\sigma_\Pi(1, 1, \ldots, 1, z_{d+1})$ is a polynomial in z_{d+1} of degree at most d. The evaluation $\sigma_\Pi(1, 1, 1, \ldots, 1)$ of this polynomial at $z_{d+1} = 1$ is not zero, because $\sigma_\Pi(1, 1, 1, \ldots, 1) = \#(\Pi \cap \mathbb{Z}^{d+1})$, and the origin is a lattice point in Π.

Finally, if we specialize $\mathbf{z}^{\mathbf{w}_k}$ to $z_1 = z_2 = \cdots = z_d = 1$, we obtain z_{d+1}^1, so that

$$\sigma_{\mathrm{cone}(\Delta)}(1, 1, \ldots, 1, z_{d+1}) = \frac{\sigma_\Pi(1, 1, \ldots, 1, z_{d+1})}{(1 - z_{d+1})^{d+1}}.$$

The left-hand side is $\mathrm{Ehr}_\Delta(z_{d+1}) = 1 + \sum_{t \geq 1} L_\Delta(t) z_{d+1}^t$ by Lemma 3.10. $\quad\square$

3.5 The Ehrhart Series of an Integral Polytope

We can actually take our proof of Ehrhart's theorem one step further by studying the polynomial $\sigma_\Pi(1, 1, \ldots, 1, z_{d+1})$. As mentioned above, the coefficient of z_{d+1}^k simply counts the integer points in the parallelepiped Π cut with the hyperplane $x_{d+1} = k$. Let's record this.

Corollary 3.11. *Suppose Δ is an integral d-simplex with vertices $\mathbf{v}_1, \mathbf{v}_2, \ldots, \mathbf{v}_{d+1}$, and let $\mathbf{w}_j = (\mathbf{v}_j, 1)$. Then*

$$\mathrm{Ehr}_\Delta(z) = 1 + \sum_{t \geq 1} L_\Delta(t) z^t = \frac{h_d^* z^d + h_{d-1}^* z^{d-1} + \cdots + h_1^* z + h_0^*}{(1 - z)^{d+1}},$$

where h_k^* equals the number of integer points in

$$\{\lambda_1 \mathbf{w}_1 + \lambda_2 \mathbf{w}_2 + \cdots + \lambda_{d+1} \mathbf{w}_{d+1} : 0 \le \lambda_1, \lambda_2, \ldots, \lambda_{d+1} < 1\}$$

with last coordinate equal to k.   □

This result can actually be used to compute Ehr_Δ, and therefore the Ehrhart polynomial, of an integral simplex Δ in low dimensions very quickly (a fact that the reader may discover in some of the exercises). We remark, however, that things are not as simple for arbitrary integral polytopes. Not only is triangulation a nontrivial task in general, but one would also have to deal with overcounting where simplices of a triangulation meet.

Corollary 3.11 implies that the numerator of the Ehrhart series of an integral simplex has nonnegative coefficients, since they count something. Although it is not known whether the coefficients of the Ehrhart series of a general polytope count something, the nonnegativity property magically survives.

Theorem 3.12 (Stanley's nonnegativity theorem). *Suppose $\mathcal{P}$ is an integral convex d-polytope with Ehrhart series*

$$\mathrm{Ehr}_{\mathcal{P}}(z) = \frac{h_d^* z^d + h_{d-1}^* z^{d-1} + \cdots + h_0^*}{(1-z)^{d+1}}.$$

Then $h_0^*, h_1^*, \ldots, h_d^*$ are nonnegative integers.

We call the numerator $h_{\mathcal{P}}^*(z)$ of the Ehrhart series $\mathrm{Ehr}_{\mathcal{P}}(z)$ the h^***-polynomial** of $\mathcal{P}$.[1] Theorem 3.12 says that it always has nonnegative integer coefficients; we will have (much) more to say about $h_{\mathcal{P}}^*(z)$ in Chapter 10.

Proof. Triangulate $\mathrm{cone}(\mathcal{P}) \subset \mathbb{R}^{d+1}$ into the simplicial cones $\mathcal{K}_1, \mathcal{K}_2, \ldots, \mathcal{K}_m$. Exercise 3.19 ensures that there exists a vector $\mathbf{v} \in \mathbb{R}^{d+1}$ such that

$$\mathrm{cone}(\mathcal{P}) \cap \mathbb{Z}^{d+1} = (\mathbf{v} + \mathrm{cone}(\mathcal{P})) \cap \mathbb{Z}^{d+1}$$

(that is, we neither lose nor gain any lattice points in shifting $\mathrm{cone}(\mathcal{P})$ by $\mathbf{v}$) and neither the facets of $\mathbf{v} + \mathrm{cone}(\mathcal{P})$ nor the translated triangulation hyperplanes contain any lattice points. This implies that every lattice point in $\mathbf{v} + \mathrm{cone}(\mathcal{P})$ belongs to exactly one simplicial cone $\mathbf{v} + \mathcal{K}_j$:

$$\mathrm{cone}(\mathcal{P}) \cap \mathbb{Z}^{d+1} = (\mathbf{v} + \mathrm{cone}(\mathcal{P})) \cap \mathbb{Z}^{d+1} = \bigcup_{j=1}^{m} \left((\mathbf{v} + \mathcal{K}_j) \cap \mathbb{Z}^{d+1} \right), \quad (3.5)$$

and this union is a *disjoint* union. If we translate the last identity into generating-function language, it becomes

---

[1] Other names for $h_{\mathcal{P}}^*(z)$ used in the literature are δ-**vector/polynomial** and **Ehrhart** h-**vector/polynomial** of $\mathcal{P}$.

$$\sigma_{\text{cone}(\mathcal{P})}\left(z_1, z_2, \ldots, z_{d+1}\right) = \sum_{j=1}^{m} \sigma_{\mathbf{v}+\mathcal{K}_j}\left(z_1, z_2, \ldots, z_{d+1}\right).$$

But now we recall that the Ehrhart series of $\mathcal{P}$ is just a special evaluation of $\sigma_{\text{cone}(\mathcal{P})}$ (Lemma 3.10):

$$\text{Ehr}_{\mathcal{P}}(z) = \sigma_{\text{cone}(\mathcal{P})}\left(1, 1, \ldots, 1, z\right) = \sum_{j=1}^{m} \sigma_{\mathbf{v}+\mathcal{K}_j}\left(1, 1, \ldots, 1, z\right). \qquad (3.6)$$

It suffices to show that the rational generating functions $\sigma_{\mathbf{v}+\mathcal{K}_j}\left(1, 1, \ldots, 1, z\right)$ for the simplicial cones $\mathbf{v} + \mathcal{K}_j$ have nonnegative integer numerator. But this fact follows from evaluating the rational function in Corollary 3.6 at $(1, 1, \ldots, 1, z)$.   □

This proof shows a little more: Since the origin is in precisely *one* simplicial cone on the right-hand side of (3.5), we get on the right-hand side of (3.6) precisely *one* term that contributes $1/(1-z)^{d+1}$ to $\text{Ehr}_{\mathcal{P}}$; all other terms contribute to higher powers of the numerator polynomial of $\text{Ehr}_{\mathcal{P}}$. That is, the constant term h_0^* equals 1. The reader might feel that we are chasing our tail at this point, since we assumed from the very beginning that the constant term of the infinite series $\text{Ehr}_{\mathcal{P}}$ is 1, and hence h_0^* has to be 1, as a quick look at the expansion of the rational function representing $\text{Ehr}_{\mathcal{P}}$ shows. The above argument shows merely that this convention is geometrically sound. Let's record this:

Lemma 3.13. *Suppose $\mathcal{P}$ is an integral convex d-polytope with Ehrhart series*

$$\text{Ehr}_{\mathcal{P}}(z) = \frac{h_d^* z^d + h_{d-1}^* z^{d-1} + \cdots + h_0^*}{(1-z)^{d+1}}.$$

Then $h_0^* = 1$.   □

For a general integral polytope $\mathcal{P}$, the reader has probably already discovered how to extract the Ehrhart polynomial of $\mathcal{P}$ from its Ehrhart series:

Lemma 3.14. *Suppose $\mathcal{P}$ is an integral convex d-polytope with Ehrhart series*

$$\text{Ehr}_{\mathcal{P}}(z) = 1 + \sum_{t \geq 1} L_{\mathcal{P}}(t)\, z^t = \frac{h_d^* z^d + h_{d-1}^* z^{d-1} + \cdots + h_1^* z + 1}{(1-z)^{d+1}}.$$

Then

$$L_{\mathcal{P}}(t) = \binom{t+d}{d} + h_1^* \binom{t+d-1}{d} + \cdots + h_{d-1}^* \binom{t+1}{d} + h_d^* \binom{t}{d}.$$

Proof. Expand into a binomial series:

$$\mathrm{Ehr}_{\mathcal{P}}(z) = \frac{h_d^* z^d + h_{d-1}^* z^{d-1} + \cdots + h_1^* z + 1}{(1-z)^{d+1}}$$

$$= \left(h_d^* z^d + h_{d-1}^* z^{d-1} + \cdots + h_1^* z + 1\right) \sum_{t \geq 0} \binom{t+d}{d} z^t$$

$$= h_d^* \sum_{t \geq 0} \binom{t+d}{d} z^{t+d} + h_{d-1}^* \sum_{t \geq 0} \binom{t+d}{d} z^{t+d-1} + \cdots$$

$$+ h_1^* \sum_{t \geq 0} \binom{t+d}{d} z^{t+1} + \sum_{t \geq 0} \binom{t+d}{d} z^t$$

$$= h_d^* \sum_{t \geq d} \binom{t}{d} z^t + h_{d-1}^* \sum_{t > d-1} \binom{t+1}{d} z^t + \cdots$$

$$+ h_1^* \sum_{t \geq 1} \binom{t+d-1}{d} z^t + \sum_{t \geq 0} \binom{t+d}{d} z^t.$$

In all infinite sums on the right-hand side, we can begin the index t with 0 without changing the sums, by the definition (2.1) of the binomial coefficient. Hence

$$\mathrm{Ehr}_{\mathcal{P}}(z)$$
$$= \sum_{t \geq 0} \left(h_d^* \binom{t}{d} + h_{d-1}^* \binom{t+1}{d} + \cdots + h_1^* \binom{t+d-1}{d} + \binom{t+d}{d} \right) z^t.$$

$\square$

The representation of the polynomial $L_{\mathcal{P}}(t)$ in terms of the coefficients of $\mathrm{Ehr}_{\mathcal{P}}$ can be interpreted as the Ehrhart polynomial expressed in the basis $\binom{t}{d}, \binom{t+1}{d}, \ldots, \binom{t+d}{d}$ (see Exercise 3.14). This representation is very useful, as the following results show.

Corollary 3.15. *If $\mathcal{P}$ is an integral convex d-polytope, then the constant term of the Ehrhart polynomial $L_{\mathcal{P}}$ is 1.*

Proof. Use the expansion of Lemma 3.14. The constant term is

$$L_{\mathcal{P}}(0) = \binom{d}{d} + h_1^* \binom{d-1}{d} + \cdots + h_{d-1}^* \binom{1}{d} + h_d^* \binom{0}{d} = \binom{d}{d} = 1. \qquad \square$$

This proof is exciting, because it marks the first instance in which we extend the domain of an Ehrhart polynomial beyond the positive integers, for which the lattice-point enumerator was initially defined. More precisely, Ehrhart's theorem (Theorem 3.8) implies that the counting function

$$L_{\mathcal{P}}(t) = \#\left(t\mathcal{P} \cap \mathbb{Z}^d\right),$$

originally defined for positive integers t, can be extended to all real or even complex arguments t (as a polynomial). A natural question arises: are there nice *interpretations* of $L_{\mathcal{P}}(t)$ for arguments t that are not positive integers? Corollary 3.15 gives such an interpretation for $t = 0$. In Chapter 4, we will give interpretations of $L_{\mathcal{P}}(t)$ for *negative* integers t.

Corollary 3.16. *Suppose $\mathcal{P}$ is an integral convex d-polytope with Ehrhart series*

$$\mathrm{Ehr}_{\mathcal{P}}(z) = \frac{h_d^* \, z^d + h_{d-1}^* \, z^{d-1} + \cdots + h_1^* \, z + 1}{(1-z)^{d+1}}.$$

Then $h_1^* = L_{\mathcal{P}}(1) - d - 1 = \# \left(\mathcal{P} \cap \mathbb{Z}^d \right) - d - 1$.

Proof. Use the expansion of Lemma 3.14 with $t = 1$:

$$L_{\mathcal{P}}(1) = \binom{d+1}{d} + h_1^* \binom{d}{d} + \cdots + h_{d-1}^* \binom{2}{d} + h_d^* \binom{1}{d} = d + 1 + h_1^*. \qquad \square$$

The proof of Corollary 3.16 suggests that there are also formulas for $h_2^*, h_3^*, \ldots$ in terms of the evaluations $L_{\mathcal{P}}(1), L_{\mathcal{P}}(2), \ldots$, and we invite the reader to find them (Exercise 3.15).

A final corollary to Theorem 3.12 and Lemma 3.14 states how large the denominators of the Ehrhart coefficients can be:

Corollary 3.17. *Suppose $\mathcal{P}$ is an integral polytope with Ehrhart polynomial $L_{\mathcal{P}}(t) = c_d \, t^d + c_{d-1} \, t^{d-1} + \cdots + c_1 \, t + 1$. Then all coefficients satisfy $d! \, c_k \in \mathbb{Z}$.*

Proof. By Lemma 3.14,

$$L_{\mathcal{P}}(t) = \binom{t+d}{d} + h_1^* \binom{t+d-1}{d} + \cdots + h_{d-1}^* \binom{t+1}{d} + h_d^* \binom{t}{d},$$

where the h_k^* are integers by Corollary 3.11 and our proof of Theorem 3.12. Hence multiplying out this expression yields a polynomial in t whose coefficients can be written as rational numbers with denominator $d!$. $\qquad \square$

We finish this section with a general result that gives relations between negative integer roots of a polynomial and its generating function. This theorem will become handy in Chapter 4, in which we find an interpretation for the evaluation of an Ehrhart polynomial at negative integers.

Theorem 3.18. *Suppose p is a degree-d polynomial with the rational generating function*

$$\sum_{t \geq 0} p(t) \, z^t = \frac{h_d \, z^d + h_{d-1} \, z^{d-1} + \cdots + h_1 \, z + h_0}{(1-z)^{d+1}}.$$

Then $h_d = h_{d-1} = \cdots = h_{k+1} = 0$ and $h_k \neq 0$ if and only if $p(-1) = p(-2) = \cdots = p\left(-(d-k)\right) = 0$ and $p\left(-(d-k+1)\right) \neq 0$.

Proof. Suppose $h_d = h_{d-1} = \cdots = h_{k+1} = 0$ and $h_k \neq 0$. Then the proof of Lemma 3.14 gives

$$p(t) = h_0 \binom{t+d}{d} + \cdots + h_{k-1} \binom{t+d-k+1}{d} + h_k \binom{t+d-k}{d}.$$

All the binomial coefficients are zero for $t = -1, -2, \ldots, -d+k$, so those are roots of p. On the other hand, all binomial coefficients but the last one are zero for $t = -d+k-1$, and since $h_k \neq 0$, $-d+k-1$ is not a root of p.

Conversely, suppose $p(-1) = p(-2) = \cdots = p(-(d-k)) = 0$ and $p(-(d-k+1)) \neq 0$. The first root -1 of p gives

$$0 = p(-1) = h_0 \binom{d-1}{d} + h_1 \binom{d-2}{d} + \cdots + h_{d-1} \binom{0}{d} + h_d \binom{-1}{d} = h_d \binom{-1}{d},$$

so we must have $h_d = 0$. The next root -2 forces $h_{d-1} = 0$, and so on, up to the root $-d+k$, which forces $h_{k+1} = 0$. It remains to show that $h_k \neq 0$. But if h_k were zero, then by a similar line of reasoning as in the first part of the proof, we would have $p(-d+k-1) = 0$, a contradiction. □

3.6 From the Discrete to the Continuous Volume of a Polytope

Given a geometric object $S \subset \mathbb{R}^d$, its **volume**, defined by the integral $\mathrm{vol}\, S := \int_S d\mathbf{x}$, is one of the fundamental data of S. By the definition of the integral, say in the Riemannian sense, we can think of computing $\mathrm{vol}\, S$ by approximating S with d-dimensional boxes that get smaller and smaller. To be precise, if we take the boxes with side length $\frac{1}{t}$, then they each have volume $\frac{1}{t^d}$. We might further think of the boxes as filling out the space between grid points in the lattice $\left(\frac{1}{t}\mathbb{Z}\right)^d$. This means that volume computation can be approximated by counting boxes, or equivalently, lattice points in $\left(\frac{1}{t}\mathbb{Z}\right)^d$:

$$\mathrm{vol}\, S = \lim_{t \to \infty} \frac{1}{t^d} \cdot \# \left(S \cap \left(\frac{1}{t}\mathbb{Z}\right)^d \right).$$

It is a short step to counting integer points in dilates of S, because

$$\# \left(S \cap \left(\frac{1}{t}\mathbb{Z}\right)^d \right) = \# \left(tS \cap \mathbb{Z}^d \right).$$

Let's summarize:

Lemma 3.19. *Suppose $S \subset \mathbb{R}^d$ is d-dimensional. Then*

$$\operatorname{vol} S = \lim_{t \to \infty} \frac{1}{t^d} \cdot \# \left(tS \cap \mathbb{Z}^d \right). \qquad \square$$

We emphasize here that S is d-dimensional, because otherwise (since S could be lower-dimensional although living in d-space), by our current definition, $\operatorname{vol} S = 0$. We will extend our volume definition in Chapter 5 to give nonzero *relative* volume to objects that are not full-dimensional.

Part of the magic of Ehrhart's theorem lies in the fact that for an integral d-polytope $\mathcal{P}$, we do not have to take a limit to compute $\operatorname{vol} \mathcal{P}$; we need to compute "only" the $d + 1$ coefficients of a polynomial.

Corollary 3.20. *Suppose $\mathcal{P} \subset \mathbb{R}^d$ is an integral convex d-polytope with Ehrhart polynomial $c_d\, t^d + c_{d-1}\, t^{d-1} + \cdots + c_1\, t + 1$. Then $c_d = \operatorname{vol} \mathcal{P}$.*

Proof. By Lemma 3.19,

$$\operatorname{vol} \mathcal{P} = \lim_{t \to \infty} \frac{c_d\, t^d + c_{d-1}\, t^{d-1} + \cdots + c_1\, t + 1}{t^d} = c_d. \qquad \square$$

On the one hand, this should not come as a surprise, because the number of integer points in some object should grow roughly like the volume of the object as we make it bigger and bigger. On the other hand, the fact that we can compute the volume as one term of a polynomial should be very surprising: the polynomial is a counting function and as such is something *discrete*, yet by computing it (and its leading term), we derive some *continuous* data. Even more, we can—at least theoretically—compute this continuous datum (the volume) of the object by calculating a few values of the polynomial and then interpolating; this can be described as a completely discrete operation!

We finish this section by showing how to retrieve the continuous volume of an integral polytope from its Ehrhart series.

Corollary 3.21. *Suppose $\mathcal{P} \subset \mathbb{R}^d$ is an integral convex d-polytope, and*

$$\operatorname{Ehr}_{\mathcal{P}}(z) = \frac{h_d^*\, z^d + h_{d-1}^*\, z^{d-1} + \cdots + h_1^*\, z + 1}{(1 - z)^{d+1}}.$$

Then $\operatorname{vol} \mathcal{P} = \dfrac{1}{d!} \left(h_d^* + h_{d-1}^* + \cdots + h_1^* + 1 \right).$

Proof. Use the expansion of Lemma 3.14. The leading coefficient is

$$\frac{1}{d!} \left(h_d^* + h_{d-1}^* + \cdots + h_1^* + 1 \right). \qquad \square$$

3.7 Interpolation

We now use the polynomial behavior of the discrete volume $L_{\mathcal{P}}$ of an integral polytope $\mathcal{P}$ to compute the continuous volume $\mathrm{vol}\,\mathcal{P}$ and the discrete volume $L_{\mathcal{P}}$ from finite data.

Two points uniquely determine a line. There exists a unique quadratic whose graph passes through any three given noncollinear points. More generally, a degree-d polynomial p is determined by $d+1$ points $(x, p(x)) \in \mathbb{R}^2$ in general position. Namely, evaluating $p(x) = c_d x^d + c_{d-1} x^{d-1} + \cdots + c_0$ at distinct inputs $x_1, x_2, \ldots, x_{d+1}$ gives

$$\begin{pmatrix} p(x_1) \\ p(x_2) \\ \vdots \\ p(x_{d+1}) \end{pmatrix} = \mathbf{V} \begin{pmatrix} c_d \\ c_{d-1} \\ \vdots \\ c_0 \end{pmatrix}, \tag{3.7}$$

where

$$\mathbf{V} = \begin{pmatrix} x_1^d & x_1^{d-1} & \cdots & x_1 & 1 \\ x_2^d & x_2^{d-1} & \cdots & x_2 & 1 \\ \vdots & \vdots & & \vdots & \vdots \\ x_{d+1}^d & x_{d+1}^{d-1} & \cdots & x_{d+1} & 1 \end{pmatrix},$$

so that

$$\begin{pmatrix} c_d \\ c_{d-1} \\ \vdots \\ c_0 \end{pmatrix} = \mathbf{V}^{-1} \begin{pmatrix} p(x_1) \\ p(x_2) \\ \vdots \\ p(x_{d+1}) \end{pmatrix}. \tag{3.8}$$

(Exercise 3.21 makes sure that $\mathbf{V}$ is invertible.) The identity (3.8) gives the famous *Lagrange interpolation formula*.

This gives us an efficient way to compute $L_{\mathcal{P}}$, at least when $\dim \mathcal{P}$ is not too large. The continuous volume of $\mathcal{P}$ will follow instantly, since it is the leading coefficient c_d of $L_{\mathcal{P}}$. In the case of an Ehrhart polynomial $L_{\mathcal{P}}$, we know that $L_{\mathcal{P}}(0) = 1$, so that (3.7) simplifies to

$$\begin{pmatrix} L_{\mathcal{P}}(x_1) - 1 \\ L_{\mathcal{P}}(x_2) - 1 \\ \vdots \\ L_{\mathcal{P}}(x_d) - 1 \end{pmatrix} = \begin{pmatrix} x_1^d & x_1^{d-1} & \cdots & x_1 \\ x_2^d & x_2^{d-1} & \cdots & x_2 \\ \vdots & \vdots & & \vdots \\ x_d^d & x_d^{d-1} & \cdots & x_d \end{pmatrix} \begin{pmatrix} c_d \\ c_{d-1} \\ \vdots \\ c_1 \end{pmatrix}.$$

Example 3.22 (Reeve's tetrahedron). Let $\mathcal{T}_h$ be the tetrahedron with vertices $(0, 0, 0)$, $(1, 0, 0)$, $(0, 1, 0)$, and $(1, 1, h)$, where h is a positive integer (see Figure 3.7).

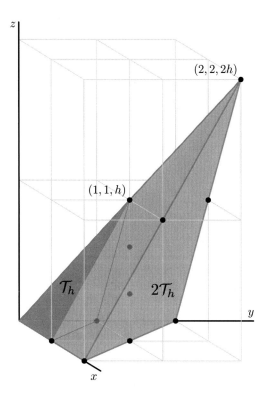

Fig. 3.7 Reeve's tetrahedron $\mathcal{T}_h$ (and $2\mathcal{T}_h$).

To interpolate the Ehrhart polynomial $L_{\mathcal{T}_h}(t)$ from its values at various points, we use Figure 3.7 to deduce the following:

$$4 = L_{\mathcal{T}_h}(1) = \text{vol}\,(\mathcal{T}_h) + c_2 + c_1 + 1\,,$$
$$h + 9 = L_{\mathcal{T}_h}(2) = \text{vol}\,(\mathcal{T}_h) \cdot 2^3 + c_2 \cdot 2^2 + c_1 \cdot 2 + 1\,.$$

Using the volume formula for a pyramid, we know that

$$\text{vol}\,(\mathcal{T}_h) = \frac{1}{3}(\text{base area})(\text{height}) = \frac{h}{6}\,.$$

Thus $h + 1 = h + 2c_2 - 1$, which gives us $c_2 = 1$ and $c_1 = 2 - \frac{h}{6}$. Therefore,

$$L_{\mathcal{T}_h}(t) = \frac{h}{6}\,t^3 + t^2 + \left(2 - \frac{h}{6}\right)t + 1\,. \qquad \square$$

3.8 Rational Polytopes and Ehrhart Quasipolynomials

We do not have to change much to study lattice-point enumeration for *rational* polytopes, and most of this section will consist of exercises for the reader. The structural result paralleling Theorem 3.8 is as follows.

Theorem 3.23 (Ehrhart's theorem for rational polytopes). *If $\mathcal{P}$ is a rational convex d-polytope, then $L_{\mathcal{P}}(t)$ is a quasipolynomial in t of degree d. Its period divides the least common multiple of the denominators of the coordinates of the vertices of $\mathcal{P}$.*

We will call the least common multiple of the denominators of the co-ordinates of the vertices of $\mathcal{P}$ the **denominator** of $\mathcal{P}$. Theorem 3.23, also due to Ehrhart, extends Theorem 3.8, because the denominator of an integral polytope $\mathcal{P}$ is 1. Exercises 3.26 and 3.27 show that the word *divides* in Theorem 3.23 is far from being replaceable by *equals*.

We set out along the path toward a proof of Theorem 3.23 by stating the analogue of Lemma 3.9 for quasipolynomials (see Exercise 3.24):

Lemma 3.24. *If*

$$\sum_{t \geq 0} f(t) \, z^t = \frac{g(z)}{h(z)} \, ,$$

then f is a quasipolynomial of degree d with period dividing p if and only if g and h are polynomials such that $\deg(g) < \deg(h)$, all roots of h are p^{th} roots of unity of multiplicity at most $d + 1$, and there is a root of multiplicity equal to $d + 1$ (all of this assuming that $\frac{g}{h}$ has been reduced to lowest terms). $\qquad\square$

Our goal is now evident: we will prove that if $\mathcal{P}$ is a rational convex d-polytope with denominator p, then

$$\mathrm{Ehr}_{\mathcal{P}}(z) = 1 + \sum_{t \geq 1} L_{\mathcal{P}}(t) \, z^t = \frac{g(z)}{\left(1 - z^p\right)^{d+1}} \, ,$$

for some polynomial g of degree less than $p(d + 1)$. As in Section 3.4, we will have to prove this only for the case of a rational *simplex*. So suppose the d-simplex Δ has vertices $\mathbf{v}_1, \mathbf{v}_2, \ldots, \mathbf{v}_{d+1} \in \mathbb{Q}^d$, and the denominator of Δ is p. Again we will cone over Δ: let

$$\mathbf{w}_1 = (\mathbf{v}_1, 1) \, , \mathbf{w}_2 = (\mathbf{v}_2, 1) \, , \ldots, \mathbf{w}_{d+1} = (\mathbf{v}_{d+1}, 1) \, ;$$

then

$$\mathrm{cone}(\Delta) = \{\lambda_1 \mathbf{w}_1 + \lambda_2 \mathbf{w}_2 + \cdots + \lambda_{d+1} \mathbf{w}_{d+1} : \lambda_1, \lambda_2, \ldots, \lambda_{d+1} \geq 0\} \subset \mathbb{R}^{d+1}.$$

To be able to use Theorem 3.5, we first have to ensure that we have a description of $\mathrm{cone}(\Delta)$ with integral generators. But since the denominator of

Δ is p, we can replace each generator $\mathbf{w}_k$ by $p\mathbf{w}_k \in \mathbb{Z}^{d+1}$, and we are ready to apply Theorem 3.5. From this point, the proof of Theorem 3.23 proceeds exactly like that of Theorem 3.8, and we invite the reader to finish it up (Exercise 3.25).

Although the proofs of Theorem 3.23 and Theorem 3.8 are almost identical, the arithmetic structure of Ehrhart quasipolynomials is much more subtle and less well known than that of Ehrhart polynomials.

3.9 Reflections on the Coin-Exchange Problem and the Gallery of Chapter 2

At this point, we encourage the reader to look back at the first two chapters in light of the basic Ehrhart-theory results. Theorem 1.5 and its higher-dimensional analogue give a special set of Ehrhart quasipolynomials. On the other hand, in Chapter 2 we encountered many integral polytopes. Ehrhart's theorem (Theorem 3.8) explains why their lattice-point enumeration functions were all polynomials.

Notes

1. Triangulations of polytopes and manifolds are an active source of research with many interesting open problems; for further study, we highly recommend [98].

2. Eugène Ehrhart laid the foundation for the central theme of this book in the 1960s, starting with the proof of Theorem 3.8 in 1962 [109]. The proof we give here follows Ehrhart's original lines of thought. An interesting fact is that he did his most beautiful work as a teacher at a *lycée* in Strasbourg (France), receiving his doctorate at age 60 on the urging of some colleagues.

3. Given any d linearly independent vectors $\mathbf{w}_1, \mathbf{w}_2, \ldots, \mathbf{w}_d \in \mathbb{R}^d$, the **lattice** generated by them is the set of all integer linear combinations of $\mathbf{w}_1, \mathbf{w}_2, \ldots, \mathbf{w}_d$. Alternatively, one can define a lattice as a discrete subgroup of $\mathbb{R}^d$, and these two notions can be shown to be equivalent. One might wonder whether replacing the lattice $\mathbb{Z}^d$ by an arbitrary lattice $\mathcal{L}$ throughout the statements of the theorems—requiring now that the vertices of a polytope be in $\mathcal{L}$—gives us any different results. The fact that the theorems of this chapter remain the same follows from the observation that every d-dimensional lattice can be mapped to $\mathbb{Z}^d$ by an invertible linear transformation.

4. Richard Stanley developed much of the theory of Ehrhart (quasi)polyno-mials, initially from a commutative-algebra point of view. Theorem 3.12 is due to him [226]. The proof we give here appeared in [46]. We will give several extensions of Theorem 3.12 in Chapter 10.

5. The tetrahedron $\mathcal{T}_h$ of Example 3.22 was used by John Reeve to show that Pick's theorem does not hold in $\mathbb{R}^3$ (see Exercise 3.23) [202]. Incidentally, the formula for $L_{\mathcal{T}_h}$ also proves that the coefficients of an Ehrhart polynomial are not always positive.

6. There are several interesting questions (some of which are still open) regarding the periods of Ehrhart quasipolynomials. Some particularly nice examples about what can happen with periods were given by Tyrrell McAllister and Kevin Woods [169].

7. Most of the results remain true if we replace *convex polytope* by *polytopal complex*, which is a finite union of polytopes. One important exception is Corollary 3.15: the constant term of an "Ehrhart polynomial" of an integral polytopal complex C is the *Euler characteristic* of C.

8. The reader might wonder why we do not discuss polytopes with *irrational* vertices. The answer is simple: nobody has yet found a theory that would parallel the results in this chapter, even in dimension 2. One notable exception is [19], in which irrational extensions of Brion's theorem are given; we will study the rational case of Brion's theorem in Chapter 11. On the other hand, there has been recent activity to study Ehrhart quasipolynomials of rational polytopes with *real* dilation parameters [14, 164]. Ehrhart theory has been extended to functions other than strict lattice-point counting; one instance is described in Chapter 13.

Exercises

3.1. Show that the number of triangulations of a polygon $\mathcal{P}$ that uses only the n vertices of $\mathcal{P}$ equals $\frac{1}{n-1}\binom{2n-4}{n-2}$.[2]

3.2. To each permutation $\pi \in S_d$ on d elements, we associate the simplex

$$\Delta_\pi := \operatorname{conv}\left\{\mathbf{0}, \mathbf{e}_{\pi(1)}, \mathbf{e}_{\pi(1)} + \mathbf{e}_{\pi(2)}, \ldots, \mathbf{e}_{\pi(1)} + \mathbf{e}_{\pi(2)} + \cdots + \mathbf{e}_{\pi(d)}\right\},$$

where $\mathbf{e}_1, \mathbf{e}_2, \ldots, \mathbf{e}_d$ denote the unit vectors in $\mathbb{R}^d$.

(a) Prove that $\{\Delta_\pi : \pi \in S_d\}$ is a triangulation of the unit d-cube $[0, 1]^d$.

---

[2] The integer $\frac{1}{n+1}\binom{2n}{n}$ is known as the nth **Catalan number**.

(b) Prove that all Δ_π are congruent to each other, that is, each one can be obtained from any other by reflections, translations, and rotations.

(c) Show that for all $\pi \in S_d$, $L_{\Delta_\pi}(t) = \binom{d+t}{d}$.

3.3. ♣ Given the polytope $\mathcal{P} \subseteq \mathbb{R}^d$, construct a lifted polytope $\mathcal{Q}$ as in (3.1).

(a) Let $\mathcal{F}$ be a lower face of $\mathcal{Q}$, defined by the supporting hyperplane $H = \{\mathbf{x} \in \mathbb{R}^{d+1} : \mathbf{a} \cdot \mathbf{x} = b\}$. Show that $a_{d+1} \neq 0$. (This means that the hyperplane is not horizontal.)

(b) Let $\mathcal{F}$ be a face of $\mathcal{Q}$, defined by the supporting hyperplane $H = \{\mathbf{x} \in \mathbb{R}^{d+1} : \mathbf{a} \cdot \mathbf{x} = b\}$ such that $\mathcal{Q} \subseteq \{\mathbf{x} \in \mathbb{R}^{d+1} : \mathbf{a} \cdot \mathbf{x} \geq b\}$. Prove that $\mathcal{F}$ is a lower face if and only if $a_{d+1} > 0$.

(c) Let $\mathcal{F}_1$ and $\mathcal{F}_2$ be lower faces of $\mathcal{Q}$. Prove that $\pi(\mathcal{F}_1) \cap \pi(\mathcal{F}_2)$ is a face common to both $\pi(\mathcal{F}_1)$ and $\pi(\mathcal{F}_2)$.

3.4. Prove that every triangulation of a polygon $\mathcal{P}$ that uses only the vertices of $\mathcal{P}$ is regular. Give an example of a (3-dimensional) polytope that has a nonregular triangulation. (*Hint:* Begin by proving that the triangulation of the 2-dimensional point configuration pictured in Figure 3.8 is not regular.[3])

Fig. 3.8 A nonregular triangulation of a point configuration.

3.5. ♣ Suppose T is a triangulation of a pointed cone. Prove that the intersection of two simplicial cones in T is again a simplicial cone.

3.6. Find the generating function $\sigma_\mathcal{K}(\mathbf{z})$ for the following cones:

(a) $\mathcal{K} = \{\lambda_1(0,1) + \lambda_2(1,0) : \lambda_1, \lambda_2 \geq 0\}$;

(b) $\mathcal{K} = \{\lambda_1(0,1) + \lambda_2(1,1) : \lambda_1, \lambda_2 \geq 0\}$;

(c) $\mathcal{K} = \{(3,4) + \lambda_1(0,1) + \lambda_2(2,1) : \lambda_1, \lambda_2 \geq 0\}$.

3.7. Fix two relatively prime positive integers a and b, and let

$$\mathcal{K} = \{\lambda_1(0,1) + \lambda_2(a,b) : \lambda_1, \lambda_2 \geq 0\}.$$

Show that

$$\sigma_\mathcal{K}(z_1, z_2) = \frac{z_1 + \sum_{k=1}^{a-1} z_1^k z_2^{\lfloor \frac{kb}{a} \rfloor + 1}}{(1 - z_2)\left(1 - z_1^a z_2^b\right)}.$$

---

[3] A **triangulation** of a point configuration $\{\mathbf{x}_1, \mathbf{x}_2, \ldots, \mathbf{x}_n\}$ is a triangulation of conv $\{\mathbf{x}_1, \mathbf{x}_2, \ldots, \mathbf{x}_n\}$ using $\mathbf{x}_1, \mathbf{x}_2, \ldots, \mathbf{x}_n$ as vertices.

3.8. ♣ Let $S \subseteq \mathbb{R}^m$ and $T \subseteq \mathbb{R}^n$. Show that $\sigma_{S \times T}(z_1, z_2, \ldots, z_{m+n}) = \sigma_S(z_1, z_2, \ldots, z_m) \sigma_T(z_{m+1}, z_{m+2}, \ldots, z_{m+n})$.

3.9. ♣ Let $\mathcal{K}$ be a rational d-cone, and let $\mathbf{m} \in \mathbb{Z}^d$. Show that $\sigma_{\mathbf{m}+\mathcal{K}}(\mathbf{z}) = \mathbf{z}^\mathbf{m} \sigma_\mathcal{K}(\mathbf{z})$.

3.10. ♣ For a set $S \subset \mathbb{R}^d$, let $-S := \{-\mathbf{x} : \mathbf{x} \in S\}$. Prove that

$$\sigma_{-S}(z_1, z_2, \ldots, z_d) = \sigma_S\left(\frac{1}{z_1}, \frac{1}{z_2}, \ldots, \frac{1}{z_d}\right).$$

3.11. Given a pointed cone $\mathcal{K} \subset \mathbb{R}^d$ with apex at the origin, let $S := \mathcal{K} \cap \mathbb{Z}^d$. Show that if $\mathbf{x}, \mathbf{y} \in S$, then $\mathbf{x} + \mathbf{y} \in S$. (In algebraic terms, S is a *semigroup*, since $\mathbf{0} \in S$ and associativity of the addition in S follows trivially from associativity in $\mathbb{R}^d$.)

3.12. Suppose $\mathcal{K} \subset \mathbb{R}^d$ is a pointed cone with apex $\mathbf{v}$, H is a supporting hyperplane of $\mathcal{K}$ with $H \cap \mathcal{K} = \{\mathbf{v}\}$, and $\mathbf{w} \in \mathbb{R}^d$ is such that $\mathbf{v} + \mathbf{w} \in \mathcal{K}$. Show that $(H + \mathbf{w}) \cap \mathcal{K}$ is a convex polytope.

3.13. ♣ Prove Lemma 3.9: If

$$\sum_{t \geq 0} f(t) z^t = \frac{g(z)}{(1-z)^{d+1}},$$

then f is a polynomial of degree d if and only if g is a polynomial of degree at most d and $g(1) \neq 0$.

3.14. Prove that $\binom{x+n}{n}, \binom{x+n-1}{n}, \ldots, \binom{x}{n}$ is a basis for the vector space Pol_n of polynomials (in the variable x) of degree less than or equal to n.

3.15. For a polynomial $p(t) = c_d t^d + c_{d-1} t^{d-1} + \cdots + c_0$, let $H_p(z)$ be defined by

$$\sum_{t \geq 0} p(t) z^t = \frac{H_p(z)}{(1-z)^{d+1}}.$$

Consider the map $\phi_d : \text{Pol}_d \to \text{Pol}_d$ given by $p \mapsto H_p$.

(a) Show that ϕ_d is a linear transformation.
(b) Compute the matrix describing ϕ_d for $d = 0, 1, 2, \ldots$.
(c) Deduce formulas for $h_2^*, h_3^*, \ldots$, similar to the one in Corollary 3.16.

3.16. Compute the Ehrhart polynomials and the Ehrhart series of the simplices with the following vertices:

(a) $(0,0,0)$, $(1,0,0)$, $(0,2,0)$, and $(0,0,3)$;
(b) $(0,0,0,0)$, $(1,0,0,0)$, $(0,2,0,0)$, $(0,0,3,0)$, and $(0,0,0,4)$.

3.17. Define the **hypersimplex** $\Delta(d, k)$ as the convex hull of

$$\{\mathbf{e}_{j_1} + \mathbf{e}_{j_2} + \cdots + \mathbf{e}_{j_k} : 1 \le j_1 < j_2 < \cdots < j_k \le d\},$$

where $\mathbf{e}_1, \mathbf{e}_2, \ldots, \mathbf{e}_d$ are the standard basis vectors in $\mathbb{R}^d$. For example, $\Delta(d, 1)$ and $\Delta(d, d-1)$ are regular $(d-1)$-simplices. Compute the Ehrhart polynomial and the Ehrhart series of $\Delta(d, k)$.

3.18. ♣ Suppose H is the hyperplane given by

$$H = \{\mathbf{x} \in \mathbb{R}^d : a_1 x_1 + a_2 x_2 + \cdots + a_d x_d = 0\}$$

for some $a_1, a_2, \ldots, a_d \in \mathbb{Z}$, which we may assume to have no common factor. Prove that there exists $\mathbf{v} \in \mathbb{Z}^d$ such that $\bigcup_{n \in \mathbb{Z}} ((n\mathbf{v} + H) \cap \mathbb{Z}^d) = \mathbb{Z}^d$. (This implies, in particular, that the points in $\mathbb{Z}^d \setminus H$ are all at least some minimal distance from H; this minimal distance is essentially given by the dot product of $\mathbf{v}$ with $(a_1, a_2, \ldots, a_d)$.)

3.19. ♣ A hyperplane H is **rational** if it can be written in the form

$$H = \{\mathbf{x} \in \mathbb{R}^d : a_1 x_1 + a_2 x_2 + \cdots + a_d x_d = b\}$$

for some $a_1, a_2, \ldots, a_d, b \in \mathbb{Z}$. A **hyperplane arrangement** in $\mathbb{R}^d$ is a finite set of hyperplanes in $\mathbb{R}^d$. Prove that a rational hyperplane arrangement $\mathcal{H}$ can be translated so that no hyperplane contains any integer points.

3.20. The conclusion of the previous exercise can be strengthened: Prove that a rational hyperplane arrangement $\mathcal{H}$ can be translated such that no hyperplane contains any *rational* points.

3.21. ♣ Show that

$$\det \begin{pmatrix} x_1^d & x_1^{d-1} & \cdots & x_1 & 1 \\ x_2^d & x_2^{d-1} & \cdots & x_2 & 1 \\ \vdots & \vdots & & \vdots & \vdots \\ x_{d+1}^d & x_{d+1}^{d-1} & \cdots & x_{d+1} & 1 \end{pmatrix} = \prod_{1 \le j < k \le d+1} (x_j - x_k).$$

Conclude that for distinct numbers $x_1, x_2, \ldots, x_{d+1}$, the matrix

$$\mathbf{V} = \begin{pmatrix} x_1^d & x_1^{d-1} & \cdots & x_1 & 1 \\ x_2^d & x_2^{d-1} & \cdots & x_2 & 1 \\ \vdots & \vdots & & \vdots & \vdots \\ x_{d+1}^d & x_{d+1}^{d-1} & \cdots & x_{d+1} & 1 \end{pmatrix}$$

is not singular. (**V** is known as the **Vandermonde matrix**.)

3.22. Let $\mathcal{P}$ be an integral d-polytope. Show that

$$\operatorname{vol}\mathcal{P} = \frac{1}{d!}\left((-1)^d + \sum_{k=1}^{d}\binom{d}{k}(-1)^{d-k}L_{\mathcal{P}}(k)\right).$$

3.23. As in Example 3.22, let $\mathcal{T}_n$ be the tetrahedron with vertices $(0,0,0)$, $(1,0,0)$, $(0,1,0)$, and $(1,1,n)$, where n is a positive integer. Show that the volume of $\mathcal{T}_n$ is unbounded as $n \to \infty$, yet for all n, $\mathcal{T}_n$ has no interior and precisely four boundary lattice points. This example proves that Pick's theorem does not hold for a 3-dimensional integral polytope $\mathcal{P}$, in the sense that there is no linear relationship among $\operatorname{vol}\mathcal{P}$, $L_{\mathcal{P}}(1)$, and $L_{\mathcal{P}^\circ}(1)$.

3.24. ♣ Prove Lemma 3.24: If

$$\sum_{t\geq 0} f(t)\, z^t = \frac{g(z)}{h(z)},$$

then f is a quasipolynomial of degree d with period dividing p if and only if g and h are polynomials such that $\deg(g) < \deg(h)$, all roots of h are p^{th} roots of unity of multiplicity at most $d+1$, and there is a root of multiplicity equal to $d+1$ (all of this assuming that $\frac{g}{h}$ has been reduced to lowest terms).

3.25. ♣ Provide the details for the proof of Theorem 3.23: If $\mathcal{P}$ is a rational convex d-polytope, then $L_{\mathcal{P}}(t)$ is a quasipolynomial in t of degree d. Its period divides the least common multiple of the denominators of the coordinates of the vertices of $\mathcal{P}$.

3.26. Let $\mathcal{T}$ be the rational triangle with vertices $(0,0)$, $\left(1, \frac{p-1}{p}\right)$, and $(p,0)$, where p is a fixed integer ≥ 2. Show that $L_{\mathcal{T}}(t) = \frac{p-1}{2}t^2 + \frac{p+1}{2}t + 1$; in particular, $L_{\mathcal{T}}$ is a *polynomial*.

3.27. Prove that for every $d \geq 2$ and $p \geq 1$, there exists a d-polytope $\mathcal{P}$ whose Ehrhart quasipolynomial is a *polynomial* (i.e., it has period 1), yet $\mathcal{P}$ has a vertex with denominator p.

3.28. Prove that the period of the Ehrhart quasipolynomial of a 1-dimensional polytope is *always* equal to the least common multiple of the denominators of its vertices.

3.29. Let $\mathcal{T}$ be the triangle with vertices $\left(-\frac{1}{2}, -\frac{1}{2}\right)$, $\left(\frac{1}{2}, -\frac{1}{2}\right)$, and $\left(0, \frac{3}{2}\right)$. Show that $L_{\mathcal{T}}(t) = t^2 + c(t)\, t + 1$, where

$$c(t) = \begin{cases} 1 & \text{if } t \text{ is even,} \\ 0 & \text{if } t \text{ is odd.} \end{cases}$$

(This shows that the periods of the "coefficients" of an Ehrhart quasipolynomial do not necessarily increase with decreasing power.) Find the Ehrhart series of $\mathcal{T}$.

3.30. Prove the following extension of Theorem 3.12: Suppose $\mathcal{P}$ is a rational d-polytope with denominator p. Then

$$\text{Ehr}_{\mathcal{P}}(z) = \frac{h^*(z)}{(1 - z^p)^{d+1}},$$

where $h^*(z)$ is a polynomial with nonnegative integral coefficients.

3.31. Find and prove a statement that extends Lemma 3.14 to Ehrhart quasipolynomials.

3.32. ♣ Prove the following extension of Corollary 3.15 to rational polytopes. Namely, the Ehrhart quasipolynomial $L_{\mathcal{P}}$ of the rational convex polytope $\mathcal{P} \subset \mathbb{R}^d$ satisfies $L_{\mathcal{P}}(0) = 1$.

3.33. Prove the following analogue of Corollary 3.17 for rational polytopes: Suppose $\mathcal{P}$ is a rational polytope with Ehrhart quasipolynomial $L_{\mathcal{P}}(t) = c_d(t)\, t^d + c_{d-1}(t)\, t^{d-1} + \cdots + c_1(t)\, t + c_0(t)$. Then for all $t \in \mathbb{Z}$ and $0 \le k \le d$, we have $d!\, c_k(t) \in \mathbb{Z}$.

3.34. ♣ Prove that Corollary 3.20 also holds for rational polytopes: Suppose $\mathcal{P} \subset \mathbb{R}^d$ is a rational convex d-polytope with Ehrhart quasipolynomial $c_d(t)\, t^d + c_{d-1}(t)\, t^{d-1} + \cdots + c_0(t)$. Then $c_d(t)$ equals the volume of $\mathcal{P}$; in particular, $c_d(t)$ is constant.

3.35. Suppose $\mathcal{P}$ is a rational convex polytope. Show that as rational functions,

$$\text{Ehr}_{2\mathcal{P}}(z) = \tfrac{1}{2}\left(\text{Ehr}_{\mathcal{P}}\left(\sqrt{z}\right) + \text{Ehr}_{\mathcal{P}}\left(-\sqrt{z}\right) \right).$$

3.36. Suppose f and g are quasipolynomials. Prove that the **convolution**

$$F(t) := \sum_{s=0}^{t} f(s)\, g(t - s)$$

is also a quasipolynomial. What can you say about the degree and the period of F, given the degrees and periods of f and g?

3.37. Given two positive integers a and b, let

$$f(t) := \begin{cases} 1 & \text{if } a|t, \\ 0 & \text{otherwise,} \end{cases} \qquad \text{and} \qquad g(t) := \begin{cases} 1 & \text{if } b|t, \\ 0 & \text{otherwise.} \end{cases}$$

Form the convolution of f and g. What function is it?

3.38. Suppose $\mathcal{P} \subset \mathbb{R}^m$ and $\mathcal{Q} \subset \mathbb{R}^n$ are rational polytopes. Prove that the convolution of $L_{\mathcal{P}}$ and $L_{\mathcal{Q}}$ equals the Ehrhart quasipolynomial of the polytope given by the convex hull of $\mathcal{P} \times \{\mathbf{0}_n\} \times \{0\}$ and $\{\mathbf{0}_m\} \times \mathcal{Q} \times \{1\}$. Here $\mathbf{0}_d$ denotes the origin in $\mathbb{R}^d$.

3.39. We define the **unimodular group** $\mathrm{SL}_d(\mathbb{Z})$ as the set of all $d \times d$ matrices with integer entries and determinant ± 1.

(a) Show that each element of $\mathrm{SL}_d(\mathbb{Z})$ acts on the integer lattice $\mathbb{Z}^d$ in a bijective fashion.
(b) Let $\mathcal{P}$ be an integral polytope, and let $\mathcal{Q} := \mathbf{A}(\mathcal{P})$, where $\mathbf{A} \in \mathrm{SL}_d(\mathbb{Z})$, so that $\mathcal{P}$ and $\mathcal{Q}$ are unimodular images of each other. Show that $L_{\mathcal{P}}(t) = L_{\mathcal{Q}}(t)$.

3.40. Search on the Internet for the program LattE: Lattice-Point Enumeration [94, 154]. You can download it for free. Experiment.

Open Problems

3.41. How many triangulations are there for a given polytope?

3.42. What is the minimal number of simplices needed to triangulate the unit d-cube? (These numbers are known for $d \le 7$.)

3.43. Classify the polynomials of a fixed degree d that are Ehrhart polynomials. This has been done completely for $d = 2$ [214] and is partially known for $d = 3$ and 4 [36, Section 3]. (See also Open Problem 10.21.)

3.44. Study the roots of Ehrhart polynomials of integral polytopes in a fixed dimension [36, 55, 62, 131]. Study the roots of the numerators of Ehrhart series.

3.45. Come up with an efficient algorithm that computes the period of an Ehrhart quasipolynomial. (See [253], in which Woods describes an efficient algorithm that checks whether a given integer is a period of an Ehrhart quasipolynomial.)

3.46. Suppose $\mathcal{P}$ and $\mathcal{Q}$ are integral polytopes with the same Ehrhart polynomial, that is, $L_{\mathcal{P}}(t) = L_{\mathcal{Q}}(t)$. What additional conditions on $\mathcal{P}$ and $\mathcal{Q}$ do we need to ensure that integer translates of $\mathcal{P}$ and $\mathcal{Q}$ are unimodular images of each other? That is, when is $\mathcal{Q} = \mathbf{A}(\mathcal{P}) + \mathbf{m}$ for some $\mathbf{A} \in \mathrm{SL}_d(\mathbb{Z})$ and $\mathbf{m} \in \mathbb{Z}^d$?

3.47. Find an "Ehrhart theory" for irrational polytopes.

Chapter 4
Reciprocity

In mathematics you don't understand things. You just get used to them.

John von Neumann (1903–1957)

While Exercise 1.4(i) gave us the elementary identity

$$\left\lfloor \frac{t-1}{a} \right\rfloor = -\left\lfloor \frac{-t}{a} \right\rfloor - 1 \qquad (4.1)$$

for $t \in \mathbb{Z}$ and $a \in \mathbb{Z}_{>0}$, this fact is a special instance of a more general theme. Namely, (4.1) marks the simplest (1-dimensional) case of a *reciprocity theorem* that is central to Ehrhart theory. Let $\mathcal{I} := \left[0, \frac{1}{a}\right] \subset \mathbb{R}$, a rational 1-polytope

Fig. 4.1 Lattice points in $t\mathcal{I}$.

(see Figure 4.1). Its discrete volume is (recalling Exercise 1.3)

$$L_{\mathcal{I}}(t) = \left\lfloor \frac{t}{a} \right\rfloor + 1.$$

The lattice-point enumerator for the *interior* $\mathcal{I}^\circ = \left(0, \frac{1}{a}\right)$, on the other hand, is

$$L_{\mathcal{I}^\circ}(t) = \left\lceil \frac{t-1}{a} \right\rceil \qquad (4.2)$$

(see Exercise 4.1). We remind the reader that both $L_{\mathcal{I}}(t)$ and $L_{\mathcal{I}^\circ}(t)$, by definition, are functions in the positive-integer variable t. However, we can see in the above formulas that they are given by quasipolynomials, and (4.1) gives an algebraic relation between these two quasipolynomials:

$$L_{\mathcal{I}^\circ}(t) = -L_{\mathcal{I}}(-t).$$

This chapter is devoted to proving that a similar identity holds for rational polytopes in any dimension:

Theorem 4.1 (Ehrhart–Macdonald reciprocity). *Suppose $\mathcal{P}$ is a convex rational polytope. Then the evaluation of the quasipolynomial $L_{\mathcal{P}}$ at negative integers yields*

$$L_{\mathcal{P}}(-t) = (-1)^{\dim \mathcal{P}} L_{\mathcal{P}^\circ}(t).$$

This theorem belongs to a class of famous *reciprocity theorems*. A common theme in combinatorics is to begin with an interesting object P, and

1. define a counting function $f(t)$ attached to P that makes physical sense for positive integer values of t;
2. recognize the function f as a polynomial in t;
3. substitute negative integral values of t into the counting function f, and recognize $f(-t)$ as a counting function of a new mathematical object Q.

For us, P is a polytope, and Q is its interior.

4.1 Generating Functions for Somewhat Irrational Cones

Our approach to proving Theorem 4.1 parallels the steps of Chapter 3: we deduce Theorem 4.1 from an identity for rational cones. We begin with a reciprocity theorem for simplicial cones.

Theorem 4.2. *Fix linearly independent vectors $\mathbf{w}_1, \mathbf{w}_2, \ldots, \mathbf{w}_d \in \mathbb{Z}^d$, and let $\mathcal{K} = \{\lambda_1 \mathbf{w}_1 + \lambda_2 \mathbf{w}_2 + \cdots + \lambda_d \mathbf{w}_d : \lambda_1, \ldots, \lambda_d \geq 0\}$, the simplicial cone generated by the $\mathbf{w}_j$. Then for those $\mathbf{v} \in \mathbb{R}^d$ for which the boundary of the shifted simplicial cone $\mathbf{v} + \mathcal{K}$ contains no integer point,*

$$\sigma_{\mathbf{v}+\mathcal{K}}\left(\frac{1}{z_1}, \frac{1}{z_2}, \ldots, \frac{1}{z_d}\right) = (-1)^d \sigma_{-\mathbf{v}+\mathcal{K}}(z_1, z_2, \ldots, z_d).$$

Remark. This theorem is meaningless on the level of formal power series; however, the identity holds at the level of rational functions. We will establish that $\sigma_{\mathbf{v}+\mathcal{K}}$ is a rational function in the process of proving the theorem.

Proof. As in the proofs of Theorem 3.5 and Corollary 3.6,

$$\sigma_{\mathbf{v}+\mathcal{K}}(\mathbf{z}) = \frac{\sigma_{\mathbf{v}+\Pi}(\mathbf{z})}{(1-\mathbf{z}^{\mathbf{w}_1})(1-\mathbf{z}^{\mathbf{w}_2})\cdots(1-\mathbf{z}^{\mathbf{w}_d})},$$

where Π is the open parallelepiped

$$\Pi = \{\lambda_1\mathbf{w}_1 + \lambda_2\mathbf{w}_2 + \cdots + \lambda_d\mathbf{w}_d : 0 < \lambda_1, \lambda_2, \ldots, \lambda_d < 1\}. \qquad (4.3)$$

This also proves that $\sigma_{\mathbf{v}+\mathcal{K}}$ is a rational function. Note that by assumption, $\mathbf{v}+\Pi$ contains no integer points on its boundary. Naturally,

$$\sigma_{-\mathbf{v}+\mathcal{K}}(\mathbf{z}) = \frac{\sigma_{-\mathbf{v}+\Pi}(\mathbf{z})}{(1-\mathbf{z}^{\mathbf{w}_1})(1-\mathbf{z}^{\mathbf{w}_2})\cdots(1-\mathbf{z}^{\mathbf{w}_d})},$$

so we need to relate the parallelepipeds $\mathbf{v}+\Pi$ and $-\mathbf{v}+\Pi$. This relation is illustrated in Figure 4.2 for the case $d = 2$; the identity for general d is (see Exercise 4.2)

$$\mathbf{v}+\Pi = -(-\mathbf{v}+\Pi) + \mathbf{w}_1 + \mathbf{w}_2 + \cdots + \mathbf{w}_d. \qquad (4.4)$$

Now we translate the geometry of (4.4) into generating functions:

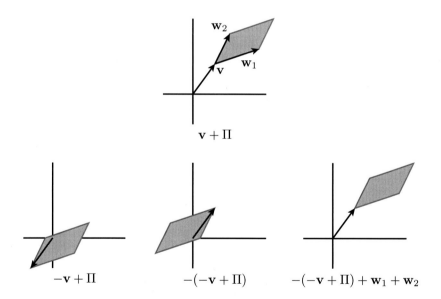

Fig. 4.2 From $-\mathbf{v}+\Pi$ to $\mathbf{v}+\Pi$.

$$\sigma_{\mathbf{v}+\Pi}(\mathbf{z}) = \sigma_{-(-\mathbf{v}+\Pi)}(\mathbf{z}) \, \mathbf{z}^{\mathbf{w}_1} \mathbf{z}^{\mathbf{w}_2} \cdots \mathbf{z}^{\mathbf{w}_d}$$

$$= \sigma_{-\mathbf{v}+\Pi} \left(\frac{1}{z_1}, \frac{1}{z_2}, \ldots, \frac{1}{z_d} \right) \mathbf{z}^{\mathbf{w}_1} \mathbf{z}^{\mathbf{w}_2} \cdots \mathbf{z}^{\mathbf{w}_d}$$

(the last equation follows from Exercise 3.10). Let's abbreviate the vector $\left(\frac{1}{z_1}, \frac{1}{z_2}, \ldots, \frac{1}{z_d} \right)$ by $\frac{1}{\mathbf{z}}$. Then the last identity is equivalent to

$$\sigma_{\mathbf{v}+\Pi} \left(\frac{1}{\mathbf{z}} \right) = \sigma_{-\mathbf{v}+\Pi}(\mathbf{z}) \, \mathbf{z}^{-\mathbf{w}_1} \mathbf{z}^{-\mathbf{w}_2} \cdots \mathbf{z}^{-\mathbf{w}_d},$$

whence

$$\sigma_{\mathbf{v}+\mathcal{K}} \left(\frac{1}{\mathbf{z}} \right) = \frac{\sigma_{\mathbf{v}+\Pi} \left(\frac{1}{\mathbf{z}} \right)}{\left(1 - \mathbf{z}^{-\mathbf{w}_1} \right) \left(1 - \mathbf{z}^{-\mathbf{w}_2} \right) \cdots \left(1 - \mathbf{z}^{-\mathbf{w}_d} \right)}$$

$$= \frac{\sigma_{-\mathbf{v}+\Pi}(\mathbf{z}) \, \mathbf{z}^{-\mathbf{w}_1} \mathbf{z}^{-\mathbf{w}_2} \cdots \mathbf{z}^{-\mathbf{w}_d}}{\left(1 - \mathbf{z}^{-\mathbf{w}_1} \right) \left(1 - \mathbf{z}^{-\mathbf{w}_2} \right) \cdots \left(1 - \mathbf{z}^{-\mathbf{w}_d} \right)}$$

$$= \frac{\sigma_{-\mathbf{v}+\Pi}(\mathbf{z})}{\left(\mathbf{z}^{\mathbf{w}_1} - 1 \right) \left(\mathbf{z}^{\mathbf{w}_2} - 1 \right) \cdots \left(\mathbf{z}^{\mathbf{w}_d} - 1 \right)}$$

$$= (-1)^d \frac{\sigma_{-\mathbf{v}+\Pi}(\mathbf{z})}{\left(1 - \mathbf{z}^{\mathbf{w}_1} \right) \left(1 - \mathbf{z}^{\mathbf{w}_2} \right) \cdots \left(1 - \mathbf{z}^{\mathbf{w}_d} \right)}$$

$$= (-1)^d \, \sigma_{-\mathbf{v}+\mathcal{K}}(\mathbf{z}) . \qquad \square$$

4.2 Stanley's Reciprocity Theorem for Rational Cones

For the general reciprocity theorem for cones, we patch the simplicial cones of a triangulation together, in a manner very similar to what we did in our proof of Theorem 3.12.

Theorem 4.3 (Stanley reciprocity). *Suppose $\mathcal{K}$ is a rational d-cone with the origin as apex. Then*

$$\sigma_{\mathcal{K}} \left(\frac{1}{z_1}, \frac{1}{z_2}, \ldots, \frac{1}{z_d} \right) = (-1)^d \, \sigma_{\mathcal{K}^\circ} \left(z_1, z_2, \ldots, z_d \right) .$$

Proof. Triangulate $\mathcal{K}$ into simplicial cones $\mathcal{K}_1, \mathcal{K}_2, \ldots, \mathcal{K}_m$. Exercise 3.19 ensures that there exists a vector $\mathbf{v} \in \mathbb{R}^d$ such that the shifted cone $\mathbf{v} + \mathcal{K}$ contains exactly the *interior* lattice points of $\mathcal{K}$,

$$\mathcal{K}^\circ \cap \mathbb{Z}^d = (\mathbf{v} + \mathcal{K}) \cap \mathbb{Z}^d, \qquad (4.5)$$

and there are no boundary lattice points on any of the triangulation cones:

$$\partial \left(\mathbf{v} + \mathcal{K}_j \right) \cap \mathbb{Z}^d = \varnothing \qquad \text{for all } j = 1, \ldots, m, \qquad (4.6)$$

as well as

$$\partial \left(-\mathbf{v} + \mathcal{K}_j\right) \cap \mathbb{Z}^d = \varnothing \qquad \text{for all } j = 1, \ldots, m. \tag{4.7}$$

We invite the reader (Exercise 4.4) to realize that (4.5)–(4.7) imply

$$\mathcal{K} \cap \mathbb{Z}^d = (-\mathbf{v} + \mathcal{K}) \cap \mathbb{Z}^d. \tag{4.8}$$

Now by Theorem 4.2,

$$\sigma_{\mathcal{K}} \left(\tfrac{1}{\mathbf{z}}\right) = \sigma_{-\mathbf{v}+\mathcal{K}} \left(\tfrac{1}{\mathbf{z}}\right) = \sum_{j=1}^{m} \sigma_{-\mathbf{v}+\mathcal{K}_j} \left(\tfrac{1}{\mathbf{z}}\right) = \sum_{j=1}^{m} (-1)^d \, \sigma_{\mathbf{v}+\mathcal{K}_j} (\mathbf{z})$$
$$= (-1)^d \, \sigma_{\mathbf{v}+\mathcal{K}} (\mathbf{z}) = (-1)^d \, \sigma_{\mathcal{K}^\circ} (\mathbf{z}) .$$

Note that the second and fourth equalities are true because of the validity of (4.7) and (4.6), respectively. $\qquad\square$

4.3 Ehrhart–Macdonald Reciprocity for Rational Polytopes

In preparation for the proof of Theorem 4.1, we define the **Ehrhart series** for the *interior* of the rational polytope $\mathcal{P}$ as

$$\mathrm{Ehr}_{\mathcal{P}^\circ} (z) := \sum_{t \geq 1} L_{\mathcal{P}^\circ} (t) \, z^t.$$

Our convention of beginning the series with $t = 1$ stems from the fact that this generating function is a special evaluation of the integer-point transform of the open cone $(\mathrm{cone}(\mathcal{P}))^\circ$: much in sync with Lemma 3.10,

$$\mathrm{Ehr}_{\mathcal{P}^\circ} (z) = \sigma_{(\mathrm{cone}(\mathcal{P}))^\circ} (1, 1, \ldots, 1, z) . \tag{4.9}$$

We are now ready to prove the Ehrhart-series analogue of Theorem 4.1.

Theorem 4.4. *Suppose $\mathcal{P}$ is a convex rational polytope. Then the evaluation of the rational function $\mathrm{Ehr}_{\mathcal{P}}$ at $\frac{1}{z}$ yields*

$$\mathrm{Ehr}_{\mathcal{P}} \left(\frac{1}{z}\right) = (-1)^{\dim \mathcal{P} + 1} \, \mathrm{Ehr}_{\mathcal{P}^\circ} (z) .$$

Proof. Suppose $\mathcal{P}$ is a d-polytope. We recall Lemma 3.10, which states that the generating function of the Ehrhart polynomial of $\mathcal{P}$ is an evaluation of the generating function of $\mathrm{cone}(\mathcal{P})$:

$$\mathrm{Ehr}_{\mathcal{P}}(z) = \sum_{t \geq 0} L_{\mathcal{P}}(t) \, z^t = \sigma_{\mathrm{cone}(\mathcal{P})} (1, 1, \ldots, 1, z) .$$

Equation (4.9) gives the analogous evaluation of $\sigma_{(\text{cone}(\mathcal{P}))^\circ}$ that yields $\text{Ehr}_{\mathcal{P}^\circ}$. Now we apply Theorem 4.3 to the $(d+1)$-cone $\mathcal{K} = \text{cone}(\mathcal{P})$:

$$\sigma_{(\text{cone}(\mathcal{P}))^\circ}\left(1, 1, \ldots, 1, z\right) = (-1)^{d+1}\, \sigma_{\text{cone}(\mathcal{P})}\left(1, 1, \ldots, 1, \tfrac{1}{z}\right). \qquad \square$$

Theorem 4.1 now follows like a breeze.

Proof of Ehrhart–Macdonald reciprocity (Theorem 4.1). We first apply Exercise 4.7 to the Ehrhart series of $\mathcal{P}$: namely, as rational functions,

$$\text{Ehr}_{\mathcal{P}}\left(\frac{1}{z}\right) = \sum_{t \le 0} L_{\mathcal{P}}(-t)\, z^t = -\sum_{t \ge 1} L_{\mathcal{P}}(-t)\, z^t.$$

Now we combine this identity with Theorem 4.4 to obtain

$$\sum_{t \ge 1} L_{\mathcal{P}^\circ}(t)\, z^t = (-1)^{d+1} \text{Ehr}_{\mathcal{P}}\left(\frac{1}{z}\right) = (-1)^d \sum_{t \ge 1} L_{\mathcal{P}}(-t)\, z^t.$$

Comparing the coefficients of the two power series yields the reciprocity theorem. $\qquad \square$

With Ehrhart–Macdonald reciprocity, we can now restate Theorem 3.18 in terms of Ehrhart polynomials:

Theorem 4.5. *Suppose $\mathcal{P}$ is an integral d-polytope with Ehrhart series*

$$\text{Ehr}_{\mathcal{P}}(z) = \frac{h_d^* z^d + h_{d-1}^* z^{d-1} + \cdots + h_1^* z + 1}{(1-z)^{d+1}}.$$

Then $h_d^* = h_{d-1}^* = \cdots = h_{k+1}^* = 0$ and $h_k^* \ne 0$ if and only if $(d-k+1)\mathcal{P}$ is the smallest integer dilate of $\mathcal{P}$ that contains an interior lattice point.

Proof. Theorem 3.18 says that h_k^* is the highest nonzero coefficient if and only if $L_{\mathcal{P}}(-1) = L_{\mathcal{P}}(-2) = \cdots = L_{\mathcal{P}}\left(-(d-k)\right) = 0$ and $L_{\mathcal{P}}\left(-(d-k+1)\right) \ne 0$. Now use Ehrhart–Macdonald reciprocity (Theorem 4.1). $\qquad \square$

The largest k for which $h_k^* \ne 0$ is called the **degree** of $\mathcal{P}$. The above theorem says that the degree of $\mathcal{P}$ is k precisely if $(d-k+1)\mathcal{P}$ is the smallest integer dilate of $\mathcal{P}$ that contains an interior lattice point.

4.4 The Ehrhart Series of Reflexive Polytopes

As an application of Theorem 4.4, we now study a special class of integral polytopes whose Ehrhart series have an additional symmetry structure. We call a polytope $\mathcal{P}$ **reflexive** if it is integral and has the hyperplane description

$$\mathcal{P} = \{ \mathbf{x} \in \mathbb{R}^d : \mathbf{A}\mathbf{x} \le \mathbf{1} \},$$

where $\mathbf{A}$ is an integral matrix. (Here $\mathbf{1}$ denotes a vector all of whose coordinates are 1.) The following theorem gives a characterization of reflexive polytopes through their Ehrhart series.

Theorem 4.6 (Hibi's palindromic theorem). *Suppose $\mathcal{P}$ is an integral d-polytope that contains the origin in its interior and that has the Ehrhart series*

$$\mathrm{Ehr}_{\mathcal{P}}(z) = \frac{h_d^* z^d + h_{d-1}^* z^{d-1} + \cdots + h_1^* z + h_0^*}{(1-z)^{d+1}}.$$

Then $\mathcal{P}$ is reflexive if and only if $h_k^* = h_{d-k}^*$ for all $0 \le k \le \frac{d}{2}$.

The two main ingredients for the proof of this result are Theorem 4.4 and the following:

Lemma 4.7. *Suppose $a_1, a_2, \ldots, a_d, b \in \mathbb{Z}$ satisfy $\gcd(a_1, a_2, \ldots, a_d, b) = 1$ and $b > 1$. Then there exist positive integers c and t such that $tb < c < (t+1)b$ and $\{(m_1, m_2, \ldots, m_d) \in \mathbb{Z}^d : a_1 m_1 + a_2 m_2 + \cdots + a_d m_d = c\} \ne \varnothing$.*

Proof. Let $g = \gcd(a_1, a_2, \ldots, a_d)$; by our assumption, $\gcd(g, b) = 1$, so one can find integers k and t such that

$$kg - tb = 1. \tag{4.10}$$

Furthermore, we can choose k and t in such a way that $t > 0$. Let $c = kg$; Equation (4.10) and the condition $b > 1$ imply that $tb < c < (t+1)b$. Finally, since $g = \gcd(a_1, a_2, \ldots, a_d)$, there exist $m_1, m_2, \ldots, m_d \in \mathbb{Z}$ such that

$$a_1 m_1 + a_2 m_2 + \cdots + a_d m_d = kg = c. \qquad \square$$

Proof of Theorem 4.6. We recall that $\mathcal{P}$ is reflexive if and only if

$$\mathcal{P} = \{ \mathbf{x} \in \mathbb{R}^d : \mathbf{A}\mathbf{x} \le \mathbf{1} \} \quad \text{for some integral matrix } \mathbf{A}. \tag{4.11}$$

We claim that $\mathcal{P}$ has such a hyperplane description if and only if

$$(t+1)\mathcal{P}^\circ \cap \mathbb{Z}^d = t\mathcal{P} \cap \mathbb{Z}^d \quad \text{for all } t \in \mathbb{Z}_{\ge 0}. \tag{4.12}$$

This condition means that the only lattice points that we gain in passing from $t\mathcal{P}$ to $(t+1)\mathcal{P}$ are those on the boundary of $(t+1)\mathcal{P}$. The fact that (4.11) implies (4.12) is the content of Exercise 4.13. Conversely, if $\mathcal{P}$ satisfies (4.12), then there are no lattice points between tH and $(t+1)H$ for any facet hyperplane H of $\mathcal{P}$ (Exercise 4.14). That is, if a facet hyperplane is given by $H = \{ \mathbf{x} \in \mathbb{R}^d : a_1 x_1 + a_2 x_2 + \cdots + a_d x_d = b \}$, where we may assume $\gcd(a_1, a_2, \ldots, a_d, b) = 1$, then

$$\{ \mathbf{x} \in \mathbb{Z}^d : tb < a_1 x_1 + a_2 x_2 + \cdots + a_d x_d < (t+1)b \} = \varnothing.$$

But then Lemma 4.7 implies that $b = 1$, and so $\mathcal{P}$ has a hyperplane description of the form (4.11).

Thus we have established that $\mathcal{P}$ is reflexive if and only if it satisfies (4.12). Now by Theorem 4.4,

$$\text{Ehr}_{\mathcal{P}^\circ}(z) = (-1)^{d+1} \text{Ehr}_{\mathcal{P}}\left(\frac{1}{z}\right) = \frac{h_0^* z^{d+1} + h_1^* z^d + \cdots + h_{d-1}^* z^2 + h_d^* z}{(1-z)^{d+1}}.$$

By condition (4.12), $\mathcal{P}$ is reflexive if and only if this rational function is equal to

$$\sum_{t \geq 1} L_{\mathcal{P}}(t-1) z^t = z \sum_{t \geq 0} L_{\mathcal{P}}(t) z^t = z\, \text{Ehr}_{\mathcal{P}}(z)$$

$$= \frac{h_d^* z^{d+1} + h_{d-1}^* z^d + \cdots + h_1^* z^2 + h_0^* z}{(1-z)^{d+1}},$$

that is, if and only if $h_k^* = h_{d-k}^*$ for all $0 \leq k \leq \frac{d}{2}$. $\qquad\square$

4.5 More "Reflections" on the Coin-Exchange Problem and the Gallery of Chapter 2

We have already encountered special cases of Ehrhart–Macdonald reciprocity several times. Note that Theorem 4.1 allows us to conclude that counting the number of *interior* lattice points in a rational polytope is tantamount to counting lattice points in its *closure*. Exercises 1.33, 2.1, and 2.7, as well as part (b) of each theorem in the gallery of Chapter 2, confirm that

$$L_{\mathcal{P}}(-t) = (-1)^{\dim \mathcal{P}} L_{\mathcal{P}^\circ}(t).$$

Notes

1. Ehrhart–Macdonald reciprocity (Theorem 4.1) had been conjectured (and proved in several special cases) by Eugène Ehrhart for about a decade before I. G. Macdonald found a general proof in 1971 [166]. One can actually relax the condition of Ehrhart–Macdonald reciprocity: it holds for polytopal complexes that are homeomorphic to a d-manifold. The proof we give here (including the proof of Theorem 4.3) appeared in [46]. An attractive alternative approach was given by Steven Sam [209].

2. Theorem 4.3 is due to Richard Stanley [225], who proved more general versions of this theorem. The reader might recall that the rational function representing the Ehrhart series of a rational cone can be thought of as its meromorphic continuation. Stanley reciprocity (Theorem 4.3) gives a functional identity for such meromorphic continuations.

3. The term *reflexive polytope* was coined by Victor Batyrev, who found exciting applications of these polytopes to mirror symmetry in physical string theory [25]. Batyrev proved that the toric variety $X_{\mathcal{P}}$ defined by a reflexive polytope $\mathcal{P}$ is *Fano*, and that every generic hypersurface of $X_{\mathcal{P}}$ is *Calabi–Yau*. That the Ehrhart series of a reflexive polytope exhibits an unexpected symmetry (Theorem 4.6) was discovered by Takayuki Hibi [135]. Matthew Fiset and Alexander Kasprzyk proved a version of Theorem 4.6 for *rational* polytopes [114]. The number of reflexive polytopes in dimension d is known for $d \leq 4$ [156, 157]; for example, there are precisely 16 reflexive polytopes in dimension 2, up to symmetries (see also [1, Sequence A090045]).

4. Reflexive polytopes bear some pleasant surprises beyond those mentioned in Section 4.4. For example, Benjamin Braun [61] proved that if $\mathcal{P}$ is reflexive and $\mathcal{Q}$ is an integral polytope that contains the origin in its interior, where $\mathcal{P}$ and $\mathcal{Q}$ live in orthogonal subspaces of $\mathbb{R}^d$, then the **free sum** $\mathcal{P} \oplus \mathcal{Q} := \mathrm{conv}\,(\mathcal{P} \cup \mathcal{Q})$ has Ehrhart series

$$\mathrm{Ehr}_{\mathcal{P}\oplus\mathcal{Q}}(z) = (1 - z)\,\mathrm{Ehr}_{\mathcal{P}}(z)\,\mathrm{Ehr}_{\mathcal{Q}}(z)\,, \qquad (4.13)$$

an analogue of sorts to Exercise 2.4; see Exercise 4.17. Another striking result is that the sum of the numbers of lattice points on the boundaries of a reflexive *polygon* and its dual is always 12 [137, 194]. A similar result holds in dimension 3 (with 12 replaced by 24) [26], but no elementary proof of the latter fact is known [34, Section 4].

5. There is an equivalent definition for reflexive polytopes: $\mathcal{P}$ is reflexive if and only if both $\mathcal{P}$ and its dual $\mathcal{P}^*$ are integral polytopes. The **dual polytope** of $\mathcal{P}$ (often also called the **polar polytope**) is defined as

$$\mathcal{P}^* := \left\{ \mathbf{x} \in \mathbb{R}^d : \mathbf{x} \cdot \mathbf{y} \leq 1 \text{ for all } \mathbf{y} \in \mathcal{P} \right\}.$$

The concept of (polar) duality is not confined to polytopes but can be defined for any nonempty subset of $\mathbb{R}^d$. Duality is a crucial chapter in the theory of polytopes, and one of its applications is the equivalence of the vertex and hyperplane description of a polytope. For more about (polar) duality, the reader might consult [20, Chapter IV].

6. The cross-polytopes $\diamond$ from Section 2.5 form a special class of reflexive polytopes. We mentioned in the notes of Chapter 2 that the roots of the Ehrhart polynomials $L_{\diamond}$ all have real part $-\frac{1}{2}$ [75, 148]. Christian Bey, Martin Henk, and Jörg Wills proved in [55] that if all complex roots of $L_{\mathcal{P}}(t)$, for some integral polytope $\mathcal{P}$, have real part $-\frac{1}{2}$, then $\mathcal{P}$ is the unimodular image of a reflexive polytope.

7. An integral polytope $\mathcal{P} \subset \mathbb{R}^d$ for which there exists a positive integer k such that $(k-1)\mathcal{P}° \cap \mathbb{Z}^d = \varnothing$, $\#(k\mathcal{P} \cap \mathbb{Z}^d) = 1$ and $t\mathcal{P}° \cap \mathbb{Z}^d = (t-k)\mathcal{P} \cap \mathbb{Z}^d$ for all integers $t > k$ is called **Gorenstein** of index k. Thus reflexive polytopes are Gorenstein of index 1. (Some of the polytopes in Chapters 2 and 6 are Gorenstein.) Theorem 4.6 can be extended to the statement that an integral polytope is Gorenstein if and only if its nonzero h_j^*'s are symmetric (not necessarily with center of symmetry $\frac{d}{2}$ as in the case of reflexive polytopes); see Exercise 4.8.

Exercises

4.1. ♣ Prove (4.2): for $a, t \in \mathbb{Z}_{>0}$, $L_{(0,\frac{1}{a})}(t) = \left| \frac{t-1}{a} \right|$.

4.2. ♣ Explain (4.4): if $\mathbf{w}_1, \mathbf{w}_2, \ldots, \mathbf{w}_d \in \mathbb{R}^d$ are linearly independent and

$$\Pi = \{\lambda_1 \mathbf{w}_1 + \lambda_2 \mathbf{w}_2 + \cdots + \lambda_d \mathbf{w}_d : 0 < \lambda_1, \lambda_2, \ldots, \lambda_d < 1\},$$

then $\mathbf{v} + \Pi = -(-\mathbf{v} + \Pi) + \mathbf{w}_1 + \mathbf{w}_2 + \cdots + \mathbf{w}_d$.

4.3. Revisit Exercise 2.15, but now give a one-line proof that the Bernoulli polynomials satisfy $B_d(1-x) = (-1)^d B_d(x)$.

4.4. ♣ Prove that (4.5)–(4.7) imply (4.8); that is, if $\mathcal{K}$ is a rational pointed d-cone with the origin as apex and $\mathbf{v} \in \mathbb{R}^d$ is such that

$$\mathcal{K}° \cap \mathbb{Z}^d = (\mathbf{v} + \mathcal{K}) \cap \mathbb{Z}^d,$$
$$\partial(\mathbf{v} + \mathcal{K}_j) \cap \mathbb{Z}^d = \varnothing \qquad \text{for all } j = 1, \ldots, m,$$

and

$$\partial(-\mathbf{v} + \mathcal{K}_j) \cap \mathbb{Z}^d = \varnothing \qquad \text{for all } j = 1, \ldots, m,$$

then

$$\mathcal{K} \cap \mathbb{Z}^d = (-\mathbf{v} + \mathcal{K}) \cap \mathbb{Z}^d.$$

4.5. Prove the following generalization of Theorem 4.3 to rational pointed cones with arbitrary apex: Suppose $\mathcal{K}$ is a rational pointed d-cone with the origin as apex, and $\mathbf{v} \in \mathbb{R}^d$. Then the integer-point transform $\sigma_{\mathbf{v}+\mathcal{K}}(\mathbf{z})$ of the pointed d-cone $\mathbf{v} + \mathcal{K}$ is a rational function that satisfies

$$\sigma_{\mathbf{v}+\mathcal{K}}\left(\frac{1}{\mathbf{z}}\right) = (-1)^d \sigma_{(-\mathbf{v}+\mathcal{K})^\circ}(\mathbf{z}).$$

4.6. Generalize Theorem 4.3 by showing that we do not need to assume that $\mathcal{K}$ is full dimensional.

4.7. ♣ Suppose $Q\colon \mathbb{Z} \to \mathbb{C}$ is a quasipolynomial. We know that $R_Q^+(z) := \sum_{t \geq 0} Q(t)\, z^t$ evaluates to a rational function.

(a) Prove that $R_Q^-(z) := \sum_{t < 0} Q(t)\, z^t$ also evaluates to a rational function.
(b) Let $Q(t) = 1$. Prove that as rational functions, $R_Q^+(z) + R_Q^-(z) = 0$.
(c) Suppose Q is a polynomial. Prove that as rational functions, $R_Q^+(z) + R_Q^-(z) = 0$.
(d) Suppose Q is a quasipolynomial. Prove that as rational functions, $R_Q^+(z) + R_Q^-(z) = 0$.

4.8. ♣ Suppose that $\mathcal{P}$ is an integral d-polytope for which

$$L_{\mathcal{P}^\circ}(t) = L_{\mathcal{P}}(t - k) \qquad \text{and} \qquad L_{\mathcal{P}^\circ}(1) = L_{\mathcal{P}^\circ}(2) = \cdots = L_{\mathcal{P}^\circ}(k-1) = 0$$

for some integer k (so $\mathcal{P}$ is a Gorenstein polytope, as mentioned in the notes). Prove that

$$\mathrm{Ehr}_{\mathcal{P}}\left(\frac{1}{z}\right) = (-1)^{d+1} z^k \, \mathrm{Ehr}_{\mathcal{P}}(z).$$

4.9. Suppose $\mathcal{P}$ is an integral d-polytope with Ehrhart series

$$\mathrm{Ehr}_{\mathcal{P}}(z) = \frac{h_d^* z^d + h_{d-1}^* z^{d-1} + \cdots + h_1^* z + 1}{(1 - z)^{d+1}}.$$

Prove that $h_d^* = L_{\mathcal{P}^\circ}(1)$.

4.10. Suppose $\mathcal{P}$ is an integral d-polytope. Show that the dilate $(d + 1)\mathcal{P}$ contains an interior lattice point.

4.11. Suppose $\mathcal{P}$ is an integral polytope. Denote the boundary of $\mathcal{P}$ by $\partial \mathcal{P}$. Prove that $L_{\partial \mathcal{P}}(t)$ is a polynomial that is either even or odd. Determine its constant term.

4.12. Recall the restricted partition function

$$p_{\{a_1, a_2, \ldots, a_d\}}(n) := \#\left\{(m_1, \ldots, m_d) \in \mathbb{Z}_{\geq 0}^d : m_1 a_1 + \cdots + m_d a_d = n\right\}$$

from Chapter 1. Prove that as quasipolynomials,

$$p_{\{a_1, a_2, \ldots, a_d\}}(-n - a_1 - a_2 - \cdots - a_d) = (-1)^{d-1} p_{\{a_1, a_2, \ldots, a_d\}}(n)$$

and that

$$p_{\{a_1,a_2,\ldots,a_d\}}(-1) = p_{\{a_1,a_2,\ldots,a_d\}}(-2) = \cdots$$
$$= p_{\{a_1,a_2,\ldots,a_d\}}(-a_1 - a_2 - \cdots - a_d + 1) = 0.$$

4.13. ♣ Prove that (4.11) implies (4.12), that is, show that if the polytope $\mathcal{P}$ is given by $\mathcal{P} = \{\mathbf{x} \in \mathbb{R}^d : \mathbf{A}\mathbf{x} \leq \mathbf{1}\}$ for an integral matrix $\mathbf{A}$, then

$$(t+1)\mathcal{P}^\circ \cap \mathbb{Z}^d = t\mathcal{P} \cap \mathbb{Z}^d$$

for all $t \in \mathbb{Z}_{\geq 0}$.

4.14. ♣ Suppose $\mathcal{P}$ is an integral polytope that satisfies (4.12): $\mathcal{P}^\circ \cap \mathbb{Z}^d = \{\mathbf{0}\}$ and for all $t \in \mathbb{Z}_{>0}$, $(t+1)\mathcal{P}^\circ \cap \mathbb{Z}^d = t\mathcal{P} \cap \mathbb{Z}^d$. Then for every $t \in \mathbb{Z}$, there are no lattice points between tH and $(t+1)H$ for any facet hyperplane H of $\mathcal{P}$.

4.15. Compute the dual polytope of the d-dimensional cross-polytope ◇.

4.16. Suppose $\mathcal{P} \subset \mathbb{R}^d$ is an integral d-polytope with $\mathbf{0} \in \mathcal{P}^\circ$. Show that its dual $\mathcal{P}^*$ is integral if and only if $\mathcal{P}$ is of the form $\{\mathbf{x} \in \mathbb{R}^d : \mathbf{A}\mathbf{x} \leq \mathbf{1}\}$ for some integral matrix $\mathbf{A}$.

4.17. Prove (4.13): if $\mathcal{P}$ is reflexive and $\mathcal{Q}$ is an integral polytope that contains the origin in its interior, where $\mathcal{P}$ and $\mathcal{Q}$ live in orthogonal subspaces of $\mathbb{R}^d$, then the free sum $\mathcal{P} \oplus \mathcal{Q} := \mathrm{conv}\,(\mathcal{P} \cup \mathcal{Q})$ has Ehrhart series

$$\mathrm{Ehr}_{\mathcal{P} \oplus \mathcal{Q}}(z) = (1-z)\,\mathrm{Ehr}_{\mathcal{P}}(z)\,\mathrm{Ehr}_{\mathcal{Q}}(z).$$

Open Problems

4.18. Suppose $\mathcal{P}$ is a 3-dimensional reflexive polytope. Denote by e^* the edge in the dual polytope $\mathcal{P}^*$ that corresponds to the edge e in $\mathcal{P}$. Give an elementary proof that

$$\sum_{e \text{ edge of } \mathcal{P}} \mathrm{length}\,(e) \cdot \mathrm{length}\,(e^*) = 24.$$

4.19. Find the number of reflexive polytopes in dimension $d \geq 5$.

4.20. Prove *Mahler's conjecture*: if $\mathcal{K}$ is a d-dimensional centrally symmetric convex set, then

$$\mathrm{vol}(\mathcal{K})\,\mathrm{vol}(K^*) \geq \frac{4^d}{d!}.$$

(The right-hand side is the quantity $\mathrm{vol}(\mathcal{K})\,\mathrm{vol}(K^*)$ for $\mathcal{K} = [-1,1]^d$. See [125, Chapter 9] for further background and references.)

Chapter 5
Face Numbers and the Dehn–Sommerville Relations in Ehrhartian Terms

"Data! Data! Data!" he cried, impatiently. "I can't make bricks without clay."

Sherlock Holmes (*The Adventure of the Copper Beeches*, by Arthur Conan Doyle, 1859–1930)

Our goal in this chapter is twofold, or rather, there is one goal in two different guises. The first is to prove a set of fascinating identities, which give linear relations among the face numbers of a polytope. They are called *Dehn–Sommerville relations*, in honor of their discoverers Max Wilhelm Dehn (1878–1952)[1] and Duncan MacLaren Young Sommerville (1879–1934).[2] Our second goal is to unify the Dehn–Sommerville relations (Theorem 5.1 below) with Ehrhart–Macdonald reciprocity (Theorem 4.1).

5.1 Face It!

We denote the number of k-dimensional faces of $\mathcal{P}$ by the symbol f_k. As k varies from 0 to d, the **face numbers** f_k encode intrinsic information about the polytope $\mathcal{P}$. The d-polytope $\mathcal{P}$ is **simple** if each vertex of $\mathcal{P}$ lies on precisely d edges of $\mathcal{P}$.

Theorem 5.1 (Dehn–Sommerville relations). *If $\mathcal{P}$ is a simple d-polytope and $0 \le k \le d$, then*

$$f_k = \sum_{j=0}^{k} (-1)^j \binom{d-j}{d-k} f_j .$$

---

[1] For more information about Dehn, see
http://www-history.mcs.st-andrews.ac.uk/Biographies/Dehn.html.

[2] For more information about Sommerville, see
http://www-history.mcs.st-andrews.ac.uk/Biographies/Sommerville.html.

This theorem takes on a particularly nice form for $k = d$, namely the famous *Euler relation*, which holds for every polytope (not just simple ones).

Theorem 5.2 (Euler relation). *If $\mathcal{P}$ is a convex d-polytope, then*

$$\sum_{j=0}^{d}(-1)^j f_j = 1\,.$$

This identity is less trivial than it might look. We give a quick proof for *rational* polytopes, for which we can use Ehrhart–Macdonald reciprocity (Theorem 4.1).

Proof of Theorem 5.2, assuming that $\mathcal{P}$ is rational. Let us count the integer points in $t\mathcal{P}$ according to the (relatively) open faces that contain them:[3]

$$L_{\mathcal{P}}(t) = \sum_{\mathcal{F}\subseteq\mathcal{P}} L_{\mathcal{F}^\circ}(t) = \sum_{\mathcal{F}\subseteq\mathcal{P}} (-1)^{\dim\mathcal{F}} L_{\mathcal{F}}(-t)\,. \tag{5.1}$$

Here and in the remainder of this chapter, the sums are over all *nonempty* faces. (Alternatively, we could agree that $L_\varnothing(t) = 0$.) Note that the first equality in (5.1), while innocent-looking, uses the not entirely trivial Exercise 5.3. The constant term of $L_{\mathcal{F}}(t)$ is 1 for every face $\mathcal{F}$ (by Exercise 3.32). Hence the constant terms of (5.1) give

$$1 = \sum_{\mathcal{F}\subseteq\mathcal{P}} (-1)^{\dim\mathcal{F}} = \sum_{j=0}^{d}(-1)^j f_j\,,$$

which proves our claim.   □

There is a natural structure on the faces of a polytope $\mathcal{P}$ induced by the containment relation $\mathcal{F} \subseteq \mathcal{G}$. This relation gives a partial ordering on the set of all faces of $\mathcal{P}$, called the **face lattice** of $\mathcal{P}$.[4] A useful way to illustrate this partially ordered set is through a directed graph whose nodes correspond to the faces of $\mathcal{P}$, and we have an edge from $\mathcal{F}$ to $\mathcal{G}$ if

$$\mathcal{F} \subset \mathcal{G} \qquad \text{and} \qquad \dim\mathcal{G} = \dim\mathcal{F} + 1\,.$$

We draw this directed graph in such a way that the edge directions are upward; Figure 5.1 shows the face lattice for a triangle. Exercise 2.6 implies that the face lattice of every simplex is a **Boolean lattice**, which is the partially ordered set formed by all subsets of a finite set, where the partial ordering is again subset containment.

We already mentioned that we will unify the Dehn–Sommerville relations (Theorem 5.1) with Ehrhart–Macdonald reciprocity (Theorem 4.1). It is for

---

[3] Note that the relative interior of a vertex is the vertex itself.

[4] The use of the word *lattice* here is disjoint from our previous definition of the word.

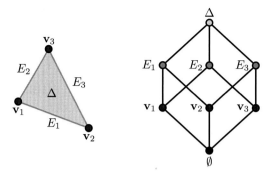

Fig. 5.1 The face lattice of a triangle.

this reason that we will prove Theorem 5.1 only for *rational* polytopes. To combine the notions of face numbers and lattice-point enumeration, we define

$$F_k(t) := \sum_{\substack{\mathcal{F} \subseteq \mathcal{P} \\ \dim \mathcal{F} = k}} L_{\mathcal{F}}(t),$$

the sum being taken over all k-faces of $\mathcal{P}$. By Ehrhart's theorem (Theorem 3.23), F_k is a quasipolynomial. Since $L_{\mathcal{F}}(0) = 1$ for all $\mathcal{F}$,

$$F_k(0) = f_k,$$

the number of k-faces of $\mathcal{P}$. We also remark that the leading coefficient of F_k measures the relative volume of the k-**skeleton** of $\mathcal{P}$, that is, the union of all k-faces; see Section 5.4 for a precise definition of relative volume.

Our common extension of Theorems 5.1 and 4.1 is the subject of the next section.

5.2 Dehn–Sommerville Extended

Theorem 5.3. *If $\mathcal{P}$ is a simple rational d-polytope and $0 \le k \le d$, then*

$$F_k(t) = \sum_{j=0}^{k} (-1)^j \binom{d-j}{d-k} F_j(-t).$$

The classical Dehn–Sommerville equations (Theorem 5.1)—again, only for rational polytopes—are obtained from the constant terms of the counting functions on both sides of the identity. On the other hand, for $k = d$, Theorem 5.3 gives (with t replaced by $-t$)

$$L_{\mathcal{P}}(-t) = F_d(-t) = \sum_{j=0}^{d}(-1)^j F_j(t) = (-1)^d \sum_{j=0}^{d}(-1)^{d-j}F_j(t).$$

The sum on the right-hand side is an inclusion–exclusion formula for the number of lattice points in the interior of $t\mathcal{P}$ (count all the points in $t\mathcal{P}$, subtract the ones on the facets, add back what you have overcounted, etc.), so in a sense, we recover Ehrhart–Macdonald reciprocity.

Proof. Suppose $\mathcal{F}$ is a k-face of $\mathcal{P}$. Then again by counting the integer points in $\mathcal{F}$ according to relatively open faces of $\mathcal{F}$ (Exercise 5.3),

$$L_{\mathcal{F}}(t) = \sum_{\mathcal{G} \subset \mathcal{F}} L_{\mathcal{G}^\circ}(t),$$

or by the Ehrhart–Macdonald reciprocity (Theorem 4.1),

$$L_{\mathcal{F}}(t) = \sum_{\mathcal{G} \subseteq \mathcal{F}}(-1)^{\dim \mathcal{G}} L_{\mathcal{G}}(-t) = \sum_{j=0}^{k}(-1)^j \sum_{\substack{\mathcal{G} \subseteq \mathcal{F} \\ \dim \mathcal{G}=j}} L_{\mathcal{G}}(-t). \qquad (5.2)$$

Now sum both left- and right-hand sides over all k-faces and rearrange the sum on the right-hand side:

$$
\begin{aligned}
F_k(t) &= \sum_{\substack{\mathcal{F} \subset \mathcal{P} \\ \dim \mathcal{F}=k}} \sum_{j=0}^{k}(-1)^j \sum_{\substack{\mathcal{G} \subset \mathcal{F} \\ \dim \mathcal{G}=j}} L_{\mathcal{G}}(-t) \\
&= \sum_{j=0}^{k}(-1)^j \sum_{\substack{\mathcal{F} \subset \mathcal{P} \\ \dim \mathcal{F}=k}} \sum_{\substack{\mathcal{G} \subseteq \mathcal{F} \\ \dim \mathcal{G}=j}} L_{\mathcal{G}}(-t) \\
&= \sum_{j=0}^{k}(-1)^j \sum_{\substack{\mathcal{G} \subseteq \mathcal{P} \\ \dim \mathcal{G}=j}} f_k(\mathcal{P}/\mathcal{G}) L_{\mathcal{G}}(-t) \\
&= \sum_{j=0}^{k}(-1)^j \sum_{\substack{\mathcal{G} \subseteq \mathcal{P} \\ \dim \mathcal{G}=j}} \binom{d-j}{d-k} L_{\mathcal{G}}(-t) \\
&= \sum_{j=0}^{k}(-1)^j \binom{d-j}{d-k} F_j(-t).
\end{aligned}
$$

Here $f_k(\mathcal{P}/\mathcal{G})$ denotes the number of k-faces of $\mathcal{P}$ containing a given j-face $\mathcal{G}$ of $\mathcal{P}$. Since $\mathcal{P}$ is simple, this number equals $\binom{d-j}{d-k}$ (see Exercise 5.5). $\square$

5.3 Applications to the Coefficients of an Ehrhart Polynomial

We will now apply Theorem 5.3 to the computation of the Ehrhart polynomial of an integral d-polytope $\mathcal{P}$. The only face-lattice-point enumerator involving the face $\mathcal{P}$ is $F_d(t)$, for which Theorem 5.3 specializes to

$$L_{\mathcal{P}}(t) = F_d(t) = \sum_{j=0}^{d} (-1)^j F_j(-t) \, .$$

In fact, we do not have to assume that $\mathcal{P}$ is simple, since this identity simply counts integer points by faces. (Recall that $(-1)^j F_j(-t)$ counts the integer points in the t-dilates of the interior of the j-faces.)[5] The last term on the right-hand side is

$$(-1)^d F_d(-t) = (-1)^d L_{\mathcal{P}}(-t) = L_{\mathcal{P}^\circ}(t)$$

by Ehrhart–Macdonald reciprocity. Shifting this term to the left gives

$$L_{\mathcal{P}}(t) - L_{\mathcal{P}^\circ}(t) = \sum_{j=0}^{d-1} (-1)^j F_j(-t) \, . \tag{5.3}$$

The difference on the left-hand side of this identity has a natural interpretation: it counts the integer points on the *boundary* of $t\mathcal{P}$. (And in fact, the right-hand side is once more an inclusion–exclusion formula for this number.) Let us write $L_{\mathcal{P}}(t) = c_d \, t^d + c_{d-1} \, t^{d-1} + \cdots + c_0$. Then $L_{\mathcal{P}^\circ}(t) = c_d \, t^d - c_{d-1} \, t^{d-1} + \cdots + (-1)^d c_0$, so that

$$L_{\mathcal{P}}(t) - L_{\mathcal{P}^\circ}(t) = 2c_{d-1} \, t^{d-1} + 2c_{d-3} \, t^{d-3} + \cdots \, ,$$

where this sum ends with $2c_0$ if d is odd and $2c_1 t$ if d is even (this should look familiar; see Exercise 4.11). Combining this expression with (5.3) yields the following useful result.

Theorem 5.4. *Suppose $L_{\mathcal{P}}(t) = c_d \, t^d + c_{d-1} \, t^{d-1} + \cdots + c_0$ is the Ehrhart polynomial of $\mathcal{P}$. Then*

$$c_{d-1} \, t^{d-1} + c_{d-3} \, t^{d-3} + \cdots = \frac{1}{2} \sum_{j=0}^{d-1} (-1)^j F_j(-t) \, . \qquad \Box$$

We can make the statement of this theorem more precise (but also messier) by writing

---

[5] So one might argue that we did not need the Dehn–Sommerville machinery for the computations in the current section. This argument is correct, although Theorem 5.3 is a strong motivation.

$$F_j(t) = \sum_{\substack{\mathcal{F} \subseteq \mathcal{P} \\ \dim \mathcal{F} = j}} L_{\mathcal{F}}(t) = c_{j,j}\, t^j + c_{j,j-1} t^{j-1} + \cdots + c_{j,0}\,.$$

Then collecting the coefficients of t^k in Theorem 5.4 yields the following relations.

Corollary 5.5. *If k and d are of different parities, then*

$$c_k = \frac{1}{2} \sum_{j=0}^{d-1} (-1)^{j+k} c_{j,k}\,. \qquad\qquad \square$$

If k and d have the same parity, then the left-hand side has to be replaced by 0.

The first coefficient c_k in the Ehrhart polynomial of a d-polytope $\mathcal{P}$ satisfying the parity condition is c_{d-1}. In this case, Corollary 5.5 tells us that c_{d-1} equals $\frac{1}{2}$ times the sum of the leading coefficients of the Ehrhart polynomials of the facets of $\mathcal{P}$.

The next interesting coefficient is c_{d-3}. For example, if $\dim \mathcal{P} = 4$, we can use Corollary 5.5 to compute c_1 entirely from (the linear coefficients of) the Ehrhart polynomials of the faces of dimension ≤ 3.

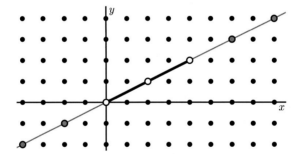

Fig. 5.2 The line segment from $(0,0)$ to $(4,2)$ and its affine sublattice.

5.4 Relative Volume

It is time to return to continuous volume. Recall Lemma 3.19: if $S \subset \mathbb{R}^d$ is d-dimensional, then $\mathrm{vol}\, S = \lim_{t \to \infty} \frac{1}{t^d} \cdot \#\left(tS \cap \mathbb{Z}^d\right)$. Back in Chapter 3, we stressed the importance of S being d-dimensional, because otherwise (i.e., S is lower-dimensional although living in d-space), by our definition, $\mathrm{vol}\, S = 0$. However, the case that $S \subset \mathbb{R}^d$ is *not* of dimension d is often very interesting; an example is the polytope $\mathcal{P}$ that we encountered in connection with the coin-exchange problem in Chapter 1. We still would like to compute the volume of such objects, in the relative sense. This makes for a slight complication. Let us say $S \subset \mathbb{R}^d$ is of dimension $m < d$, and let span $S = \{\mathbf{x} + \lambda(\mathbf{y} - \mathbf{x}) : \mathbf{x}, \mathbf{y} \in S, \lambda \in \mathbb{R}\}$, the affine span of S. If we follow the same procedure as above (counting boxes or grid points), we compute the volume relative to the sublattice (span S) $\cap\ \mathbb{Z}^d$; we call this the **relative volume** of S.

For example, the line segment L from $(0, 0)$ to $(4, 2)$ in $\mathbb{R}^2$ has the relative volume 2, because in span $L = \{(x, y) \in \mathbb{R}^2 : y = \frac{1}{2}x\}$, L is covered by two segments of "unit length" in this affine subspace, as pictured in Figure 5.2. A 3-dimensional instance that should be reminiscent of Chapter 1 is illustrated in Figure 5.3.

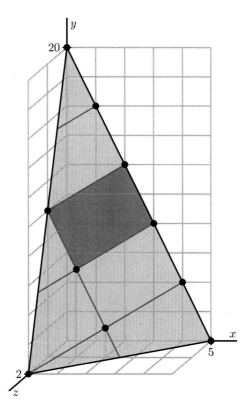

Fig. 5.3 The triangle defined by $\frac{x}{5} + \frac{y}{20} + \frac{z}{2} = 1, x \geq 0, y \geq 0, z \geq 0$. The shaded region is a fundamental domain for the sublattice that lies on the affine span of the triangle.

If $S \subseteq \mathbb{R}^d$ has full dimension d, the relative volume coincides with the "full-dimensional" volume. Henceforth, when we write vol S we refer to the relative volume of S. With this convention, we can rewrite Lemma 3.19 to accommodate a set $S \subset \mathbb{R}^d$ that is m-dimensional: its relative volume can be computed as

$$\operatorname{vol} S = \lim_{t \to \infty} \frac{1}{t^m} \cdot \# \left(tS \cap \mathbb{Z}^d \right) .$$

In the case that $\# \left(tS \cap \mathbb{Z}^d \right)$ has the special form of a polynomial—for example, if S is an integral polytope—we can further simplify this theorem. Suppose $\mathcal{P} \subset \mathbb{R}^d$ is an integral m-polytope with Ehrhart polynomial

$$L_{\mathcal{P}}(t) = c_m \, t^m + c_{m-1} \, t^{m-1} + \cdots + c_1 \, t + 1 .$$

Then according to the above discussion, and much in sync with Lemma 3.19,

$$\operatorname{vol} \mathcal{P} = \lim_{t \to \infty} \frac{1}{t^m} L_{\mathcal{P}}(t) = \lim_{t \to \infty} \frac{c_m \, t^m + c_{m-1} \, t^{m-1} + \cdots + c_1 \, t + 1}{t^m} = c_m .$$

The relative volume of $\mathcal{P}$ is the leading term of the corresponding counting function $L_{\mathcal{P}}$.

For example, in the previous section we found that Corollary 5.5 implies that the second leading coefficient c_{d-1} of the Ehrhart polynomial of the d-polytope $\mathcal{P}$ equals $\frac{1}{2}$ times the sum of the leading terms of the Ehrhart polynomials of the facets of $\mathcal{P}$. The leading term for one facet is simply the relative volume of that facet:

Theorem 5.6. *Suppose* $L_{\mathcal{P}}(t) = c_d \, t^d + c_{d-1} \, t^{d-1} + \cdots + c_0$ *is the Ehrhart polynomial of the integral polytope* $\mathcal{P}$. *Then*

$$c_{d-1} = \frac{1}{2} \sum_{\mathcal{F} \text{ a facet of } \mathcal{P}} \operatorname{vol} \mathcal{F} . \qquad \square$$

Notes

1. The Dehn–Sommerville relations (Theorem 5.1) first surfaced in the work of Max Dehn, who proved them in 1905 for dimension 5 [100]. (The Dehn–Sommerville relations are not that complicated for $d \leq 4$; see Exercise 5.4.) Some two decades later, D. M. Y. Sommerville proved the general case [223]. Theorem 5.1 was neither well known nor much used in the first half of the twentieth century, achieving renown only after its rediscovery by Victor Klee [150] and its appearance in Branko Grünbaum's famous and widely read book [126]. The Dehn–Sommerville relations will play a prominent role in Chapter 10.

2. The Euler relation (Theorem 5.2) is easy to prove directly for $d = 3$ (this case is attributed to Euler), but for higher dimension, one has to be somewhat careful, as we already remarked in the text. The classical proof for general d was found in 1852 by Ludwig Schläfli [211], although it (like numerous later proofs) assumes that the boundary of a convex polytope can be built up inductively in a "good" way. This nontrivial fact—which is called *shelling* of a polytope—was proved by Heinz Bruggesser and Peter Mani in 1971 [71]. (One instance of a shelling will be the subject of Exercise 10.16.) Shellability is nicely discussed in [258, Lecture 8]. There are short proofs of the Euler relation that do not use the shelling of a polytope (see, for example, [162, 184, 248]).

3. The reader might suspect that proving Theorems 5.1 and 5.2 for *rational* polytopes suffices for the general case, since it seems that we can transform a polytope with irrational vertices slightly to one with only rational vertices without changing the face structure of the polytope. This is true in our everyday world, but it fails in dimension ≥ 4 (see [206] for dimension 4 and [258, pp. 172–173] for general dimension).

4. Theorem 5.3 is due to Peter McMullen [171], who, in fact, proved this result in somewhat greater generality. Another generalization of Theorem 5.3 can be found in [84].

Exercises

5.1. Consider a simple 3-polytope with at least five facets. Two players play the following game: each player, in turn, signs his or her name on a previously unsigned face. The winner is the player who first succeeds in signing three facets that share a common vertex. Show that the player who signs first will always win by playing as well as possible.[6]

5.2. Show that for the d-cube, $f_k = 2^{d-k} \binom{d}{k}$.

5.3. ♣ Given a polytope $\mathcal{P}$, prove that

$$\mathcal{P} = \bigcup_{\mathcal{F} \subseteq \mathcal{P}} \mathcal{F}^\circ$$

as a disjoint union over all faces of $\mathcal{P}$, and deduce from this (5.1).

5.4. Give an elementary proof of the Dehn–Sommerville relations (Theorem 5.1) for $d \leq 4$.

5.5. ♣ Let $\mathcal{P}$ be a simple d-polytope. Prove that the number of k-faces of $\mathcal{P}$ containing a given j-face of $\mathcal{P}$ equals $\binom{d-j}{d-k}$.

5.6. ♣ Show directly, without using Theorem 5.2, that for a d-simplex,

---

[6] This was one of the 2002 Putnam contest problems.

(a) $f_k = \binom{d+1}{k+1}$;

(b) $\displaystyle\sum_{k=0}^{d} (-1)^k f_k = 1$.

5.7. ♣ Prove Theorem 5.1 directly (and hence not requiring $\mathcal{P}$ to be an integral polytope). (*Hint:* Orient yourself along the proof of Theorem 5.3, but start with the Euler relation (Theorem 5.2) for a given face $\mathcal{F}$ instead of (5.2).)

5.8. Let $\mathcal{F}$ be a face of a simple polytope $\mathcal{P}$. Prove that for $0 \le k \le d$,

$$\sum_{\mathcal{G} \supseteq \mathcal{F}} (-1)^{\dim \mathcal{G}} \binom{\dim \mathcal{G}}{k} = (-1)^d \binom{\dim \mathcal{F}}{d-k}.$$

5.9. ♣ The **dual polytope** of $\mathcal{P}$ (often also called the **polar polytope**) is defined as

$$\mathcal{P}^* := \{\mathbf{x} \in \mathbb{R}^d : \mathbf{x} \cdot \mathbf{y} \le 1 \text{ for all } \mathbf{y} \in \mathcal{P}\}.$$

(For $\mathcal{P}^*$ to be a polytope, we need to require that the origin be in the interior of $\mathcal{P}$; see also the notes at the end of Chapter 4.)

(a) Prove that the face lattice of $\mathcal{P}^*$ equals the face lattice of $\mathcal{P}$ turned upside down.

(b) Show that $\mathcal{P}^*$ is simple if and only if $\mathcal{P}$ is **simplicial**, that is, all nontrivial faces of $\mathcal{P}$ are simplices.

(c) Prove the Dehn–Sommerville relations for a simplicial d-polytope: for $0 \le k < d$,

$$f_k = \sum_{j=k}^{d-1} (-1)^{d-j-1} \binom{j+1}{k+1} f_j.$$

This identity also holds for $k = -1$ if we define $f_{-1} := 1$. (*Hint:* Try a proof that is dual to the one hinted at in Exercise 5.7.)

5.10. Show that the equations in Theorem 5.3 are equivalent to the following identities. If $\mathcal{P}$ is a simple lattice d-polytope and $k \le d$, then

$$\sum_{j=0}^{k} (-1)^{k-j} \binom{d-j}{k-j} F_{d-j}(-t) = \sum_{i=k}^{d} (-1)^{i-k} \binom{i}{k} F_i(t).$$

5.11. Prove that the equations in the previous exercise imply the following identities, which compare the number of lattice points in faces and relative interiors of faces of the simple polytope $\mathcal{P}$:

$$\sum_{j=0}^{k} (-1)^j \binom{d-j}{k-j} \sum_{\substack{\mathcal{F} \subseteq \mathcal{P} \\ \dim \mathcal{F} = d-j}} \#\left(\mathcal{F} \cap \mathbb{Z}^d\right) = \sum_{i=k}^{d} \binom{i}{k} \sum_{\substack{\mathcal{G} \subseteq \mathcal{P} \\ \dim \mathcal{G} = i}} \#\left(\mathcal{G}^\circ \cap \mathbb{Z}^d\right),$$

where $k = 0, \ldots, d = \dim \mathcal{P}$. For example, for $k = 0$,

$$\#(\mathcal{P} \cap \mathbb{Z}^d) = \sum_{\mathcal{G} \subseteq \mathcal{P}} \# \left(\mathcal{G}^\circ \cap \mathbb{Z}^d \right),$$

and for $k = d$, we obtain the inclusion–exclusion formula

$$\#(\mathcal{P}^\circ \cap \mathbb{Z}^d) = \sum_{j=0}^{d} (-1)^{d-j} \sum_{\substack{\mathcal{F} \subseteq \mathcal{P} \\ \dim \mathcal{F} = j}} \# \left(\mathcal{F} \cap \mathbb{Z}^d \right).$$

5.12. Another nice reformulation of Theorem 5.3 is the following generalized reciprocity law. For a simple integral d-polytope $\mathcal{P}$, define the *generalized Ehrhart polynomial*

$$E_k(t) := \sum_{j=0}^{k} (-1)^j \binom{d-j}{k-j} \sum_{\substack{\mathcal{F} \subseteq \mathcal{P} \\ \dim \mathcal{F} = d-j}} L_{\mathcal{F}}(t),$$

for $0 \le k \le d$. Prove the following generalized reciprocity law: for $0 \le k \le d$,

$$E_k(-t) = (-1)^d \, E_{d-k}(t),$$

which implies Ehrhart–Macdonald reciprocity (Theorem 4.1) for simple integral polytopes when $k = 0$.

5.13. What happens when $\mathcal{P}$ is *not* simple? Give an example for which Theorem 5.3 fails.

5.14. This exercise shows that sometimes, we may use the pigeonhole principle in a surprising way to find geometric structure.

(a) Given any collection of five points on the unit sphere in $\mathbb{R}^3$, show that there is a closed half-space containing at least four of them.
(b) Suppose we are given a collection of five vectors in $\mathbb{R}^3$ such that no two of those vectors lie in a hyperplane. Show that we can always find four of them that form the edges of a 3-dimensional *pointed* cone.

5.15. Let $\mathcal{P} \subset \mathbb{R}^3$ be a 3-dimensional integral polytope. Show that the sum of the relative areas of the facets of $\mathcal{P}$ equals the number of integer points on the boundary of $\mathcal{P}$ minus 2.

5.16. Give an alternative proof of Theorem 5.6 by considering $L_{\mathcal{P}}(t) - L_{\mathcal{P}^\circ}(t)$ as the lattice-point enumerator of the boundary of $\mathcal{P}$.

Chapter 6
Magic Squares

The peculiar interest of magic squares and all lusus numerorum *in general lies in the fact that they possess the charm of mystery.*

W. S. Andrews

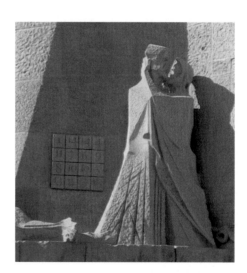

Fig. 6.1 Magic square at the *Temple de la Sagrada Família* (Barcelona, Spain).

Equipped with a solid base of theoretical results, we are now ready to return to computations. We use Ehrhart theory to assist us in enumerating *magic squares*.

Loosely speaking, a magic square is an $n \times n$ array of integers (usually required to be positive, often restricted to the numbers $1, 2, \ldots, n^2$, usually required to have distinct entries) whose sum along every row, column, and main diagonal is the same number, called the *magic sum*. Magic squares have turned up time and again, some in mathematical contexts, others in philosophical or religious contexts. According to legend, the first magic square (the ancient *Luo Shu square*) was discovered in China sometime before the first century B.C.E. on the back of a turtle emerging from a river. It looked like this:

4	9	2
3	5	7
8	1	6

Our task in this chapter is to develop a theory for counting certain classes of magic squares, which we now introduce.

6.1 It's a Kind of Magic

One should notice that the Luo Shu square has the distinct entries $1, 2, \ldots, 9$, so these entries are distinct positive integers drawn from a particular set. Such requirements are too restrictive for our purposes. We define a **semimagic square** to be a square matrix whose entries are nonnegative integers and whose rows and columns sum to the same number. A **magic square** is a semimagic square whose main diagonals also add up to the line sum. Here is one example each of a semimagic and a magic square:

3	0	0
0	1	2
0	2	1

1	2	0
0	1	2
2	0	1

We caution the reader about clashing definitions in the literature. For example, some people reserve the term *magic square* for what we will call a **traditional magic square**, a magic square of order n whose entries are the distinct integers $1, 2, \ldots, n^2$. (The Luo Shu square is an example of a traditional magic square.) Others are slightly less restrictive and use the term *magic square* for a magic square with *distinct* entries. We stress that we do not make this requirement in this chapter.

Our goal is to count semimagic and magic squares. In the traditional case, this is in some sense not very interesting:[1] for each order, there is a fixed number of traditional magic squares. For example, there are 7040 traditional 4×4 magic squares.

The situation becomes more interesting if we drop the condition of traditionality and study the number of magic squares as a function of the line sum.

---

[1] It is, nevertheless, an incredibly hard problem to count all traditional magic squares of a given size n. At present, these numbers are known only for $n \leq 5$ [1, Sequence A006052].

We denote the total numbers of semimagic and magic squares of order n and line sum t by $H_n(t)$ and $M_n(t)$, respectively.

$\heartsuit$	$t - \heartsuit$
$t - \heartsuit$	$\heartsuit$

$\dfrac{t}{2}$	$\dfrac{t}{2}$
$\dfrac{t}{2}$	$\dfrac{t}{2}$

Fig. 6.2 Semimagic and magic squares for $n = 2$.

Example 6.1. We illustrate these notions for the case $n = 2$, which is not very complicated. Here a semimagic square is determined once we know one entry, say the upper left one, denoted by $\heartsuit$ in Figure 6.2. Because of the upper row sum, the upper right entry has to be $t - \heartsuit$, as does the lower left entry (because of the left column sum). But then the lower right entry has to be $t - (t - \heartsuit) = \heartsuit$ (for two reasons: the lower row sum and the right column sum). The entry $\heartsuit$ can be any integer between 0 and t. Since there are $t + 1$ such integers,

$$H_2(t) = t + 1. \tag{6.1}$$

In the magic case, we have also to think of the diagonals. Looking back at our semimagic square in Figure 6.2, we see that the first diagonal gives $2 \cdot \heartsuit = t$, or $\heartsuit = \frac{t}{2}$. In this case, $t - \heartsuit = \frac{t}{2}$, and so a 2×2 magic square has to have identical entries in each position. Because we require the entries to be integers, this is possible only if t is even, in which case we obtain precisely 1 solution, the square on the right in Figure 6.2. That is,

$$M_2(t) = \begin{cases} 1 & \text{if } t \text{ is even,} \\ 0 & \text{if } t \text{ is odd.} \end{cases}$$

These easy results already hint at something: the counting function H_n is of a different character from that of the function M_n.   □

6.2 Semimagic Squares: Integer Points in the Birkhoff–von Neumann Polytope

Just as the Frobenius problem was intrinsically connected to questions about integer points on line segments, triangles, and higher-dimensional simplices, magic squares and their relatives have a life in the world of geometry. The most famous example is connected to semimagic squares.

A semimagic $n \times n$ square has n^2 nonnegative entries that sum to the same number along every row and along every column. Consider, therefore, the polytope

$$
\mathcal{B}_n := \left\{ \begin{pmatrix} x_{11} & \cdots & x_{1n} \\ \vdots & & \vdots \\ x_{n1} & \cdots & x_{nn} \end{pmatrix} \in \mathbb{R}^{n^2} : x_{jk} \geq 0, \begin{array}{l} \sum_j x_{jk} = 1 \text{ for all } 1 \leq k \leq n \\ \sum_k x_{jk} = 1 \text{ for all } 1 \leq j \leq n \end{array} \right\},
$$

(6.2)

consisting of nonnegative real matrices, in which all rows and columns sum to 1. The polytope $\mathcal{B}_n$ is called the nth **Birkhoff–von Neumann polytope**, in honor of Garrett Birkhoff (1911–1996)[2] and John von Neumann (1903–1957).[3] Because the matrices contained in the Birkhoff–von Neumann polytope appear frequently in probability and statistics (the line sum 1 representing probability 1), $\mathcal{B}_n$ is often described as the set of all $n \times n$ *doubly stochastic matrices*.

Geometrically, $\mathcal{B}_n$ is a subset of $\mathbb{R}^{n^2}$ and as such is difficult to picture once n exceeds 1.[4] However, we can get a glimpse of $\mathcal{B}_2 \subset \mathbb{R}^4$ when we think about what form points in $\mathcal{B}_2$ can possibly attain. Very much in sync with Figure 6.2, such a point is determined by its upper left entry $\heartsuit$ and looks like this:

$$
\begin{pmatrix} \heartsuit & 1 - \heartsuit \\ 1 - \heartsuit & \heartsuit \end{pmatrix}.
$$

The entry $\heartsuit$ is a real number between 0 and 1, which suggests that $\mathcal{B}_2$ should look like a line segment in 4-space. Indeed, the vertices of $\mathcal{B}_2$ should be given by $\heartsuit = 0$ and $\heartsuit = 1$, that is, by the points

$$
\begin{pmatrix} 1 & 0 \\ 0 & 1 \end{pmatrix}, \begin{pmatrix} 0 & 1 \\ 1 & 0 \end{pmatrix} \in \mathcal{B}_2.
$$

These results generalize: $\mathcal{B}_n$ is an $(n-1)^2$-polytope (see Exercise 6.3) whose vertices (Exercise 6.5) are the **permutation matrices**, namely, those $n \times n$ matrices that have precisely one 1 in each row and column (every other entry being zero). For dimensional reasons, we can talk about the continuous volume of $\mathcal{B}_n$ only in the relative sense, following the definition of Section 5.4.

The connection of the semimagic counting function $H_n(t)$ to the Birkhoff–von Neumann polytope $\mathcal{B}_n$ becomes clear in the light of the lattice-point enumerator for $\mathcal{B}_n$: the counting function $H_n(t)$ enumerates precisely the integer points in $t\mathcal{B}_n$, that is,

---

[2] For more information about Birkhoff, see
http://www-history.mcs.st-andrews.ac.uk/Biographies/Birkhoff_Garrett.html.

[3] For more information about von Neumann, see
http://www-history.mcs.st-andrews.ac.uk/Biographies/Von_Neumann.html.

[4] And the case $n = 1$ is not terribly interesting: $\mathcal{B}_1 = \{1\}$ is a point.

$$H_n(t) = \# \left(t\mathcal{B}_n \cap \mathbb{Z}^{n^2} \right) = L_{\mathcal{B}_n}(t) \,.$$

We can say more after noticing that permutation matrices are *integer* points in $\mathcal{B}_n$, and so Ehrhart's theorem (Theorem 3.8) applies:

Theorem 6.2. $H_n(t)$ *is a polynomial in* t *of degree* $(n-1)^2$. $\square$

The fact that H_n is a polynomial—apart from being mathematically appealing—has the same nice computational consequence that we exploited in Section 3.7: we can calculate this counting function by interpolation. For example, to compute H_2, a linear polynomial, we need to know only two values. In fact, since we know that the constant term of H_2 is 1 (by Corollary 3.15), we need only one value. It is not hard to convince even a lay person that $H_2(1) = 2$ (which two semimagic squares are those?), from which we interpolate

$$H_2(t) = t + 1 \,.$$

To interpolate the polynomial H_3, we need to know four values aside from $H_3(0) = 1$. In fact, we do not even have to know that many values, because Ehrhart–Macdonald reciprocity (Theorem 4.1) assists us in computations. To see this, let $H_n^\circ(t)$ denote the number of $n \times n$ square matrices with *positive* integer entries summing up to t along each row and column. A moment's thought (Exercise 6.6) reveals that

$$H_n^\circ(t) = H_n(t - n) \,. \tag{6.3}$$

But there is a second relationship between H_n and H_n°, namely, $H_n^\circ(t)$ counts, by definition, the integer points in the *relative interior* of the Birkhoff–von Neumann polytope $\mathcal{B}_n$, that is, $H_n^\circ(t) = L_{\mathcal{B}_n^\circ}(t)$. Ehrhart–Macdonald reciprocity (Theorem 4.1) now gives

$$H_n^\circ(-t) = (-1)^{(n-1)^2} H_n(t) \,.$$

Combining this identity with (6.3) gives us a symmetry identity for the counting function for semimagic squares:

Theorem 6.3. *The polynomial* H_n *satisfies*

$$H_n(-n - t) = (-1)^{(n-1)^2} H_n(t)$$

and

$$H_n(-1) = H_n(-2) = \cdots = H_n(-n + 1) = 0 \,.   \square$$

The roots of H_n at the first $n-1$ negative integers follow from (Exercise 6.7)

$$H_n^\circ(1) = H_n^\circ(2) = \cdots = H_n^\circ(n - 1) = 0 \,.$$

Theorem 6.3 gives the degree of $\mathcal{B}_n$, and it implies that the numerator of the Ehrhart series of the Birkhoff–von Neumann polytope is palindromic:

Corollary 6.4. *The Ehrhart series of the Birkhoff–von Neumann polytope* $\mathcal{B}_n$ *has the form*

$$\mathrm{Ehr}_{\mathcal{B}_n}(z) = \frac{h^*_{(n-1)(n-2)} \, z^{(n-1)(n-2)} + \cdots + h^*_0}{(1-z)^{(n-1)^2+1}},$$

where $h^*_0, h^*_1, \ldots, h^*_{(n-1)(n-2)} \in \mathbb{Z}_{\geq 0}$ *satisfy* $h^*_k = h^*_{(n-1)(n-2)-k}$ *for* $0 \leq k \leq \frac{(n-1)(n-2)}{2}$.

Proof. Denote the Ehrhart series of $\mathcal{B}_n$ by

$$\mathrm{Ehr}_{\mathcal{B}_n}(z) = \frac{h^*_{(n-1)^2} \, z^{(n-1)^2} + \cdots + h^*_0}{(1-z)^{(n-1)^2+1}}.$$

The fact that $h^*_{(n-1)^2} = \cdots = h^*_{(n-1)^2-(n-2)} = 0$ follows from the second part of Theorem 6.3 and Theorem 4.5. The palindromicity of the numerator coefficients follows from the first part of Theorem 6.3 and Exercise 4.8: it implies

$$\mathrm{Ehr}_{\mathcal{B}_n}\left(\frac{1}{z}\right) = (-1)^{(n-1)^2+1} z^n \, \mathrm{Ehr}_{\mathcal{B}_n}(z),$$

which yields $h^*_k = h^*_{(n-1)(n-2)-k}$ on simplifying both sides of the equation. □

Let's return to the interpolation of H_3: Theorem 6.3 gives, in addition to $H_3(0) = 1$, the values

$$H_3(-3) = 1 \qquad \text{and} \qquad H_3(-1) = H_3(-2) = 0.$$

These four values together with $H_3(1) = 6$ (see Exercise 6.1) suffice to interpolate the quartic polynomial H_3, and one computes

$$H_3(t) = \frac{1}{8} t^4 + \frac{3}{4} t^3 + \frac{15}{8} t^2 + \frac{9}{4} t + 1. \tag{6.4}$$

This interpolation example suggests the use of a computer; we let it calculate enough values of H_n and then simply interpolate. As far as computations are concerned, however, we should not get too excited about the fact that we computed H_2 and H_3 so effortlessly. In general, the polynomial H_n has degree $(n-1)^2$, so we need to compute $(n-1)^2 + 1$ values of H_n to be able to interpolate. Of those, we know n (the constant term and the roots given by Theorem 6.3), so $n^2 - 3n + 2$ values of H_n remain to be computed. Ehrhart–Macdonald reciprocity reduces the number of values to be computed to $\frac{n^2-3n+2}{2}$. That is still a large number, as anyone can testify who has tried to get a computer to enumerate all semimagic 7×7 squares with line sum 15. Nevertheless, it is a fun fact that we can compute H_n for small n by interpolation. It is amusing to test one's computer against the constant-term computation we will outline below, and we invite the reader to try both.

For small n, interpolation is clearly superior to a constant-term computation in the spirit of Chapter 1. The turning point is somewhere around $n = 5$: the computer needs more and more time to compute the values $H_n(t)$ as t increases. Methods superior to interpolation are needed.

6.3 Magic Generating Functions and Constant-Term Identities

Now we will construct a generating function for H_n, for which we will use Theorem 2.13. The semimagic counting function H_n is the Ehrhart polynomial of the nth Birkhoff–von Neumann polytope $\mathcal{B}_n$, which, in turn, is defined as a set of matrices by (6.2). First we rewrite the definition of $\mathcal{B}_n$ to fit the general description (2.23) of a polytope. If we consider the points in $\mathcal{B}_n$ as column vectors in $\mathbb{R}^{n^2}$ (rather than as matrices in $\mathbb{R}^{n \times n}$), then

$$\mathcal{B}_n = \left\{ \mathbf{x} \in \mathbb{R}^{n^2}_{\geq 0} : \mathbf{A}\,\mathbf{x} = \mathbf{b} \right\},$$

where

$$\mathbf{A} = \begin{pmatrix} 1 \cdots 1 & & & \\ & 1 \cdots 1 & & \\ & & \ddots & \\ & & & 1 \cdots 1 \\ 1 & 1 & & 1 \\ \ddots & \ddots & \cdots & \ddots \\ 1 & 1 & & 1 \end{pmatrix} \in \mathbb{Z}^{2n \times n^2} \qquad (6.5)$$

(here we do not show the zero entries) and

$$\mathbf{b} = \begin{pmatrix} 1 \\ 1 \\ \vdots \\ 1 \end{pmatrix} \in \mathbb{Z}^{2n}.$$

From this description of $\mathcal{B}_n$, we can easily build the generating function for H_n. According to Theorem 2.13, for a general rational polytope $\mathcal{P} = \{\mathbf{x} \in \mathbb{R}^d_{\geq 0} : \mathbf{A}\,\mathbf{x} = \mathbf{b}\}$,

$$L_{\mathcal{P}}(t) = \operatorname{const}\left(\frac{1}{(1 - \mathbf{z}^{\mathbf{c}_1})(1 - \mathbf{z}^{\mathbf{c}_2}) \cdots (1 - \mathbf{z}^{\mathbf{c}_d})\,\mathbf{z}^{t\mathbf{b}}} \right),$$

where $\mathbf{c}_1, \mathbf{c}_2, \ldots, \mathbf{c}_d$ denote the columns of $\mathbf{A}$. In our special case, the columns of $\mathbf{A}$ are of a simple form: they contain exactly two 1's and elsewhere 0's. We need one generating-function variable for each row of $\mathbf{A}$. To keep things as clear as possible, we use $z_1, z_2, \ldots, z_n$ for the first n rows of $\mathbf{A}$ (representing the row constraints of $\mathcal{B}_n$) and $w_1, w_2, \ldots, w_n$ for the last n rows of $\mathbf{A}$ (representing the column constraints of $\mathcal{B}_n$). With this notation, Theorem 2.13 applied to $\mathcal{B}_n$ gives the following starting point for our computations:

Theorem 6.5. *The number* $H_n(t)$ *of semimagic* $n \times n$ *squares with line sum* t *satisfies*

$$H_n(t) = \text{const} \left(\frac{1}{\prod_{1 \le j,k \le n} (1 - z_j w_k) \left(\prod_{1 \le j \le n} z_j \prod_{1 \le k \le n} w_k \right)^t} \right). \qquad \square$$

This identity is of both theoretical and practical use. One can use it to compute H_3 and even H_4 by hand. For now, we work on refining it further, exemplified by the case $n = 2$.

We first note that in the formula for H_2, the variables w_1 and w_2 are separated, in the sense that we can write this formula as a product of two factors, one involving only w_1 and the other involving only w_2:

$$H_2(t) = \text{const} \left(\frac{1}{z_1^t z_2^t} \frac{1}{(1 - z_1 w_1)(1 - z_2 w_1) w_1^t} \frac{1}{(1 - z_1 w_2)(1 - z_2 w_2) w_2^t} \right).$$

Now we put an ordering on the constant-term computation: let's first compute the constant term with respect to w_2, then the one with respect to w_1. (We do not order the computation with respect to z_1 and z_2 just yet.) Since z_1, z_2, and w_1 are considered constants when we do constant-term computations with respect to w_2, we can simplify:

$$H_2(t) = \text{const}_{z_1, z_2} \left(\frac{1}{z_1^t z_2^t} \text{const}_{w_1} \left(\frac{1}{(1 - z_1 w_1)(1 - z_2 w_1) w_1^t} \right. \right.$$
$$\left. \left. \times \text{const}_{w_2} \left(\frac{1}{(1 - z_1 w_2)(1 - z_2 w_2) w_2^t} \right) \right) \right).$$

Now we can see the effect of the separate appearance of w_1 and w_2: the constant-term identity *factors*. This is very similar to the factoring that can appear in computations of integrals in several variables. Let's rewrite our identity to emphasize the factoring:

$$H_2(t) = \text{const}_{z_1, z_2} \left(\frac{1}{z_1^t z_2^t} \text{const}_{w_1} \left(\frac{1}{(1 - z_1 w_1)(1 - z_2 w_1) w_1^t} \right) \right.$$
$$\left. \times \text{const}_{w_2} \left(\frac{1}{(1 - z_1 w_2)(1 - z_2 w_2) w_2^t} \right) \right).$$

But now the expressions in the last two sets of parentheses are identical, except that in one case, the constant-term variable is called w_1, and in the other case, w_2. Since these are just "dummy" variables, we can call them w, and combine:

$$H_2(t) = \text{const}_{z_1,z_2}\left(\frac{1}{z_1^t z_2^t}\left(\text{const}_w \frac{1}{(1-z_1 w)(1-z_2 w)w^t}\right)^2\right).$$

(Note the square!) Naturally, all of this factoring works in the general case, and we invite the reader to prove it (Exercise 6.8):

$$H_n(t) = \text{const}_{z_1,\ldots,z_n}\left((z_1\cdots z_n)^{-t}\left(\text{const}_w \frac{1}{(1-z_1 w)\cdots(1-z_n w)w^t}\right)^n\right).$$

(6.6)

We can go further, namely, we can compute the innermost constant term

$$\text{const}_w \frac{1}{(1-z_1 w)\cdots(1-z_n w)w^t}.$$

It should come as no surprise that we shall use a partial fraction expansion to do so. The w-poles of the rational function are at $w = \frac{1}{z_1}$, $w = \frac{1}{z_2}$, ..., $w = \frac{1}{z_n}$, $w = 0$, and so we expand

$$\frac{1}{(1-z_1 w)\cdots(1-z_n w)w^t} = \frac{A_1}{w - \frac{1}{z_1}} + \frac{A_2}{w - \frac{1}{z_2}} + \cdots + \frac{A_n}{w - \frac{1}{z_n}} + \sum_{k=1}^{t}\frac{B_k}{w^k}. \quad (6.7)$$

Just as in Chapter 1, we can forget about the B_k-terms, since they do not contribute to the constant term, that is,

$$\text{const}_w \frac{1}{(1-z_1 w)\cdots(1-z_n w)w^t}$$

$$= \text{const}_w\left(\frac{A_1}{w - \frac{1}{z_1}} + \frac{A_2}{w - \frac{1}{z_2}} + \cdots + \frac{A_n}{w - \frac{1}{z_n}}\right)$$

$$= -A_1 z_1 - A_2 z_2 - \cdots - A_n z_n.$$

We invite the reader to show (Exercise 6.9) that

$$A_k = -\frac{z_k^{t-1}}{\left(1 - \frac{z_1}{z_k}\right)\cdots\left(1 - \frac{z_{k-1}}{z_k}\right)\left(1 - \frac{z_{k+1}}{z_k}\right)\cdots\left(1 - \frac{z_n}{z_k}\right)}$$

$$= -\frac{z_k^{t+n-2}}{\prod_{j\neq k}(z_k - z_j)}. \quad (6.8)$$

Putting these coefficients back into the partial fraction expansion yields the following identity.

Theorem 6.6. *The number $H_n(t)$ of semimagic $n \times n$ squares with line sum t satisfies*

$$H_n(t) = \text{const}\left((z_1 \cdots z_n)^{-t}\left(\sum_{k=1}^{n} \frac{z_k^{t+n-1}}{\prod_{j \neq k}(z_k - z_j)}\right)^n\right). \qquad \square$$

Amidst all this generality, we almost forgot to compute H_2 with our partial fraction approach. The last theorem says that

$$H_2(t) = \text{const}\left((z_1 z_2)^{-t}\left(\frac{z_1^{t+1}}{z_1 - z_2} + \frac{z_2^{t+1}}{z_2 - z_1}\right)^2\right)$$

$$= \text{const}\left(\frac{z_1^{t+2} z_2^{-t}}{(z_1 - z_2)^2} - 2\frac{z_1 z_2}{(z_1 - z_2)^2} + \frac{z_1^{-t} z_2^{t+2}}{(z_1 - z_2)^2}\right). \qquad (6.9)$$

It is now time to put more order on the constant-term computation. Let's say that we first compute the constant term with respect to z_1, and after that with respect to z_2. So we have to compute first

$$\text{const}_{z_1}\left(\frac{z_1^{t+2} z_2^{-t}}{(z_1 - z_2)^2}\right), \quad \text{const}_{z_1}\left(\frac{z_1 z_2}{(z_1 - z_2)^2}\right), \quad \text{and} \quad \text{const}_{z_1}\left(\frac{z_1^{-t} z_2^{t+2}}{(z_1 - z_2)^2}\right).$$

To obtain these constant terms, we need to expand the function $\frac{1}{(z_1-z_2)^2}$. As we know from calculus, this expansion depends on the order of the magnitudes of z_1 and z_2. For example, if $|z_1| < |z_2|$, then

$$\frac{1}{z_1 - z_2} = \frac{1}{z_2} \frac{1}{\frac{z_1}{z_2} - 1} = -\frac{1}{z_2}\sum_{k \geq 0}\left(\frac{z_1}{z_2}\right)^k = -\sum_{k \geq 0}\frac{1}{z_2^{k+1}} z_1^k,$$

and hence

$$\frac{1}{(z_1 - z_2)^2} = -\frac{d}{dz_1}\left(\frac{1}{z_1 - z_2}\right) = \sum_{k \geq 1}\frac{k}{z_2^{k+1}} z_1^{k-1} = \sum_{k \geq 0}\frac{k+1}{z_2^{k+2}} z_1^k.$$

So let's assume for the moment that $|z_1| < |z_2|$. This might sound funny, since z_1 and z_2 are *variables*. However, as such they are simply tools that enable us to compute some quantity that is independent of z_1 and z_2. In view of these ideas, we may assume any order of the magnitudes of the variables. In Exercise 6.11, we will check that indeed the order does not matter. Now,

$$\text{const}_{z_1}\left(\frac{z_1^{t+2} z_2^{-t}}{(z_1 - z_2)^2}\right) = z_2^{-t}\, \text{const}_{z_1}\left(\frac{z_1^{t+2}}{(z_1 - z_2)^2}\right)$$

$$= z_2^{-t}\, \text{const}_{z_1}\left(z_1^{t+2}\sum_{k \geq 0}\frac{k+1}{z_2^{k+2}} z_1^k\right) \qquad (6.10)$$

$$= z_2^{-t} \operatorname{const}_{z_1} \left(\sum_{k \geq 0} \frac{k+1}{z_2^{k+2}} z_1^{k+t+2} \right)$$

$$= 0 \,,$$

since there are only positive powers of z_1 (recall that $t \geq 0$). Analogously (see Exercise 6.10), one checks that

$$\operatorname{const}_{z_1} \left(\frac{z_1 z_2}{(z_1 - z_2)^2} \right) = 0 \,. \tag{6.11}$$

For the last constant term, we compute

$$\operatorname{const}_{z_1} \left(\frac{z_1^{-t} z_2^{t+2}}{(z_1 - z_2)^2} \right) = z_2^{t+2} \operatorname{const}_{z_1} \left(z_1^{-t} \sum_{k \geq 0} \frac{k+1}{z_2^{k+2}} z_1^k \right)$$

$$= z_2^{t+2} \operatorname{const}_{z_1} \left(\sum_{k \geq 0} \frac{k+1}{z_2^{k+2}} z_1^{k-t} \right).$$

The constant term on the right-hand side is the term with $k = t$, that is,

$$\operatorname{const}_{z_1} \left(\frac{z_1^{-t} z_2^{t+2}}{(z_1 - z_2)^2} \right) = z_2^{t+2} \frac{t+1}{z_2^{t+2}} = t+1 \,.$$

So of the three constant terms, only one survives, and with $\operatorname{const}_{z_2}(t+1) = t+1$ we recover what we have known since the beginning of this chapter:

$$H_2(t) = \operatorname{const} \left(\frac{z_1^{t+2} z_2^{-t}}{(z_1 - z_2)^2} - 2 \frac{z_1 z_2}{(z_1 - z_2)^2} + \frac{z_1^{-t} z_2^{t+2}}{(z_1 - z_2)^2} \right) = t+1 \,.$$

This was a lot of work for this seemingly trivial polynomial. Recall, for example, that we can get the same result by an easy interpolation. However, to compute a similar interpolation, e.g., for H_4, we likely would need to use a computer (to obtain the interpolation values). On the other hand, the constant-term computation of H_4 boils down to only five iterated constant terms, which can actually be computed by hand (see Exercise 6.14). The result is

$$H_4(t) = \frac{11}{11340} t^9 + \frac{11}{630} t^8 + \frac{19}{135} t^7 + \frac{2}{3} t^6 + \frac{1109}{540} t^5 + \frac{43}{10} t^4 + \frac{35117}{5670} t^3$$

$$+ \frac{379}{63} t^2 + \frac{65}{18} t + 1 \,.$$

6.4 The Enumeration of Magic Squares

What happens when we bring the diagonal constraints, which are not present in the semimagic case, into the magic picture? In the introduction of this chapter we have already seen an example, namely the number of 2×2 magic squares,

$$M_2(t) = \begin{cases} 1 & \text{if } t \text{ is even,} \\ 0 & \text{if } t \text{ is odd.} \end{cases}$$

This is a very simple example of a quasipolynomial. In fact, like H_n, the counting function M_n is defined by integral linear equations and inequalities, so it is the lattice-point enumerator of a rational polytope, and Theorem 3.23 gives at once the following result.

Theorem 6.7. *The counting function $M_n(t)$ is a quasipolynomial in t.* □

We invite the reader to prove that the degree of M_n is $n^2 - 2n - 1$ (Exercise 6.16).

Let's see what happens in the first nontrivial case, 3×3 magic squares. We follow our recipe and assign variables $m_1, m_2, \ldots, m_9$ to the entries of our 3×3 squares:

m_1	m_2	m_3
m_4	m_5	m_6
m_7	m_8	m_9

The magic conditions require now that $m_1, m_2, \ldots, m_9 \in \mathbb{Z}_{\geq 0}$ and

$$m_1 + m_2 + m_3 = t\,,$$
$$m_4 + m_5 + m_6 = t\,,$$
$$m_7 + m_8 + m_9 = t\,,$$
$$m_1 + m_4 + m_7 = t\,,$$
$$m_2 + m_5 + m_8 = t\,,$$
$$m_3 + m_6 + m_9 = t\,,$$
$$m_1 + m_5 + m_9 = t\,,$$
$$m_3 + m_5 + m_7 = t\,,$$

according to the row sums (the first three equations), the column sums (the next three equations), and the diagonal sums (the last two equations). By now, we are experienced in translating this system into a generating function: we need one variable for each equation, so let's take z_1, z_2, z_3 for the first three, w_1, w_2, w_3 for the next three, and y_1, y_2 for the last two equations. The function $M_3(t)$ is thus the constant term of

$$\frac{1}{(1 - z_1 w_1 y_1)(1 - z_1 w_2)(1 - z_1 w_3 y_2)(1 - z_2 w_1)(1 - z_2 w_2 y_1 y_2)(1 - z_2 w_3)}$$

$$\times \frac{1}{(1 - z_3 w_1 y_2)(1 - z_3 w_2)(1 - z_3 w_3 y_1)(z_1 z_2 z_3 w_1 w_2 w_3 y_1 y_2)^t} . \tag{6.12}$$

It does take some work, but it is instructive to compute this constant term (just try it!). The result is

$$M_3(t) = \begin{cases} \frac{2}{9} t^2 + \frac{2}{3} t + 1 & \text{if } 3|t, \\ 0 & \text{otherwise.} \end{cases} \tag{6.13}$$

As predicted by Theorem 6.7, M_3 is a quasipolynomial. It has degree 2 and period 3. This may be more apparent if we rewrite it as

$$M_3(t) = \begin{cases} \frac{2}{9} t^2 + \frac{2}{3} t + 1 & \text{if } t \equiv 0 \bmod 3, \\ 0 & \text{if } t \equiv 1 \bmod 3, \\ 0 & \text{if } t \equiv 2 \bmod 3, \end{cases}$$

and we can see the three constituents of the quasipolynomial M_3. There is an alternative way to describe M_3; namely, let

$$c_2(t) = \begin{cases} \frac{2}{9} & \text{if } t \equiv 0 \bmod 3, \\ 0 & \text{if } t \equiv 1 \bmod 3, \\ 0 & \text{if } t \equiv 2 \bmod 3, \end{cases}$$

$$c_1(t) = \begin{cases} \frac{2}{3} & \text{if } t \equiv 0 \bmod 3, \\ 0 & \text{if } t \equiv 1 \bmod 3, \\ 0 & \text{if } t \equiv 2 \bmod 3, \end{cases}$$

$$c_0(t) = \begin{cases} 1 & \text{if } t \equiv 0 \bmod 3, \\ 0 & \text{if } t \equiv 1 \bmod 3, \\ 0 & \text{if } t \equiv 2 \bmod 3. \end{cases}$$

Then the quasipolynomial M_3 can be written as

$$M_3(t) = c_2(t) t^2 + c_1(t) t + c_0(t) .$$

Notes

1. Magic squares date back to China in the first millennium B.C.E. [77]; they underwent much further development in the Islamic world late in the first millennium C.E. and in the next millennium (or sooner; the data are lacking) in India [78]. From Islam, they passed to Christian Europe in the later Middle

Ages, probably initially through the Jewish community [78, Part II, pp. 290 ff.] and later possibly Byzantium [78, Part I, p. 198], and no later than the early eighteenth century (the data are buried in barely tapped archives) to sub-Saharan Africa [256, Chapter 12]. The contents of a magic square have varied with time and writer; usually they have been the first n^2 consecutive integers, but often any arithmetic sequence or arbitrary positive numbers. In the past century, mathematicians have made some simplifications in the interest of obtaining results about the number of squares with a fixed magic sum, in particular, allowing repeated entries as in this chapter.

2. The problem of counting magic squares (other than traditional magic squares) seems not to have occurred to anyone before the twentieth century, no doubt because there was no way to approach the question previously. The first nontrivial formulas addressing the counting problem, namely (6.4) and (6.13) for H_3 and M_3, were established by Percy Macmahon in 1915 [167]. Recently, there has grown up a literature on exact formulas (see, for example, [110, 224] for semimagic squares; for magic squares, see [2, 35]; for magic squares with distinct entries, see [47, 48, 254]).

3. Another famous kind of square is a *Latin square* (see, for example, [101]). Here each row and column has n different numbers, the same n numbers in every row/column (usually taken to be the first n positive integers). There are counting problems associated with Latin squares, which can be attacked using Ehrhart theory [48] (see also [1, Sequence A002860]).

4. Recent work includes mathematical-historical research, such as the discovery of unpublished magic squares of Benjamin Franklin [3, 187]. Aside from mathematical research, magic squares and their siblings naturally continue to be an excellent source of topics for popular mathematics books (see, for example, [7, 191]).

5. The Birkhoff–von Neumann polytope $\mathcal{B}_n$ possesses fascinating combinatorial properties [56,69,70,83,257] and relates to many mathematical areas [104,151]. Its name honors Garrett Birkhoff and John von Neumann, who proved that the extremal points of $\mathcal{B}_n$ are the permutation matrices [57,251] (see Exercise 6.5). A long-standing open problem is the determination of the relative volume of $\mathcal{B}_n$, which is known only for $n \leq 10$ [1, Sequence A037302]. In fact, the last two records ($n = 9$ and 10) for computing vol $\mathcal{B}_n$ rely on the theory of counting functions that is introduced in this book, more precisely, Theorem 6.6 [42].

6. An important generalization of the Birkhoff–von Neumann polytope is given by *transportation polytopes*, which consist of *contingency tables*. They have applications to statistics and in particular to disclosure limitation procedures [97]. The Birkhoff–von Neumann polytopes are special transportation polytopes that consist of two-way contingency tables with given 1-marginals.

7. The polynomiality of H_n (Theorem 6.2) and its symmetry (Theorem 6.3) were conjectured in 1966 by Harsh Anand, Vishwa Dumir, and Hansraj Gupta [6] and proved seven years later independently by Eugène Ehrhart [110] and Richard Stanley [224]. See [72,231] for more about the history of Theorems 6.2 and 6.3 and their connections to commutative algebra. Stanley also conjectured that the numerator coefficients in Corollary 6.4 are unimodal, a fact that was proved only in 2005, by Christos Athanasiadis [13] (see also [74]). The quasipolynomiality of M_n (Theorem 6.7) and its degree are discussed in [35]. The period of M_n is in general not known. In [35], it is conjectured that it is always nontrivial for $n > 1$. The work in [2] gives some credence to this conjecture by proving that the polytope of magic $n \times n$ squares is *not* integral for $n \geq 2$.

8. We close with a story about Cornelius Agrippa's *De Occulta Philosophia*, written in 1510. In it he describes the spiritual powers of magic squares and produces some squares of orders from three up to nine. His work, although influential in the mathematical community, enjoyed only brief success, for the Counter-Reformation and the witch hunts of the Inquisition began soon thereafter: Agrippa himself was accused of being allied with the devil.

Exercises

6.1. ♣ Find and prove a formula for $H_n(1)$.

6.2. Let $(x_{ij})_{1 \leq i,j \leq 3}$ be a magic 3×3 square.

(a) Show that the center term x_{22} is the average over all x_{ij}.
(b) Show that $M_3(t) = 0$ if 3 does not divide t.

6.3. ♣ Prove that $\dim \mathcal{B}_n = (n-1)^2$.

6.4. Prove the following characterization of a vertex of a convex polytope $\mathcal{P}$: A point $\mathbf{v} \in \mathcal{P}$ is a vertex of $\mathcal{P}$ if and only if for every line L through $\mathbf{v}$ and every neighborhood N of $\mathbf{v}$, there exists a point in $L \cap N$ that is not in $\mathcal{P}$.

6.5. ♣ Prove that the vertices of $\mathcal{B}_n$ are the $n \times n$ permutation matrices.

6.6. ♣ Let $H_n^\circ(t)$ denote the number of $n \times n$ matrices with *positive* integer entries summing up to t along each row and column. Show that $H_n^\circ(t) = H_n(t-n)$ for $t > n$.

6.7. ♣ Show that $H_n^\circ(1) = H_n^\circ(2) = \cdots = H_n^\circ(n-1) = 0$.

6.8. ♣ Prove (6.6):

$$H_n(t) = \mathrm{const}_{z_1,\ldots,z_n}\left((z_1 \cdots z_n)^{-t} \left(\mathrm{const}_w \frac{1}{(1-z_1 w)\cdots(1-z_n w)w^t} \right)^n \right).$$

6.9. ♣ Compute the partial fraction coefficients (6.8).

6.10. ♣ Verify (6.11).

6.11. Repeat the constant-term computation of H_2 starting from (6.9), but now by first computing the constant term with respect to z_2, and after that with respect to z_1.

6.12. Use your favorite computer program to calculate the formula for $H_3(t)$, $H_4(t), \dots$ by interpolation.

6.13. Compute H_3 using Theorem 6.6.

6.14. Compute H_4 using Theorem 6.6.

6.15. Show that

$$\sum_{k=1}^{n} \frac{z_k^{t+n-1}}{\prod_{j \neq k}(z_k - z_j)} = \sum_{m_1+\cdots+m_n=t} z_1^{m_1} \cdots z_n^{m_n},$$

and use this identity to give an alternative proof of Theorem 6.6.

6.16. ♣ Prove that for $n \geq 3$, the degree of M_n is $n^2 - 2n - 1$.

6.17. Explain the period of M_3 by computing the vertices of the polytope of 3×3 magic squares.

6.18. ♣ Verify (6.12) and use it to compute M_3.

6.19. Compute M_3 by interpolation. (*Hint:* Use Exercises 6.2 and 6.17.)

6.20. A **symmetric semimagic square** is a semimagic square that is a symmetric matrix. Show that the number of *symmetric* semimagic $n \times n$ squares with line sum t is a quasipolynomial in t. Determine its degree and period.

Open Problems

6.21. Compute the number of *traditional* magic $n \times n$ squares for $n > 5$.

6.22. Compute $\mathrm{vol}\,\mathcal{B}_n$ for $n > 10$. Compute H_n for $n > 9$.

6.23. Prove that the period of M_n is nontrivial for $n > 1$.

6.24. The vertices of the Birkhoff–von Neumann polytope are in one-to-one correspondence with the elements of the symmetric group S_n. Consider a subgroup of S_n and take the convex hull of the corresponding permutation matrices. Compute the Ehrhart polynomials of this polytope. (The face numbers of the polytope corresponding to the subgroup A_n, the even permutations, were studied in [141].)

6.25. Consider the rational polytope $\mathcal{S}_n$ formed by all *symmetric* $n \times n$ doubly stochastic matrices (see Exercise 6.20).

(a) Write $\mathrm{Ehr}_{\mathcal{S}_n}(z)$ as a rational function with denominator $(1 - z^2)^{\binom{n}{2}+1}$. Prove that the numerator polynomial is palindromic.

(b) Prove that the numerator polynomial of $\mathrm{Ehr}_{2\mathcal{S}_n}(z)$ is palindromic.

(See [92] for partial results.)

6.26. Prove that the graph formed by the vertices and edges of every 2-way transportation polytope is Hamiltonian.

Part II
Beyond the Basics

Chapter 7
Finite Fourier Analysis

God created infinity, and man, unable to understand infinity, created finite sets.

Gian–Carlo Rota (1932–1999)

We now consider the vector space of all complex-valued periodic functions on the integers with period b. It turns out that every such function $a(n)$ on the integers can be written as a polynomial in the bth root of unity $\xi^n := e^{2\pi i n/b}$. Such a representation for $a(n)$ is called a **finite Fourier series**. Here we develop finite Fourier theory using rational functions and their partial fraction decomposition. We then define the *Fourier transform* and the *convolution* of finite Fourier series, and show how one can use these ideas to prove identities on trigonometric functions, as well as find connections to the classical *Dedekind sums*.

The more we know about roots of unity and their various sums, the deeper are the results that we can prove (see Exercise 7.20); in fact, certain statements about sums of roots of unity even imply the Riemann hypothesis! However, this chapter is elementary and draws connections to the sawtooth functions and Dedekind sums, two basic sums over roots of unity. The general philosophy here is that finite sums of rational functions of roots of unity are basic ingredients in many mathematical structures.

7.1 A Motivating Example

To ease the reader into the general theory, let's work out the finite Fourier series for a simple example first, an arithmetic function with a period of 3.

Example 7.1. Consider the following arithmetic function, of period 3:

$$n : 0, 1, 2, 3, 4, 5, \ldots$$
$$a(n) : 1, 5, 2, 1, 5, 2, \ldots$$

We first embed this sequence into a generating function as follows:

$$F(z) := 1 + 5z + 2z^2 + z^3 + 5z^4 + 2z^5 + \cdots = \sum_{n \geq 0} a(n)\, z^n.$$

Since the sequence is periodic, we can simplify $F(z)$ using a geometric series argument:

$$F(z) = \sum_{n \geq 0} a(n)\, z^n$$
$$= 1 + 5z + 2z^2 + z^3 \left(1 + 5z + 2z^2\right) + z^6 \left(1 + 5z + 2z^2\right) + \cdots$$
$$= \left(1 + 5z + 2z^2\right) \sum_{k \geq 0} z^{3k}$$
$$= \frac{1 + 5z + 2z^2}{1 - z^3}.$$

We now use the same technique that was employed in Chapter 1, namely the technique of expanding a rational function into its partial fraction decomposition. Here all the poles are simple, and located at the three cube roots of unity, so that

$$F(z) = \frac{\hat{a}(0)}{1 - z} + \frac{\hat{a}(1)}{1 - \rho z} + \frac{\hat{a}(2)}{1 - \rho^2 z}, \qquad (7.1)$$

where the constants $\hat{a}(0), \hat{a}(1), \hat{a}(2)$ remain to be found, and where $\rho := e^{2\pi i/3}$, a third root of unity. Using the geometric series for each of these terms separately, we arrive at

$$F(z) = \sum_{n \geq 0} \left(\hat{a}(0) + \hat{a}(1)\rho^n + \hat{a}(2)\rho^{2n}\right) z^n,$$

so that we have derived the finite Fourier series of our sequence $a(n)$. The only remaining piece of information that we need is the computation of the constants $\hat{a}(j)$, for $j = 0, 1, 2$. It turns out that this is also quite easy to do. We have, from (7.1) above, the identity

$$\hat{a}(0) \left(1 - \rho z\right) \left(1 - \rho^2 z\right) + \hat{a}(1) \left(1 - z\right) \left(1 - \rho^2 z\right) + \hat{a}(2) \left(1 - z\right) \left(1 - \rho z\right)$$
$$= 1 + 5z + 2z^2,$$

valid for all $z \in \mathbb{C}$. On letting $z = 1$, ρ^2, and ρ, respectively, we obtain

$$3\,\hat{a}(0) = 1 + 5 + 2\,,$$
$$3\,\hat{a}(1) = 1 + 5\rho^2 + 2\rho^4\,,$$
$$3\,\hat{a}(2) = 1 + 5\rho + 2\rho^2\,,$$

where we have used the identity $(1 - \rho)(1 - \rho^2) = 3$ (see Exercise 7.2). We can simplify a bit to get $\hat{a}(0) = \frac{8}{3}$, $\hat{a}(1) = \frac{-4-3\rho}{3}$, and $\hat{a}(2) = \frac{-1+3\rho}{3}$. Thus the finite Fourier series for our sequence is

$$a(n) = \frac{8}{3} + \left(-\frac{4}{3} - \rho\right)\rho^n + \left(-\frac{1}{3} + \rho\right)\rho^{2n}. \qquad \square$$

The object of the next section is to show that this simple process follows just as easily for every periodic function on $\mathbb{Z}$. The ensuing sections contain some applications of the finite Fourier series of periodic functions.

7.2 Finite Fourier Series for Periodic Functions on $\mathbb{Z}$

The general theory is just as easy conceptually as Example 7.1, and we now develop it. Consider a periodic sequence on $\mathbb{Z}$, defined by $\{a(n)\}_{n=0}^{\infty}$, of period b. Throughout the chapter, we fix the bth root of unity $\xi := e^{2\pi i/b}$. As before, we embed our periodic sequence $\{a(n)\}_{n=0}^{\infty}$ into a generating function,

$$F(z) := \sum_{n\geq 0} a(n)\,z^n,$$

and use the periodicity of the sequence to immediately get

$$F(z) = \left(\sum_{k=0}^{b-1} a(k)\,z^k\right) + \left(\sum_{k=0}^{b-1} a(k)\,z^k\right) z^b + \left(\sum_{k=0}^{b-1} a(k)\,z^k\right) z^{2b} + \cdots$$
$$= \frac{\sum_{k=0}^{b-1} a(k)\,z^k}{1 - z^b} = \frac{P(z)}{1 - z^b},$$

where the last step simply defines the polynomial $P(z) = \sum_{k=0}^{b-1} a(k)\,z^k$. Now we expand the rational generating function $F(z)$ into partial fractions, as before:

$$F(z) = \frac{P(z)}{1 - z^b} = \sum_{m=0}^{b-1} \frac{\hat{a}(m)}{1 - \xi^m z}.$$

As in Example 7.1, we expand each of the terms $\frac{1}{1-\xi^m z}$ as a geometric series, and substitute into the sum above to get

$$F(z) = \sum_{n \geq 0} a(n)\, z^n = \sum_{m=0}^{b-1} \frac{\hat{a}(m)}{1 - \xi^m z}$$

$$= \sum_{m=0}^{b-1} \hat{a}(m) \sum_{n \geq 0} \xi^{mn} z^n = \sum_{n \geq 0} \left(\sum_{m=0}^{b-1} \hat{a}(m)\, \xi^{mn} \right) z^n.$$

Comparing the coefficients of any fixed z^n gives us the finite Fourier series for $a(n)$, namely

$$a(n) = \sum_{m=0}^{b-1} \hat{a}(m)\, \xi^{mn}.$$

We now find a formula for the Fourier coefficients $\hat{a}(n)$, as in the example. To recapitulate,

$$P(z) = \sum_{m=0}^{b-1} \hat{a}(m) \frac{1 - z^b}{1 - \xi^m z}.$$

To solve for $P(\xi^{-n})$, we note that

$$\lim_{z \to \xi^{-n}} \frac{1 - z^b}{1 - \xi^m z} = 0 \qquad \text{if } m - n \not\equiv 0 \bmod b,$$

and

$$\lim_{z \to \xi^{-n}} \frac{1 - z^b}{1 - \xi^m z} = \lim_{z \to \xi^{-n}} \frac{b z^{b-1}}{\xi^m} = b\, \xi^{n-m} = b \qquad \text{if } m - n \equiv 0 \bmod b.$$

Thus $P(\xi^{-n}) = b\,\hat{a}(n)$, and so

$$\hat{a}(n) = \frac{1}{b} P(\xi^{-n}) = \frac{1}{b} \sum_{k=0}^{b-1} a(k)\, \xi^{-nk}.$$

We have just proved the main result of finite Fourier series, using only elementary properties of rational functions:

Theorem 7.2 (Finite Fourier series expansion and Fourier inversion). *Let $a(n)$ be any periodic function on $\mathbb{Z}$, with period b. Then we have the following finite Fourier series expansion:*

$$a(n) = \sum_{k=0}^{b-1} \hat{a}(k)\, \xi^{nk},$$

where the Fourier coefficients are

$$\hat{a}(n) = \frac{1}{b} \sum_{k=0}^{b-1} a(k)\, \xi^{-nk}, \tag{7.2}$$

with $\xi = e^{2\pi i/b}$. □

The coefficients $\hat{a}(m)$ are known as the **Fourier coefficients** of the function $a(n)$, and if $\hat{a}(m) \neq 0$, we sometimes say that the function **has frequency** m. The finite Fourier series of a periodic function provides us with surprising power and insight into its structure. We are able to analyze the function using its frequencies (only finitely many), and this window into the frequency domain becomes indispensable for computations and simplifications.

We note that the Fourier coefficients $\hat{a}(n)$ and the original sequence elements $a(n)$ are related by a linear transformation given by the matrix

$$L := \left(\xi^{(i-1)(j-1)} \right), \tag{7.3}$$

where $1 \leq i, j \leq b$, as is evident from (7.2) in the proof above. We further note that the second half of the proof, namely solving for the Fourier coefficients $\hat{a}(n)$, is just tantamount to inverting this matrix L.

One of the main building blocks of our lattice-point enumeration formulas in polytopes is the **sawtooth function**, defined by

$$((x)) := \begin{cases} \{x\} - \frac{1}{2} & \text{if } x \notin \mathbb{Z}, \\ 0 & \text{if } x \in \mathbb{Z}. \end{cases} \tag{7.4}$$

(As a reminder, $\{x\} = x - \lfloor x \rfloor$ is the fractional part of x.) The graph of this function is displayed in Figure 7.1. We have seen a closely related function before, in Chapter 1, in our study of the coin-exchange problem. Equation (1.8) gave us the finite Fourier series for essentially this function from the discrete-geometry perspective of the coin-exchange problem; however, we now compute the finite Fourier series for this periodic function directly, pretending that we do not know about its other life as a counting function.

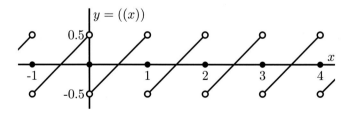

Fig. 7.1 The sawtooth function $y = ((x))$.

Lemma 7.3. *The finite Fourier series for the discrete sawtooth function $\left(\left(\frac{a}{b} \right) \right)$, a periodic function of $a \in \mathbb{Z}$ with period b, is given by*

$$\left(\!\left(\frac{a}{b}\right)\!\right) = \frac{1}{2b} \sum_{k=1}^{b-1} \frac{1+\xi^k}{1-\xi^k}\, \xi^{ak} = \frac{i}{2b} \sum_{k=1}^{b-1} \cot\frac{\pi k}{b}\, \xi^{ak}.$$

Here the second equality follows from $\frac{1+e^{2\pi i x}}{1-e^{2\pi i x}} = i\cot(\pi x)$, by the definition of the cotangent.

Proof. Using Theorem 7.2, we know that our periodic function has a finite Fourier series $\left(\!\left(\frac{a}{b}\right)\!\right) = \sum_{k=0}^{b-1} \hat{a}(k)\,\xi^{ak}$, where

$$\hat{a}(k) = \frac{1}{b} \sum_{m=0}^{b-1} \left(\!\left(\frac{m}{b}\right)\!\right) \xi^{-mk}.$$

We first compute $\hat{a}(0) = \frac{1}{b} \sum_{m=0}^{b-1} \left(\!\left(\frac{m}{b}\right)\!\right) = 0$, by Exercise 7.15. For $k \neq 0$,

$$\hat{a}(k) = \frac{1}{b} \sum_{m=1}^{b-1} \left(\frac{m}{b} - \frac{1}{2}\right) \xi^{-mk} = \frac{1}{b^2} \sum_{m=1}^{b-1} m\,\xi^{-mk} + \frac{1}{2b}$$

$$= \frac{1}{b}\left(\frac{\xi^k}{1-\xi^k} + \frac{1}{2}\right) = \frac{1}{2b}\frac{1+\xi^k}{1-\xi^k},$$

where we used Exercise 7.5 in the penultimate equality above. $\square$

We define the **Dedekind sum** by

$$s(a,b) = \sum_{k=0}^{b-1} \left(\!\left(\frac{ka}{b}\right)\!\right) \left(\!\left(\frac{k}{b}\right)\!\right),$$

for any two relatively prime integers a and $b > 0$ (although it may be easily extended to all integers—see Exercise 8.11). Note that the Dedekind sum is a periodic function of the variable a, with period b, by the periodicity of the sawtooth function. That is,

$$s(a+jb,b) = s(a,b) \qquad \text{for all } j \in \mathbb{Z}. \tag{7.5}$$

Using the finite Fourier series for the sawtooth function, we can now easily reformulate the Dedekind sums as a finite sum over the bth roots of unity or cotangents:

Lemma 7.4.

$$s(a,b) = \frac{1}{4b} \sum_{\mu=1}^{b-1} \frac{1+\xi^\mu}{1-\xi^\mu}\frac{1+\xi^{-\mu a}}{1-\xi^{-\mu a}} = \frac{1}{4b} \sum_{\mu=1}^{b-1} \cot\frac{\pi\mu}{b}\cot\frac{\pi\mu a}{b}.$$

We will see and use this alternative formulation of the Dedekind sum time and again.

Proof.

$$s(a,b) = \sum_{k=0}^{b-1} \left(\left(\frac{ka}{b} \right) \right) \left(\left(\frac{k}{b} \right) \right)$$

$$= \frac{1}{4b^2} \sum_{k=0}^{b-1} \left(\left(\sum_{\mu=1}^{b-1} \frac{1+\xi^\mu}{1-\xi^\mu} \xi^{\mu ka} \right) \left(\sum_{\nu=1}^{b-1} \frac{1+\xi^\nu}{1-\xi^\nu} \xi^{\nu k} \right) \right)$$

$$= \frac{1}{4b^2} \sum_{\mu=1}^{b-1} \sum_{\nu=1}^{b-1} \frac{1+\xi^\mu}{1-\xi^\mu} \frac{1+\xi^\nu}{1-\xi^\nu} \left(\sum_{k=0}^{b-1} \xi^{k(\nu+\mu a)} \right).$$

We note that the last sum $\sum_{k=0}^{b-1} \xi^{k(\nu+\mu a)}$ vanishes, unless $\nu \equiv -\mu a \bmod b$ (Exercise 7.6), in which case the sum equals b, and we obtain

$$s(a,b) = \frac{1}{4b} \sum_{\mu=1}^{b-1} \frac{1+\xi^\mu}{1-\xi^\mu} \frac{1+\xi^{-\mu a}}{1-\xi^{-\mu a}}.$$

Rewriting the right-hand side in terms of cotangents gives

$$s(a,b) = \frac{i^2}{4b} \sum_{\mu=1}^{b-1} \cot \frac{\pi \mu}{b} \cot \frac{-\pi \mu a}{b} = \frac{1}{4b} \sum_{\mu=1}^{b-1} \cot \frac{\pi \mu}{b} \cot \frac{\pi \mu a}{b},$$

because the cotangent is an odd function. □

7.3 The Finite Fourier Transform and Its Properties

Given a periodic function f on $\mathbb{Z}$, we have seen that f possesses a finite Fourier series, with the finite collection of Fourier coefficients that we called $\hat{f}(0), \hat{f}(1), \ldots, \hat{f}(b-1)$. We now regard f as a function on the finite set $G = \{0, 1, 2, \ldots, b-1\}$, and let V_G be the vector space of all complex-valued functions on G. Equivalently, V_G is the vector space of all complex-valued *periodic functions on* $\mathbb{Z}$ with period b.

We define the **Fourier transform** of f, denoted by $\mathbf{F}(f)$, to be the periodic function on $\mathbb{Z}$ defined by the sequence of uniquely determined values

$$\hat{f}(0), \hat{f}(1), \ldots, \hat{f}(b-1).$$

Thus

$$\mathbf{F}(f)(m) = \hat{f}(m).$$

Theorem 7.2 above gave us these coefficients as a linear combination of the values $f(k)$, with $k = 0, 1, 2, \ldots, b-1$. Thus $\mathbf{F}(f)$ is a linear transformation of the function f, thought of as a vector in V_G. In other words, we have shown that $\mathbf{F}(f)$ is a one-to-one and onto linear transformation of V_G.

The vector space V_G is a vector space of dimension b; indeed, an explicit basis can easily be given for V_G using the **delta functions** (see Exercise 7.7) defined by

$$\delta_m(x) := \begin{cases} 1 & \text{if } x = m + kb, \text{ for some integer } k, \\ 0 & \text{otherwise.} \end{cases}$$

In other words, $\delta_m(x)$ is the periodic function on $\mathbb{Z}$ that picks out the arithmetic progression $\{m + kb : k \in \mathbb{Z}\}$.

But there is another natural basis for V_G. For a fixed integer a, the roots of unity $\{e_a(x) := e^{2\pi iax/b} : x \in \mathbb{Z}\}$ can be thought of as a single function $e_a(x) \in V_G$ because of its periodicity on $\mathbb{Z}$. As we saw in Theorem 7.2, the functions $\{e_1(x), \ldots, e_b(x)\}$ give a basis for the vector space of functions V_G. A natural question now arises: how are the two bases related to each other? An initial observation is

$$\widehat{\delta_a}(n) = \frac{1}{b} e^{-2\pi ian/b},$$

which simply follows from the computation

$$\widehat{\delta_a}(n) = \frac{1}{b} \sum_{k=0}^{b-1} \delta_a(k)\, \xi^{-kn} = \frac{1}{b} \xi^{-an} = \frac{1}{b} e^{-2\pi ian/b}.$$

So to get from the first basis to the second basis, we need precisely the finite Fourier transform!

It is extremely useful to define the following **inner product** on the vector space V_G:

$$\langle f, g \rangle = \sum_{k=0}^{b-1} f(k)\, \overline{g(k)}, \tag{7.6}$$

for two functions $f, g \in V_G$. Here the bar denotes complex conjugation. The following elementary properties show that $\langle f, g \rangle$ is an inner product (see Exercise 7.8):

1. $\langle f, f \rangle \geq 0$, with equality if and only if $f = 0$, the zero function.
2. $\langle f, g \rangle = \overline{\langle g, f \rangle}$.

Equipped with this inner product, V_G can now be regarded as a metric space: We can measure *distances* between two functions, and in particular between two basis elements $e_a(x) := e^{2\pi iax/b}$ and $e_c(x) := e^{2\pi icx/b}$. Every positive definite inner product gives rise to the distance function $\sqrt{\langle f - g, f - g \rangle}$.

Lemma 7.5 (Orthogonality relations).

$$\frac{1}{b}\langle \mathbf{e}_a, \mathbf{e}_c \rangle = \delta_a(c) = \begin{cases} 1 & \text{if } b \mid (a-c), \\ 0 & \text{otherwise.} \end{cases}$$

Proof. We compute the inner product

$$\langle \mathbf{e}_a, \mathbf{e}_c \rangle = \sum_{m=0}^{b-1} \mathbf{e}_a(m)\, \overline{\mathbf{e}_c(m)} = \sum_{m=0}^{b-1} e^{2\pi i(a-c)m/b}.$$

If $b \mid (a-c)$, then each term equals 1 in the latter sum, and hence the sum equals b. This verifies the first case of the lemma.

If $b \nmid (a-c)$, then $\mathbf{e}_{a-c}(m) = e^{2\pi i m(a-c)/b}$ is a nontrivial root of unity, and we have the finite geometric series

$$\sum_{m=0}^{b-1} e^{\frac{2\pi i(a-c)m}{b}} = \frac{e^{b\frac{2\pi i(a-c)}{b}} - 1}{e^{\frac{2\pi i(a-c)}{b}} - 1} = 0,$$

verifying the second case of the lemma. $\square$

Example 7.6. We recall the sawtooth function again, since it is one of the building blocks of lattice-point enumeration, and compute its Fourier transform. Namely, we define

$$B(k) := \left(\!\left(\frac{k}{b}\right)\!\right) = \begin{cases} \{\frac{k}{b}\} - \frac{1}{2} & \text{if } \frac{k}{b} \notin \mathbb{Z}, \\ 0 & \text{if } \frac{k}{b} \in \mathbb{Z}, \end{cases}$$

a periodic function on the integers with period b. What is its finite Fourier transform? We have already seen the answer, in the course of the proof of Lemma 7.3:

$$\widehat{B}(n) = \frac{1}{2b}\frac{1+\xi^n}{1-\xi^n} = \frac{i}{2b}\cot\frac{\pi n}{b}$$

for $n \neq 0$ and $\widehat{B}(0) = 0$. As always, $\xi = e^{2\pi i/b}$. $\square$

In the next section, we delve more deeply into the behavior of this inner product, where the Parseval identity is proved.

7.4 The Parseval Identity

A nontrivial property of the inner product defined above is the following identity, linking the "norm of a function" to the "norm of its Fourier transform." It is known as the *Parseval identity*, and also goes by the name of the *Plancherel theorem*.

Theorem 7.7 (Parseval identity). *For all $f \in V_G$,*

$$\langle f, f \rangle = b \langle \hat{f}, \hat{f} \rangle.$$

Proof. Using the definition $\mathbf{e}_m(x) = \xi^{mx}$ and the relation

$$\hat{f}(x) = \frac{1}{b} \sum_{m=0}^{b-1} f(m) \, \overline{\mathbf{e}_m(x)}$$

from Theorem 7.2,

$$
\begin{aligned}
\langle \hat{f}, \hat{f} \rangle &= \left\langle \frac{1}{b} \sum_{m=0}^{b-1} f(m) \, \overline{\mathbf{e}_m}, \; \frac{1}{b} \sum_{n=0}^{b-1} f(n) \, \overline{\mathbf{e}_n} \right\rangle \\
&= \frac{1}{b^2} \sum_{k=0}^{b-1} \sum_{m=0}^{b-1} f(m) \, \overline{\mathbf{e}_m(k)} \sum_{n=0}^{b-1} \overline{f(n)} \, \mathbf{e}_n(k) \\
&= \frac{1}{b^2} \sum_{m=0}^{b-1} \sum_{n=0}^{b-1} f(m) \, \overline{f(n)} \, \langle \mathbf{e}_m, \mathbf{e}_n \rangle \\
&= \frac{1}{b} \sum_{m=0}^{b-1} \sum_{n=0}^{b-1} f(m) \, \overline{f(n)} \, \delta_m(n) \\
&= \frac{1}{b} \langle f, f \rangle,
\end{aligned}
$$

where the essential step in the proof was using the orthogonality relations (Lemma 7.5) in the fourth equality above. □

A basically identical proof yields the following stronger result, showing that the "distance between two functions" is essentially equal to the "distance between their Fourier transforms."

Theorem 7.8. *For all $f, g \in V_G$,*

$$\langle f, g \rangle = b \langle \hat{f}, \hat{g} \rangle. \qquad\qquad □$$

Example 7.9. A nice application of the generalized Parseval identity above now gives us Lemma 7.4 very quickly, the reformulation of the Dedekind sum as a sum over roots of unity. Namely, we first fix an integer a relatively prime to b and define $f(k) = \left(\!\left(\frac{k}{b}\right)\!\right)$ and $g(k) = \left(\!\left(\frac{ka}{b}\right)\!\right)$. Then, using Example 7.6, $\hat{f}(n) = \frac{i}{2b} \cot \frac{\pi n}{b}$. To find the Fourier transform of g, we need an extra twist. Since

$$\left(\!\left(\frac{ka}{b}\right)\!\right) = \frac{i}{2b} \sum_{m=1}^{b-1} \cot \frac{\pi m}{b} \, \xi^{mka},$$

we can multiply each index m by a^{-1}, the multiplicative inverse of a modulo b (recall that we require a and b to be relatively prime for the reformulation of the Dedekind sum). Since a^{-1} is relatively prime to b, this multiplication just permutes $m = 1, 2, \ldots, b-1$ modulo b, but the sum remains invariant (see Exercise 1.9):

$$\sum_{m=1}^{b-1} \cot \frac{\pi m}{b} \, \xi^{mka} = \sum_{m=1}^{b-1} \cot \frac{\pi m a^{-1}}{b} \, \xi^{ma^{-1}ka} = \sum_{m=1}^{b-1} \cot \frac{\pi m a^{-1}}{b} \, \xi^{mk},$$

that is,

$$\hat{g}(n) = \frac{i}{2b} \cot \frac{\pi n a^{-1}}{b}.$$

Hence Theorem 7.8 immediately gives us the reformulation of the Dedekind sum:

$$s(a, b) := \sum_{k=0}^{b-1} \left(\!\left(\frac{k}{b}\right)\!\right) \left(\!\left(\frac{ka}{b}\right)\!\right)$$

$$= b \sum_{m=1}^{b-1} \left(\frac{i}{2b} \cot \frac{\pi m}{b}\right) \overline{\left(\frac{i}{2b} \cot \frac{\pi m a^{-1}}{b}\right)}$$

$$= \frac{1}{4b} \sum_{m=1}^{b-1} \cot \frac{\pi m}{b} \cot \frac{\pi m a^{-1}}{b}$$

$$= \frac{1}{4b} \sum_{m=1}^{b-1} \cot \frac{\pi m a}{b} \cot \frac{\pi m}{b}.$$

For the last equality, we again used the trick of replacing m by ma. $\qquad\square$

7.5 The Convolution of Finite Fourier Series

Another basic tool in finite Fourier analysis is the convolution of two finite Fourier series. Namely, suppose f and g are periodic functions with period b. We define the **convolution** of f and g by

$$(f * g)(t) := \sum_{m=0}^{b-1} f(t - m) \, g(m).$$

Indeed, it is this convolution tool (the proof of the convolution theorem below is almost trivial) that is responsible for the fastest known algorithm for multiplying two polynomials of degree b in $O(b \log(b))$ steps (see the notes at the end of this chapter).

Theorem 7.10 (Convolution theorem for finite Fourier series). *Let* $f(t) = \frac{1}{b} \sum_{k=0}^{b-1} a_k \, \xi^{kt}$ *and* $g(t) = \frac{1}{b} \sum_{k=0}^{b-1} c_k \, \xi^{kt}$, *where* $\xi = e^{2\pi i/b}$. *Then their* *convolution satisfies*

$$(f * g)(t) = \frac{1}{b} \sum_{k=0}^{b-1} a_k c_k \, \xi^{kt}.$$

Proof. The proof is straightforward: we just compute the left-hand side, and obtain

$$\sum_{m=0}^{b-1} f(t-m)g(m) = \frac{1}{b^2} \sum_{m=0}^{b-1} \left(\sum_{k=0}^{b-1} a_k \, \xi^{k(t-m)} \right) \left(\sum_{l=0}^{b-1} c_l \, \xi^{lm} \right)$$

$$= \frac{1}{b^2} \sum_{k=0}^{b-1} \sum_{l=0}^{b-1} a_k c_l \left(\sum_{m=0}^{b-1} \xi^{kt+(l-k)m} \right)$$

$$= \frac{1}{b} \sum_{k=0}^{b-1} a_k c_k \, \xi^{kt},$$

because the sum $\sum_{m=0}^{b-1} \xi^{(l-k)m}$ vanishes, unless $l = k$ (see Exercise 7.6). In the case that $l = k$, we have $\sum_{m=0}^{b-1} \xi^{(l-k)m} = b$. $\qquad\square$

It is an easy exercise (Exercise 7.23) to show that this convolution theorem is equivalent to the following statement:

$$\mathbf{F}(f * g) = b \, \mathbf{F}(f) \, \mathbf{F}(g).$$

Note that the proof of Theorem 7.10 is essentially identical to the proof of Lemma 7.4 above; we could have proved the lemma, in fact, by applying the convolution theorem. We now show how Theorem 7.10 can be used to derive identities on trigonometric functions.

Example 7.11. We claim that

$$\sum_{k=1}^{b-1} \cot^2 \left(\frac{\pi k}{b} \right) = \frac{(b-1)(b-2)}{3}.$$

The sum suggests the use of the convolution theorem, with a function whose Fourier coefficients are $a_k = c_k = \cot \frac{\pi k}{b}$. But we already know such a function! It is just the sawtooth function $\frac{2b}{i} \left(\left(\frac{m}{b} \right) \right)$. Therefore,

$$-\frac{1}{4b} \sum_{k=1}^{b-1} \cot^2 \left(\frac{\pi k}{b} \right) \xi^{kt} = \sum_{m=1}^{b-1} \left(\left(\frac{t-m}{b} \right) \right) \left(\left(\frac{m}{b} \right) \right),$$

where the equality follows from Theorem 7.10. On setting $t = 0$, we obtain

$$\sum_{m=1}^{b-1} \left(\left(\frac{-m}{b}\right)\right) \left(\left(\frac{m}{b}\right)\right) = -\sum_{m=1}^{b-1} \left(\left(\frac{m}{b}\right)\right) \left(\left(\frac{m}{b}\right)\right)$$

$$= -\frac{1}{b^2} \sum_{m=1}^{b-1} m^2 + \frac{1}{b} \sum_{m=1}^{b-1} m - \frac{1}{4}(b-1)$$

$$= -\frac{(b-1)(b-2)}{12b},$$

as desired. We used the identity $\left(\left(\frac{-m}{b}\right)\right) = - \left(\left(\frac{m}{b}\right)\right)$ in the first equality above, and some algebra was used in the last equality. Notice, moreover, that the convolution theorem gave us more than we asked for, namely an identity for every value of t.   □

Notes

1. Finite Fourier analysis offers a wealth of applications and is, for example, one of the main tools in quantum information theory. For the reader interested in going further than the humble beginnings outlined in this chapter, we heartily recommend Audrey Terras's monograph [242].

2. The Dedekind sum is our main motivation for studying finite Fourier series, and in fact, Chapter 8 is devoted to a detailed investigation of these sums, in which the Fourier–Dedekind sums of Chapter 1 also finally reappear.

3. The reader may consult [155, p. 501] for a proof that two polynomials of degree N can be multiplied in $O(N \log N)$ steps. The proof of this fact runs along the following conceptual lines. First, let the two given polynomials of degree N be $f(x) = \sum_{n=0}^{N} a(n) x^n$ and $g(x) = \sum_{n=0}^{N} b(n) x^n$. Then we know that $\hat{f}(\xi)$ and $\hat{g}(\xi)$ are two finite Fourier series, and we abbreviate them by $\hat{f}$ and $\hat{g}$, respectively. We now note that $\widehat{bfg} = \mathbf{F}(\mathbf{F}^{-1}(\hat{f}) * \mathbf{F}^{-1}(\hat{g}))$. If we can compute the finite Fourier transform (and its inverse) quickly, then this argument shows that we can multiply polynomials quickly. It is a fact of life that we *can* compute the Fourier transform of a periodic function of period N in $O(N \log N)$ steps, by an algorithm known as the *fast Fourier transform* (again, see [155] for a complete description).

4. The continuous Fourier transform, defined by $\int_{-\infty}^{\infty} f(t) e^{-2\pi i t x} \, dt$, can be related to the finite Fourier transform in the following way. We approximate the continuous integral by discretizing a large interval $[0, a]$. Precisely, we let $\Delta := \frac{a}{b}$, and we let $t_k := k\Delta = \frac{ka}{b}$. Then

$$\int_0^a f(t)\, e^{-2\pi itx}\, dt \approx \sum_{k=1}^b f(t_k)\, e^{-2\pi it_k x}\, \Delta\,,$$

which is in essence a finite Fourier series for the function $f(\frac{a}{b}x)$ as a function of $x \in \mathbb{Z}$. Hence finite Fourier series find an application to continuous Fourier analysis as an approximation tool.

Exercises

Throughout the exercises, we fix an integer $b > 1$ and let $\xi = e^{2\pi i/b}$.

7.1. Show that $1 - x^b = \prod_{k=1}^b (1 - \xi^k x)$.

7.2. ♣ Show that $\prod_{k=1}^{b-1}(1 - \xi^k) = b$.

7.3. Consider the matrix that arose in the proof of Theorem 7.2, namely $L = (a_{ij})$, with $a_{ij} := \xi^{(i-1)(j-1)}$ and with $1 \leq i, j \leq b$. Show that the matrix $\frac{1}{\sqrt{b}}L$ is a unitary matrix (recall that a matrix U is unitary if $U^*U = I$, where U^* is the conjugate transpose of U). Thus, this exercise shows that the Fourier transform of a periodic function is always given by a unitary transformation.

7.4. Show that $\left|\det\left(\frac{1}{\sqrt{b}}L\right)\right| = 1$, where $|z|$ denotes the norm of the complex number z. (It turns out that $\det(L)$ can sometimes be a complex number, but we will not use this fact here.)

7.5. ♣ Show that for every integer a relatively prime to b,

$$\frac{1}{b}\sum_{k=1}^{b-1} k\, \xi^{-ak} = \frac{\xi^a}{1 - \xi^a}\,.$$

7.6. ♣ Let n be an integer. Show that the sum $\sum_{k=0}^{b-1} \xi^{kn}$ vanishes, unless $n \equiv 0 \pmod{b}$, in which case it is equal to b.

7.7. ♣ For an integer m, recall that we defined the delta function

$$\delta_m(x) = \begin{cases} 1 & \text{if } x = m + ab, \text{ for some integer } a, \\ 0 & \text{otherwise.} \end{cases}$$

The b functions $\delta_1(x), \ldots, \delta_b(x)$ are clearly in the vector space V_G, since they are periodic on $\mathbb{Z}$ with period b. Show that they form a basis for V_G.

7.8. ♣ Prove that for all $f, g \in V_G$:

(a) $\langle f, f \rangle \geq 0$, with equality if and only if $f = 0$, the zero function.

(b) $\langle f, g \rangle = \overline{\langle g, f \rangle}$.

7.9. Show that $\sum_{k=1}^{b-1} \frac{1}{1-\xi^k} = \frac{b-1}{2}$.

7.10. Show that $\langle \delta_a, \delta_c \rangle = \delta_a(c)$.

7.11. Prove that $(f * \delta_a)(x) = f(x-a)$.

7.12. Prove that $\delta_a * \delta_c = \delta_{(a+c) \bmod b}$.

7.13. Let $g(x) = x - a$. Prove that $\widehat{f \circ g}(x) = \hat{f}(x) \, e^{-\frac{2\pi i a x}{b}}$.

7.14. Show that the Fourier transform can have only the eigenvalues $\pm 1, \pm i$.

7.15. ♣ Prove that for every real number x, $((x)) = \sum_{k=0}^{b-1} \left(\left(\frac{x+k}{b}\right)\right)$.

7.16. If x is not an integer, show that $\sum_{n=0}^{b-1} \cot\left(\pi \frac{n+x}{b}\right) = b \cot(\pi x)$.

7.17. Show that for every integer a relatively prime to b,

$$\sum_{\xi} \frac{\xi^{a+1} - 1}{(\xi^a - 1)(\xi - 1)} = 0,$$

where the sum is taken over all b^{th} roots of unity ξ except $\xi = 1$.

7.18. We call a root of unity $e^{2\pi i a/b}$ a **primitive** b^{th} root of unity if a is relatively prime to b. Let $\Phi_b(x)$ denote the polynomial with leading coefficient 1 and of degree $\phi(b)$[1] whose roots are the $\phi(b)$ distinct primitive b^{th} roots of unity. This polynomial is known as the *cyclotomic polynomial of order b*. Show that

$$\prod_{d \mid b} \Phi_d(x) = x^b - 1,$$

where the product is taken over all positive divisors d of b.

7.19. We define the Möbius function for positive integers n by

$$\mu(n) = \begin{cases} 1 & \text{if } n = 1, \\ 0 & \text{if } n \text{ is divisible by a square}, \\ (-1)^k & \text{if } n \text{ is square-free and has } k \text{ prime divisors}. \end{cases}$$

Deduce from the previous exercise that

$$\Phi_b(x) = \prod_{d \mid b} \left(x^d - 1\right)^{\mu(b/d)}.$$

---

[1] $\phi(b) := \# \{k \in [1, b-1] : \gcd(k, b) = 1\}$ is the **Euler ϕ-function**.

7.20. Prove that for every positive integer b,

$$\sum_{\substack{1 \le a \le b \\ \gcd(a,b)=1}} e^{2\pi i a/b} = \mu(b),$$

the Möbius function.

7.21. Show that for every positive odd integer k,

$$s(1,k) = \frac{(k-1)(k-2)}{12k}.$$

7.22. Show that, if b is odd, then $\sum_{k=1}^{b-1} \tan^2\left(\frac{\pi k}{b}\right) = b(b-1)$.

7.23. ♣ Show that Theorem 7.10 is equivalent to the following statement:

$$\mathbf{F}(f * g) = b\,\mathbf{F}(f)\,\mathbf{F}(g).$$

7.24. Consider the trace of the linear transformation $L = \left(\xi^{(i-1)(j-1)}\right)$, defined in (7.3). The trace of L is $G(b) := \sum_{m=0}^{b-1} \xi^{m^2}$, known as a **Gauß sum**. Show that $|G(b)| = \sqrt{b}$ if b is an odd prime.

Chapter 8
Dedekind Sums, the Building Blocks of Lattice-Point Enumeration

If things are nice there is probably a good reason why they are nice: and if you don't know at least one reason for this good fortune, then you still have work to do.

Richard Askey

We encountered Dedekind sums in our study of finite Fourier analysis in Chapter 7, and we became intimately acquainted with their siblings in our study of the coin-exchange problem in Chapter 1. They have one shortcoming, however (which we shall remove): the definition of $s(a, b)$ requires us to sum over b terms, which is rather slow when $b = 2^{100}$, for example. Luckily, there is a magical *reciprocity law* for the Dedekind sum $s(a, b)$ that allows us to compute it in roughly $\log_2(b) = 100$ steps, in the example above. This is the kind of magic that saves the day when we try to enumerate lattice points in integral polytopes of dimension $d \leq 4$. There is an ongoing effort to extend these ideas to higher dimensions, but there is much room for improvement. In this chapter, we focus on the computational-complexity issues that arise when we try to compute Dedekind sums explicitly. In many ways, the Dedekind sums extend the notion of the greatest common divisor of two integers.

8.1 Fourier–Dedekind Sums and the Coin-Exchange Problem Revisited

Recall from Chapter 1 the Fourier–Dedekind sum (defined in (1.13))

$$
s_n\left(a_1, a_2, \ldots, a_d; b\right) = \frac{1}{b} \sum_{k=1}^{b-1} \frac{\xi_b^{kn}}{\left(1 - \xi_b^{ka_1}\right)\left(1 - \xi_b^{ka_2}\right) \cdots \left(1 - \xi_b^{ka_d}\right)},
$$

which appeared as a main player in our analysis of the Frobenius coin-exchange problem. We can now recognize the Fourier–Dedekind sums as honest finite Fourier series with period b. The Fourier–Dedekind sums unify many variations of the Dedekind sum that have appeared in the literature, and form the building blocks of Ehrhart quasipolynomials. For example, we showed in Chapter 1 that $s_n(a_1, a_2, \ldots, a_d; b)$ appears in the Ehrhart quasipolynomial of the d-simplex

$$\left\{ (x_1, \ldots, x_{d+1}) \in \mathbb{R}^{d+1}_{\geq 0} : a_1 x_1 + \cdots + a_d x_d + b x_{d+1} = 1 \right\}.$$

Example 8.1. We first notice that when $n = 0$ and $d = 2$, the Fourier–Dedekind sum reduces to a classical Dedekind sum (which—finally—explains the name): for relatively prime positive integers a and b,

$$s_0(a, 1; b) = \frac{1}{b} \sum_{k=1}^{b-1} \frac{1}{\left(1 - \xi_b^{ka}\right)\left(1 - \xi_b^{k}\right)}$$

$$= \frac{1}{b} \sum_{k=1}^{b-1} \left(\frac{1}{1 - \xi_b^{ka}} - \frac{1}{2} \right)\left(\frac{1}{1 - \xi_b^{k}} - \frac{1}{2} \right)$$

$$+ \frac{1}{2b} \sum_{k=1}^{b-1} \frac{1}{1 - \xi_b^{k}} + \frac{1}{2b} \sum_{k=1}^{b-1} \frac{1}{1 - \xi_b^{ka}} - \frac{1}{b} \sum_{k=1}^{b-1} \frac{1}{4}$$

$$= \frac{1}{4b} \sum_{k=1}^{b-1} \left(\frac{1 + \xi_b^{ka}}{1 - \xi_b^{ka}} \right)\left(\frac{1 + \xi_b^{k}}{1 - \xi_b^{k}} \right) + \frac{1}{b} \sum_{k=1}^{b-1} \frac{1}{1 - \xi_b^{k}} - \frac{b-1}{4b}.$$

In the last step, we used the fact that multiplying the index k by a does not change the middle sum. This middle sum can be further simplified by recalling (1.8):

$$\frac{1}{b} \sum_{k=1}^{b-1} \frac{1}{\left(1 - \xi_b^{k}\right) \xi_b^{kn}} = -\left\{ \frac{n}{b} \right\} + \frac{1}{2} - \frac{1}{2b},$$

whence

$$s_0(a, 1; b) = \frac{1}{4b} \sum_{k=1}^{b-1} \left(\frac{1 + \xi_b^{ka}}{1 - \xi_b^{ka}} \right)\left(\frac{1 + \xi_b^{k}}{1 - \xi_b^{k}} \right) + \frac{1}{2} - \frac{1}{2b} - \frac{b-1}{4b}$$

$$= -\frac{1}{4b} \sum_{k=1}^{b-1} \cot\left(\frac{\pi ka}{b} \right) \cot\left(\frac{\pi k}{b} \right) + \frac{b-1}{4b} \tag{8.1}$$

$$= -s(a, b) + \frac{b-1}{4b}. \qquad \Box$$

Example 8.2. The next special evaluation of a Fourier–Dedekind sum is very similar to the computation above, so we leave it to the reader to prove

(Exercise 8.7) that for a_1, a_2 relatively prime to b,

$$s_0\left(a_1, a_2; b\right) = -s\left(a_1 a_2^{-1}, b\right) + \frac{b-1}{4b}, \qquad (8.2)$$

where $a_2^{-1} a_2 \equiv 1 \bmod b$. $\qquad \square$

Returning to the general Fourier–Dedekind sum, we now prove the first of a series of *reciprocity laws*: identities for certain sums of Fourier–Dedekind sums. We first recall how these sums came up in Chapter 1, namely, from the partial fraction expansion of the function

$$\begin{aligned} f(z) &= \frac{1}{\left(1 - z^{a_1}\right) \cdots \left(1 - z^{a_d}\right) z^n} \\ &= \frac{A_1}{z} + \frac{A_2}{z^2} + \cdots + \frac{A_n}{z^n} + \frac{B_1}{z-1} + \frac{B_2}{(z-1)^2} + \cdots + \frac{B_d}{(z-1)^d} \qquad (8.3) \\ &\quad + \sum_{k=1}^{a_1-1} \frac{C_{1k}}{z - \xi_{a_1}^k} + \sum_{k=1}^{a_2-1} \frac{C_{2k}}{z - \xi_{a_2}^k} + \cdots + \sum_{k=1}^{a_d-1} \frac{C_{dk}}{z - \xi_{a_d}^k}. \end{aligned}$$

(Here we assume that $a_1, a_2, \ldots, a_d$ are pairwise relatively prime.) Theorem 1.8 states that with the help of the partial fraction coefficients $B_1, \ldots, B_d$ and Fourier–Dedekind sums, we can compute the restricted partition function for $A = \{a_1, a_2, \ldots, a_d\}$:

$$\begin{aligned} p_A(n) = &-B_1 + B_2 - \cdots + (-1)^d B_d + s_{-n}\left(a_2, a_3, \ldots, a_d; a_1\right) \\ &+ s_{-n}\left(a_1, a_3, a_4, \ldots, a_d; a_2\right) + \cdots + s_{-n}\left(a_1, a_2, \ldots, a_{d-1}; a_d\right). \end{aligned}$$

It is not hard to see that $B_1, B_2, \ldots, B_d$ are polynomials in n (Exercise 8.8), whence we call

$$\mathrm{poly}_A(n) := -B_1 + B_2 - \cdots + (-1)^d B_d$$

the **polynomial part** of the restricted partition function $p_A(n)$. These polynomials can be computed from first principles (and we give some of them in Example 8.3) but in fact, they are relatives of the Bernoulli polynomials, which we defined in (2.8) in Section 2.4: We define the **Bernoulli–Barnes polynomials** $B_k^A(x)$ through the generating function

$$\frac{z^d e^{xz}}{\left(e^{a_1 z} - 1\right)\left(e^{a_2 z} - 1\right) \cdots \left(e^{a_d z} - 1\right)} = \sum_{k \ge 0} B_k^A(x) \frac{z^k}{k!},$$

where as usual, $A = \{a_1, a_2, \ldots, a_d\}$. (The Bernoulli polynomials are the special cases $B_k^{\{1\}}$.) Then

$$\mathrm{poly}_A(n) = \frac{(-1)^d}{(d-1)!} B_d^A(-t), \qquad (8.4)$$

an identity that can be proved most easily with a pinch of complex analysis (Exercise 14.3 in Chapter 14).

Example 8.3. The first few expressions for $\mathrm{poly}_{\{a_1,\ldots,a_d\}}(n)$ are

$$\mathrm{poly}_{\{a_1\}}(n) = \frac{1}{a_1},$$

$$\mathrm{poly}_{\{a_1,a_2\}}(n) = \frac{n}{a_1 a_2} + \frac{1}{2}\left(\frac{1}{a_1} + \frac{1}{a_2}\right),$$

$$\mathrm{poly}_{\{a_1,a_2,a_3\}}(n) = \frac{n^2}{2a_1 a_2 a_3} + \frac{n}{2}\left(\frac{1}{a_1 a_2} + \frac{1}{a_1 a_3} + \frac{1}{a_2 a_3}\right) \tag{8.5}$$
$$+ \frac{1}{12}\left(\frac{3}{a_1} + \frac{3}{a_2} + \frac{3}{a_3} + \frac{a_1}{a_2 a_3} + \frac{a_2}{a_1 a_3} + \frac{a_3}{a_1 a_2}\right),$$

$$\mathrm{poly}_{\{a_1,a_2,a_3,a_4\}}(n) = \frac{n^3}{6a_1 a_2 a_3 a_4}$$
$$+ \frac{n^2}{4}\left(\frac{1}{a_1 a_2 a_3} + \frac{1}{a_1 a_2 a_4} + \frac{1}{a_1 a_3 a_4} + \frac{1}{a_2 a_3 a_4}\right)$$
$$+ \frac{n}{4}\left(\frac{1}{a_1 a_2} + \frac{1}{a_1 a_3} + \frac{1}{a_1 a_4} + \frac{1}{a_2 a_3} + \frac{1}{a_2 a_4} + \frac{1}{a_3 a_4}\right)$$
$$+ \frac{n}{12}\left(\frac{a_1}{a_2 a_3 a_4} + \frac{a_2}{a_1 a_3 a_4} + \frac{a_3}{a_1 a_2 a_4} + \frac{a_4}{a_1 a_2 a_3}\right)$$
$$+ \frac{1}{24}\left(\frac{a_1}{a_2 a_3} + \frac{a_1}{a_2 a_4} + \frac{a_1}{a_3 a_4} + \frac{a_2}{a_1 a_3} + \frac{a_2}{a_1 a_4} + \frac{a_2}{a_3 a_4}\right.$$
$$\left. + \frac{a_3}{a_1 a_2} + \frac{a_3}{a_1 a_4} + \frac{a_3}{a_2 a_4} + \frac{a_4}{a_1 a_2} + \frac{a_4}{a_1 a_3} + \frac{a_4}{a_2 a_3}\right)$$
$$+ \frac{1}{8}\left(\frac{1}{a_1} + \frac{1}{a_2} + \frac{1}{a_3} + \frac{1}{a_4}\right). \qquad \Box$$

We are about to combine the Ehrhart results of Chapter 3 with the partial fraction expansion of Chapter 1 that gave rise to the Fourier–Dedekind sums.

Theorem 8.4 (Zagier reciprocity). *For all pairwise relatively prime positive integers $a_1, a_2, \ldots, a_d$,*

$$s_0(a_2, a_3, \ldots, a_d; a_1) + s_0(a_1, a_3, a_4, \ldots, a_d; a_2) + \cdots$$
$$+ s_0(a_1, a_2, \ldots, a_{d-1}; a_d)$$
$$= 1 - \mathrm{poly}_{\{a_1, a_2, \ldots, a_d\}}(0).$$

At first sight, this reciprocity law should come as a surprise. The Fourier–Dedekind sums can be long, complicated sums, yet when combined in this fashion, they add up to a trivial rational function in $a_1, a_2, \ldots, a_d$.

Proof. We compute the constant term of the quasipolynomial $p_A(n)$:

$$p_A(0) = \text{poly}_A(0) + s_0(a_2, a_3, \ldots, a_d; a_1)$$
$$+ s_0(a_1, a_3, a_4, \ldots, a_d; a_2) + \cdots + s_0(a_1, a_2, \ldots, a_{d-1}; a_d).$$

On the other hand, Exercise 3.32 (the extension of Corollary 3.15 to Ehrhart quasipolynomials) states that $p_A(0) = 1$, whence

$$1 = \text{poly}_A(0) + s_0(a_2, a_3, \ldots, a_d; a_1)$$
$$+ s_0(a_1, a_3, a_4, \ldots, a_d; a_2) + \cdots + s_0(a_1, a_2, \ldots, a_{d-1}; a_d). \qquad \square$$

8.2 The Dedekind Sum and Its Reciprocity and Computational Complexity

We derived in (8.1) the classical Dedekind sum $s(a, b)$ as a special evaluation of the Fourier–Dedekind sum. Naturally, Theorem 8.4 takes on a particular form when we specialize this reciprocity law to the classical Dedekind sum.

Corollary 8.5 (Dedekind's reciprocity law). *For all relatively prime positive integers a and b,*

$$s(a, b) + s(b, a) = \frac{1}{12}\left(\frac{a}{b} + \frac{b}{a} + \frac{1}{ab}\right) - \frac{1}{4}.$$

Proof. A special case of Theorem 8.4 is

$$s_0(a, 1; b) + s_0(b, a; 1) + s_0(1, b; a) = 1 - \text{poly}_{\{a,1,b\}}(0)$$
$$= 1 - \frac{1}{12}\left(\frac{3}{a} + 3 + \frac{3}{b} + \frac{a}{b} + \frac{1}{ab} + \frac{b}{a}\right)$$
$$= \frac{3}{4} - \frac{1}{12}\left(\frac{a}{b} + \frac{b}{a} + \frac{1}{ab}\right) - \frac{1}{4a} - \frac{1}{4b}.$$

Now we use the fact that $s_0(b, a; 1) = 0$ and the identity (8.1):

$$s_0(a, 1; b) = -s(a, b) + \frac{1}{4} - \frac{1}{4b}. \qquad \square$$

Dedekind's reciprocity law allows us to compute the Dedekind sum $s(a, b)$ as quickly as the gcd algorithm for a and b, also known as the *Euclidean algorithm*. Let's get a better feeling for the way we can compute the Dedekind sum by working out an example. We remind the reader of another crucial property of the Dedekind sums that we already pointed out in (7.5): $s(a, b)$ remains invariant when we replace a by its residue modulo b, that is,

$$s(a, b) = s(a \bmod b, b). \qquad (8.6)$$

Example 8.6. Let $a = 100$ and $b = 147$. Now we alternately use Corollary 8.5 and the reduction identity (8.6):

$$
\begin{aligned}
s(100, 147) &= \frac{1}{12}\left(\frac{100}{147} + \frac{147}{100} + \frac{1}{14700}\right) - \frac{1}{4} - s(147, 100) \\
&= -\frac{1249}{17640} - s(47, 100) \\
&= -\frac{1249}{17640} - \left(\frac{1}{12}\left(\frac{47}{100} + \frac{100}{47} + \frac{1}{4700}\right) - \frac{1}{4} - s(100, 47)\right) \\
&= -\frac{773}{20727} + s(6, 47) \\
&= -\frac{773}{20727} + \frac{1}{12}\left(\frac{6}{47} + \frac{47}{6} + \frac{1}{282}\right) - \frac{1}{4} - s(47, 6) \\
&= \frac{166}{441} - s(5, 6) \\
&= \frac{166}{441} - \left(\frac{1}{12}\left(\frac{5}{6} + \frac{6}{5} + \frac{1}{30}\right) - \frac{1}{4} - s(6, 5)\right) \\
&= \frac{2003}{4410} + s(1, 5) \\
&= \frac{2003}{4410} - \frac{1}{4} + \frac{1}{30} + \frac{5}{12} \\
&= \frac{577}{882}.
\end{aligned}
$$

In the last step, we used Exercise 7.21: $s(1, k) = -\frac{1}{4} + \frac{1}{6k} + \frac{k}{12}$. A priori, $s(100, 147)$ takes 147 steps to compute, whereas we were able to compute it in nine steps using Dedekind's reciprocity law and (8.6). □

As a second corollary to Theorem 8.4, we mention the following three-term reciprocity law for the special Fourier–Dedekind sum $s_0(a, b; c)$. This reciprocity law could be restated in terms of the classical Dedekind sum via the identity (8.2).

Corollary 8.7. *For pairwise relatively prime positive integers* a, b, *and* c,

$$
s_0(a, b; c) + s_0(c, a; b) + s_0(b, c; a) = 1 - \frac{1}{12}\left(\frac{3}{a} + \frac{3}{b} + \frac{3}{c} + \frac{a}{bc} + \frac{b}{ca} + \frac{c}{ab}\right).
$$
□

8.3 Rademacher Reciprocity for the Fourier–Dedekind Sum

The next reciprocity law will be again for the general Fourier–Dedekind sums. It extends Theorem 8.4 beyond $n = 0$.

Theorem 8.8 (Rademacher reciprocity). *Let $a_1, a_2, \ldots, a_d$ be pairwise relatively prime positive integers. Then for $n = 1, 2, \ldots, (a_1 + \cdots + a_d - 1)$,*

$$s_n (a_2, a_3, \ldots, a_d; a_1) + s_n (a_1, a_3, a_4, \ldots, a_d; a_2) + \cdots$$
$$+ s_n (a_1, a_2, \ldots, a_{d-1}; a_d) = -\operatorname{poly}_{\{a_1, a_2, \ldots, a_d\}} (-n).$$

Proof. We recall the definition

$$p_A^\circ(n) = \# \left\{ (m_1, \ldots, m_d) \in \mathbb{Z}^d : \text{all } m_j > 0, \ m_1 a_1 + \cdots + m_d a_d = n \right\}$$

of Exercise 1.33, that is, $p_A^\circ(n)$ counts the number of partitions of n using only the elements of A as parts, *where each part is used at least once*. This counting function is, naturally, connected to p_A through Ehrhart–Macdonald reciprocity (Theorem 4.1):

$$p_A^\circ(n) = (-1)^{d-1} p_A(-n),$$

that is,

$$(-1)^{d-1} p_A^\circ(n) = \operatorname{poly}_A(-n) + s_n (a_2, a_3, \ldots, a_d; a_1)$$
$$+ s_n (a_1, a_3, a_4, \ldots, a_d; a_2) + \cdots + s_n (a_1, a_2, \ldots, a_{d-1}; a_d).$$

On the other hand, by its very definition,

$$p_A^\circ(n) = 0 \qquad \text{for} \quad n = 1, 2, \ldots, (a_1 + \cdots + a_d - 1),$$

so that for those n,

$$0 = \operatorname{poly}_A(-n) + s_n (a_2, a_3, \ldots, a_d; a_1)$$
$$+ s_n (a_1, a_3, a_4, \ldots, a_d; a_2) + \cdots + s_n (a_1, a_2, \ldots, a_{d-1}; a_d). \qquad \square$$

Just as Zagier reciprocity takes on a special form for the classical Dedekind sum, Rademacher reciprocity specializes for $d = 2$ to a reciprocity identity for the **Dedekind–Rademacher sum**

$$r_n(a, b) := \sum_{k=0}^{b-1} \left(\left(\frac{ka + n}{b} \right) \right) \left(\left(\frac{k}{b} \right) \right).$$

The classical Dedekind sum is, naturally, the specialization $r_0(a, b) = s(a, b)$. To be able to state the reciprocity law for the Dedekind–Rademacher sums, we define the function

$$\chi_a(n) := \begin{cases} 1 & \text{if } a \mid n, \\ 0 & \text{otherwise,} \end{cases}$$

which will come in handy as a bookkeeping device.

Corollary 8.9 (Reciprocity law for Dedekind–Rademacher sums).
Let a and b be relatively prime positive integers. Then for $n = 1, 2, \ldots, a + b$,

$$r_n(a,b) + r_n(b,a) = \frac{n^2}{2ab} - \frac{n}{2}\left(\frac{1}{ab} + \frac{1}{a} + \frac{1}{b}\right) + \frac{1}{12}\left(\frac{a}{b} + \frac{b}{a} + \frac{1}{ab}\right)$$

$$+ \frac{1}{2}\left(\left(\!\left(\frac{a^{-1}n}{b}\right)\!\right) + \left(\!\left(\frac{b^{-1}n}{a}\right)\!\right) + \left(\!\left(\frac{n}{a}\right)\!\right) + \left(\!\left(\frac{n}{b}\right)\!\right)\right)$$

$$+ \frac{1}{4}\left(1 + \chi_a(n) + \chi_b(n)\right),$$

where $a^{-1}a \equiv 1 \bmod b$ and $b^{-1}b \equiv 1 \bmod a$.

This identity follows almost instantly once we are able to express the Dedekind–Rademacher sum in terms of Fourier–Dedekind sums.

Lemma 8.10. *Suppose a and b are relatively prime positive integers and $n \in \mathbb{Z}$. Then*

$$r_n(a,b) = -s_n(a,1;b) + \frac{1}{2}\left(\!\left(\frac{n}{b}\right)\!\right) + \frac{1}{2}\left(\!\left(\frac{na^{-1}}{b}\right)\!\right) - \frac{1}{4b} + \frac{1}{4}\chi_b(n),$$

where $a^{-1}a \equiv 1 \bmod b$.

Proof. We begin by rewriting the finite Fourier series (1.8) for the sawtooth function $((x))$:

$$\frac{1}{b}\sum_{k=1}^{b-1}\frac{\xi_b^{kn}}{1 - \xi_b^k} = -\left\{\frac{-n}{b}\right\} + \frac{1}{2} - \frac{1}{2b} = -\left(\!\left(\frac{-n}{b}\right)\!\right) + \frac{1}{2}\chi_b(n) - \frac{1}{2b}$$

$$= \left(\!\left(\frac{n}{b}\right)\!\right) + \frac{1}{2}\chi_b(n) - \frac{1}{2b}.$$

Hence we also have

$$\frac{1}{b}\sum_{k=1}^{b-1}\frac{\xi_b^{kn}}{1 - \xi_b^{ka}} = \frac{1}{b}\sum_{k=1}^{b-1}\frac{\xi_b^{ka^{-1}n}}{1 - \xi_b^k} = \left(\!\left(\frac{a^{-1}n}{b}\right)\!\right) + \frac{1}{2}\chi_b\left(a^{-1}n\right) - \frac{1}{2b}$$

$$= \left(\!\left(\frac{a^{-1}n}{b}\right)\!\right) + \frac{1}{2}\chi_b\left(n\right) - \frac{1}{2b}.$$

Now we use the convolution theorem for finite Fourier series (Theorem 7.10) for the functions

$$f(n) := \frac{1}{b}\sum_{k=1}^{b-1}\frac{\xi_b^{kn}}{1 - \xi_b^k} \qquad \text{and} \qquad g(n) := \frac{1}{b}\sum_{k=1}^{b-1}\frac{\xi_b^{kn}}{1 - \xi_b^{ka}}.$$

It gives

$$\frac{1}{b} \sum_{k=1}^{b-1} \frac{\xi_b^{kn}}{\left(1 - \xi_b^k\right)\left(1 - \xi_b^{ka}\right)} = \sum_{m=0}^{b-1} f(n-m)\, g(m) =$$

$$\sum_{m=0}^{b-1} \left(\left(\!\left(\frac{n-m}{b}\right)\!\right) + \frac{1}{2}\chi_b(n-m) - \frac{1}{2b} \right) \left(\left(\!\left(\frac{a^{-1}m}{b}\right)\!\right) + \frac{1}{2}\chi_b(m) - \frac{1}{2b} \right).$$

We invite the reader to check (Exercise 8.12) that the sum on the right-hand side simplifies to

$$-\sum_{m=0}^{b-1} \left(\left(\!\left(\frac{am+n}{b}\right)\!\right) \right) \left(\left(\!\left(\frac{m}{b}\right)\!\right) \right) + \frac{1}{2}\left(\left(\!\left(\frac{a^{-1}n}{b}\right)\!\right) \right) + \frac{1}{2}\left(\left(\!\left(\frac{n}{b}\right)\!\right) \right) - \frac{1}{4b} + \frac{1}{4}\chi_b(n),$$

whence

$$s_n(a,1;b) = -r_n(a,b) + \frac{1}{2}\left(\left(\!\left(\frac{a^{-1}n}{b}\right)\!\right) \right) + \frac{1}{2}\left(\left(\!\left(\frac{n}{b}\right)\!\right) \right) - \frac{1}{4b} + \frac{1}{4}\chi_b(n). \qquad \square$$

Proof of Corollary 8.9. We use the following special case of Theorem 8.8:

$$s_n(a,1;b) + s_n(1,a;b) + s_n(a,b;1) = -\operatorname{poly}_{\{a,1,b\}}(-n)$$

$$= -\frac{n^2}{2ab} + \frac{n}{2}\left(\frac{1}{ab} + \frac{1}{a} + \frac{1}{b} \right) - \frac{1}{12}\left(\frac{3}{a} + \frac{3}{b} + 3 + \frac{a}{b} + \frac{b}{a} + \frac{1}{ab} \right),$$

which holds for $n = 1, 2, \ldots, a+b$. Lemma 8.10 allows us to translate this identity into one for Dedekind–Rademacher sums:

$$r_n(a,b) + r_n(b,a) = \frac{n^2}{2ab} - \frac{n}{2}\left(\frac{1}{ab} + \frac{1}{a} + \frac{1}{b} \right) + \frac{1}{12}\left(\frac{a}{b} + \frac{b}{a} + \frac{1}{ab} \right)$$

$$+ \frac{1}{2}\left(\left(\!\left(\frac{a^{-1}n}{b}\right)\!\right) + \left(\!\left(\frac{b^{-1}n}{a}\right)\!\right) + \left(\!\left(\frac{n}{a}\right)\!\right) + \left(\!\left(\frac{n}{b}\right)\!\right) \right)$$

$$+ \frac{1}{4}\left(1 + \chi_a(n) + \chi_b(n) \right). \qquad \square$$

The two-term reciprocity law allows us to compute the Dedekind–Rademacher sum as quickly as the gcd algorithm, just as obtained for the classical Dedekind sum. This fact has an interesting consequence: In Theorem 2.10 and Exercise 2.36, we showed implicitly (see Exercise 8.13) that Dedekind–Rademacher sums are the only nontrivial ingredients of the Ehrhart quasipolynomials of rational polygons. Corollary 8.9 ensures that these Ehrhart quasipolynomials can be computed almost instantly.

8.4 The Mordell–Pommersheim Tetrahedron

In this section, we return to Ehrhart polynomials and illustrate how Dedekind sums appear naturally in generating-function computations. We will study the tetrahedron that historically first gave rise to the connection of Dedekind sums and lattice-point enumeration in polytopes. It is given by

$$\mathcal{P} = \left\{ (x, y, z) \in \mathbb{R}^3 : x, y, z \geq 0, \ \frac{x}{a} + \frac{y}{b} + \frac{z}{c} \leq 1 \right\}, \tag{8.7}$$

a tetrahedron with vertices $(0,0,0)$, $(a,0,0)$, $(0,b,0)$, and $(0,0,c)$, where a, b, c are positive integers. We insert the slack variable n and interpret

$$L_{\mathcal{P}}(t) = \# \left\{ (k, l, m) \in \mathbb{Z}^3 : k, l, m \geq 0, \ \frac{k}{a} + \frac{l}{b} + \frac{m}{c} \leq t \right\}$$

$$= \# \left\{ (k, l, m, n) \in \mathbb{Z}^4 : k, l, m, n \geq 0, \ bck + acl + abm + n = abct \right\}$$

as the Taylor coefficient of z^{abct} for the function

$$\left(\sum_{k \geq 0} z^{bck} \right) \left(\sum_{l \geq 0} z^{acl} \right) \left(\sum_{m \geq 0} z^{abm} \right) \left(\sum_{n \geq 0} z^n \right)$$

$$= \frac{1}{(1 - z^{bc}) (1 - z^{ac}) (1 - z^{ab}) (1 - z)}.$$

As we have done numerous times before, we shift this coefficient to the constant term:

$$L_{\mathcal{P}}(t) = \mathrm{const} \left(\frac{1}{(1 - z^{bc}) (1 - z^{ac}) (1 - z^{ab}) (1 - z) \, z^{abct}} \right).$$

To reduce the number of poles, it is convenient to change this function slightly; the constant term of $1/(1 - z^{bc})(1 - z^{ac})(1 - z^{ab})(1 - z)$ is 1, so that

$$L_{\mathcal{P}}(t) = \mathrm{const} \left(\frac{z^{-abct} - 1}{(1 - z^{bc}) (1 - z^{ac}) (1 - z^{ab}) (1 - z)} \right) + 1.$$

This trick becomes useful in the next step, namely expanding the function into partial fractions. Strictly speaking, we cannot do that, since the numerator is not a polynomial in z. However, we can think of this rational function as a sum of two functions. The higher-order poles of both summands that we will not include in our computation below cancel each other, so we can ignore them at this stage. The only poles of

$$\frac{z^{-abct} - 1}{(1 - z^{bc}) (1 - z^{ac}) (1 - z^{ab}) (1 - z)} \tag{8.8}$$

are at the ath, b^{th}, c^{th} roots of unity and at 0. (As before, we do not have to bother with the coefficients of $z = 0$ of the partial fraction expansion.) To make life momentarily easier (the general case is the subject of Exercise 8.18), let's assume that a, b, and c are pairwise relatively prime; then all the poles besides 0 and 1 are simple. The computation of the coefficients for $z = 1$ is very similar to what we did with the restricted partition function in Chapter 1. The coefficient in the partial fraction expansion of a nontrivial root of unity, say ξ_a^k, is also computed practically as easily as in earlier examples: it is

$$-\frac{t}{a\left(1 - \xi_a^{kbc}\right)\left(1 - \xi_a^k\right)} \tag{8.9}$$

(see Exercise 8.15). Summing this fraction over $k = 1, 2, \ldots, a - 1$ gives rise to the Fourier–Dedekind sum

$$-\frac{t}{a} \sum_{k=1}^{a-1} \frac{1}{\left(1 - \xi_a^{kbc}\right)\left(1 - \xi_a^k\right)} = -t\, s_0\left(bc, 1; a\right).$$

Putting this coefficient and its siblings for the other roots of unity into the partial fraction expansion and computing the constant term yields (Exercise 8.15)

$$
\begin{aligned}
L_{\mathcal{P}}(t) = {} & \frac{abc}{6}\, t^3 + \frac{ab + ac + bc + 1}{4}\, t^2 \\
& + \left(\frac{a + b + c}{4} + \frac{1}{4}\left(\frac{1}{a} + \frac{1}{b} + \frac{1}{c}\right) + \frac{1}{12}\left(\frac{bc}{a} + \frac{ca}{b} + \frac{ab}{c} + \frac{1}{abc}\right)\right) t \\
& + \left(s_0\left(bc, 1; a\right) + s_0\left(ca, 1; b\right) + s_0\left(ab, 1; c\right)\right) t \\
& + 1\,.
\end{aligned}
$$

We recognize instantly that the Fourier–Dedekind sums in this Ehrhart polynomial are in fact classical Dedekind sums by (8.1), and so we arrive at the following celebrated result.

Theorem 8.11. *Let* $\mathcal{P}$ *be given by* (8.7) *with* a, b, *and* c *pairwise relatively prime. Then*

$$
\begin{aligned}
L_{\mathcal{P}}(t) = {} & \frac{abc}{6}\, t^3 + \frac{ab + ac + bc + 1}{4}\, t^2 + \left(\frac{3}{4} + \frac{a + b + c}{4}\right. \\
& + \left.\frac{1}{12}\left(\frac{bc}{a} + \frac{ca}{b} + \frac{ab}{c} + \frac{1}{abc}\right) - s\left(bc, a\right) - s\left(ca, b\right) - s\left(ab, c\right)\right) t + 1\,.
\end{aligned}
$$

$\square$

We finish this chapter by giving the Ehrhart series of the Mordell–Pommersheim tetrahedron $\mathcal{P}$. It follows simply from the transformation formulas (computing the Ehrhart numerator coefficients from the Ehrhart polynomial coefficients) of Corollary 3.16 and Exercise 3.15, and hence the Ehrhart series of $\mathcal{P}$ naturally contains Dedekind sums.

Corollary 8.12. *Let $\mathcal{P}$ be given by (8.7) with a, b, and c pairwise relatively prime. Then*

$$\mathrm{Ehr}_{\mathcal{P}}(z) = \frac{h_3^* z^3 + h_2^* z^2 + h_1^* z + 1}{(1-z)^4},$$

where

$$h_3^* = \frac{abc}{6} - \frac{ab + ac + bc + a + b + c}{4} - \frac{1}{2} + \frac{1}{12}\left(\frac{bc}{a} + \frac{ca}{b} + \frac{ab}{c} + \frac{1}{abc}\right)$$
$$- s\,(bc, a) - s\,(ca, b) - s\,(ab, c)$$

$$h_2^* = \frac{2abc}{3} + \frac{a + b + c}{2} + \frac{3}{2} + \frac{1}{6}\left(\frac{bc}{a} + \frac{ca}{b} + \frac{ab}{c} + \frac{1}{abc}\right)$$
$$- 2\,(s\,(bc, a) + s\,(ca, b) + s\,(ab, c))$$

$$h_1^* = \frac{abc}{6} + \frac{ab + ac + bc + a + b + c}{4} - 2 + \frac{1}{12}\left(\frac{bc}{a} + \frac{ca}{b} + \frac{ab}{c} + \frac{1}{abc}\right)$$
$$- s\,(bc, a) - s\,(ca, b) - s\,(ab, c). \qquad\qquad \square$$

It is a curious fact that the above expressions for h_1^*, h_2^*, and h_3^* are nonnegative integers due to Corollary 3.11.

Notes

1. The classical Dedekind sums came to life in the 1880s when Richard Dedekind (1831–1916)[1] studied the transformation properties of the **Dedekind η-function** [99]

$$\eta(z) := e^{\frac{\pi i z}{12}} \prod_{n \geq 1} \left(1 - e^{2\pi i n z}\right),$$

a useful computational gadget in the land of modular forms in number theory. Dedekind's reciprocity law (Corollary 8.5) follows from one of the functional transformation identities for η. Dedekind also proved that

$$12k\, s(h, k) \equiv k + 1 - 2\left(\frac{h}{k}\right) \pmod{8},$$

establishing a beautiful connection between the Dedekind sum and the Jacobi symbol $\left(\frac{h}{k}\right)$ (the reader may want to consult the lovely Carus monograph *Dedekind Sums*, by Emil Grosswald and Hans Rademacher, where the above result is proved [199, p. 34]), and then used this identity to show that the

---

[1] For more information about Dedekind, see
http://www-history.mcs.st-andrews.ac.uk/Biographies/Dedekind.html.

reciprocity law for the Dedekind sums (for which [199] contains several different proofs) is equivalent to the reciprocity law for the Jacobi symbol.

2. The Dedekind sums and their generalizations appear in various contexts besides analytic number theory and discrete geometry. Other mathematical areas in which Dedekind sums show up include topology [140, 176, 255], algebraic number theory [174, 222], and algebraic geometry [117]. They also have connections to algorithmic complexity [153] and continued fractions [17, 136, 183].

3. The reciprocity laws (Theorems 8.4 and 8.8) for the Fourier–Dedekind sums were proved in [37], though the connection to Bernoulli–Barnes polynomials was noticed only in [27]. Theorem 8.4 is equivalent to the reciprocity law for Don Zagier's *higher-dimensional Dedekind sums* [255]. Corollary 8.7 (stated in terms of the classical Dedekind sum) is originally due to Hans Rademacher [197]. Theorem 8.8 generalizes reciprocity laws by Rademacher [198] (essentially Corollary 8.9) and Ira Gessel [119].

4. The Fourier–Dedekind sums form only one set of generalizations of the classical Dedekind sums. A long, but by no means complete, list of other generalizations is [8, 9, 28, 31, 33, 39, 51–53, 80, 81, 107, 108, 119, 128, 130, 152, 174–176, 198, 241, 255].

5. The connection of Dedekind sums and lattice-point problems, namely Theorem 8.11 for $t = 1$, was first established by Louis Mordell in 1951 [181]. Some 42 years later, James Pommersheim established a proof of Theorem 8.11 as part of a much more general machinery [193]. In fact, Pommersheim's work implies that the classical Dedekind sum is the only nontrivial ingredient one needs for Ehrhart polynomials in dimensions 3 and 4.

6. We touched the question of efficient computability of Ehrhart (quasi-)polynomials in this chapter. Unfortunately, our current state of knowledge on generalized Dedekind sums does not suffice to make any general statement. However, Alexander Barvinok proved in 1994 [24] that in fixed dimension, the rational generating function of the Ehrhart quasipolynomial of a rational polytope can be efficiently computed. Barvinok's proof did not employ Dedekind sums but rather used a decomposition theorem of Brion, which is the subject of Chapter 11.

Exercises

8.1. Show that $s(a, b) = 0$ if and only if $a^2 \equiv -1 \bmod b$.

8.2. Prove that $6b\,s(a,b) \in \mathbb{Z}$. (*Hint:* Start with rewriting the Dedekind sum in terms of the greatest-integer function.)

8.3. Prove that $s(-a,b) = -s(a,b)$, for all $a \in \mathbb{Z}$ with $\gcd(a,b) = 1$.

8.4. For a fixed prime modulus p, show that we can solve $s(a,p) = s(b,p)$ in integers a,b if and only if $a \equiv b \bmod p$ or $ab \equiv 1 \bmod p$.

8.5. Let a and b be any two relatively prime positive integers. Show that the reciprocity law for the Dedekind sums implies that for $b \equiv r \bmod a$,

$$12ab\,s(a,b) = -12ab\,s(r,a) + a^2 + b^2 - 3ab + 1\,.$$

Deduce the following identities:

(a) For $b \equiv 1 \bmod a$,

$$12ab\,s(a,b) = -a^2b + b^2 + a^2 - 2b + 1\,.$$

(b) For $b \equiv 2 \bmod a$,

$$12ab\,s(a,b) = -\frac{1}{2}a^2b + a^2 + b^2 - \frac{5}{2}b + 1\,.$$

(c) For $b \equiv -1 \bmod a$,

$$12ab\,s(a,b) = a^2b + a^2 + b^2 - 6ab + 2b + 1\,.$$

8.6. Denote by f_n the sequence of Fibonacci numbers, defined as in Chapter 1 by

$$f_1 = f_2 = 1 \qquad \text{and} \qquad f_{n+2} = f_{n+1} + f_n \ \text{ for } n \geq 1\,.$$

Prove that

$$s\left(f_{2k}, f_{2k+1}\right) = 0$$

and

$$12f_{2k-1}f_{2k}\,s\left(f_{2k-1}, f_{2k}\right) = f_{2k-1}^2 + f_{2k}^2 - 3f_{2k-1}f_{2k} + 1\,.$$

8.7. ♣ Prove (8.2):

$$s_0(a_1, a_2; b) = -\,s\left(a_1 a_2^{-1}, b\right) + \frac{b-1}{4b}\,,$$

where $a_2^{-1}a_2 \equiv 1 \bmod b$.

8.8. Prove that $B_1, B_2, \ldots, B_d$ in the partial fraction expansion (8.3),

$$f(z) = \frac{1}{(1 - z^{a_1}) \cdots (1 - z^{a_d}) z^n}$$

$$= \frac{A_1}{z} + \frac{A_2}{z^2} + \cdots + \frac{A_n}{z^n} + \frac{B_1}{z - 1} + \frac{B_2}{(z - 1)^2} + \cdots + \frac{B_d}{(z - 1)^d}$$

$$+ \sum_{k=1}^{a_1 - 1} \frac{C_{1k}}{z - \xi_{a_1}^k} + \sum_{k=1}^{a_2 - 1} \frac{C_{2k}}{z - \xi_{a_2}^k} + \cdots + \sum_{k=1}^{a_d - 1} \frac{C_{dk}}{z - \xi_{a_d}^k},$$

are polynomials in n (of degree less than d) and rational functions in $a_1, \ldots, a_d$.

8.9. ♣ Verify the first few expressions for $\text{poly}_{\{a_1, \ldots, a_d\}}(n)$ in (8.5).

8.10. Show that the Dedekind–Rademacher sum satisfies $r_{-n}(a, b) = r_n(a, b)$.

8.11. The definition $s(a, b) := \sum_{k=0}^{b-1} \left(\!\left(\frac{ka}{b}\right)\!\right) \left(\!\left(\frac{k}{b}\right)\!\right)$ of the Dedekind sum may be extended to encompass all pairs of integers (a, b) with $b > 0$. Show that if $g := \gcd(a, b)$, then $s(a, b) = s(\frac{a}{g}, \frac{b}{g})$. (This exercise shows that the Dedekind sum may also be thought of as a function of the fraction $\frac{a}{b}$, defined for any two integers a and $b > 0$.)

8.12. ♣ Show that

$$\sum_{m=0}^{b-1} \left(\left(\!\left(\frac{n - m}{b}\right)\!\right) + \frac{1}{2} \chi_b(n - m) - \frac{1}{2b} \right) \left(\left(\!\left(\frac{a^{-1}m}{b}\right)\!\right) + \frac{1}{2} \chi_b(m) - \frac{1}{2b} \right)$$

$$= \sum_{m=0}^{b-1} \left(\left(\!\left(\frac{am - n}{b}\right)\!\right) \right) \left(\left(\!\left(\frac{m}{b}\right)\!\right) \right) + \frac{1}{2} \left(\left(\!\left(\frac{a^{-1}n}{b}\right)\!\right) \right) + \frac{1}{2} \left(\left(\!\left(\frac{-n}{b}\right)\!\right) \right) - \frac{1}{4b}$$

$$+ \frac{1}{4} \chi_b(n).$$

8.13. Rephrase the Ehrhart quasipolynomials for rational triangles given in Theorem 2.10 and Exercise 2.36 in terms of Dedekind–Rademacher sums.

8.14. Rephrase the restricted partition $p_{\{a,b,c\}}(n)$ from Example 1.9 in terms of Dedekind–Rademacher sums, and use an explicit formula for $p_{\{a,1,1\}}(n)$ to derive a formula for $r_n(1, a)$.

8.15. ♣ Prove Theorem 8.11 by verifying (8.9) and computing the coefficients for $z = 1$ in the partial fraction expansion of (8.8).

8.16. Let $\mathcal{P}$ be given by (8.7) with $a = k$ and $b = c = 1$. Compute $L_{\mathcal{P}}(t)$ from first principles and conclude from this an alternative proof of the formula for $s(1, k)$ given in Exercise 7.21.

8.17. Let $\mathcal{P}$ be given by (8.7) with $\gcd(a, b) = 1$ and $c = 1$. Use Dedekind's reciprocity law (Corollary 8.5) to derive from Theorem 8.11 a formula for $L_{\mathcal{P}}(t)$ not depending on Dedekind sums. Can you prove this formula from first principles?

8.18. Generalize the Ehrhart polynomial of the Mordell–Pommersheim tetrahedron given in Theorem 8.11 to the case that a, b, and c are not necessarily pairwise relatively prime.

8.19. Compute the Ehrhart polynomial of the 4-simplex

$$\left\{ (x_1, x_2, x_3, x_4) \in \mathbb{R}^4_{\geq 0} : \frac{x_1}{a} + \frac{x_2}{b} + \frac{x_3}{c} + \frac{x_4}{d} \leq 1 \right\},$$

where a, b, c, d are pairwise relatively prime positive integers. (*Hint:* You may use Corollary 5.5 to compute the linear term.)

8.20. This exercise gives an alternative proof (and extension) of Dedekind's reciprocity law (Corollary 8.5) by way of a polynomial analogue of the Dedekind sum, the **Carlitz polynomial**

$$c(u, v; a, b) := \sum_{k=1}^{b-1} u^{\left\lfloor \frac{ka}{b} \right\rfloor} v^{k-1}.$$

Here u and v are indeterminates, and a and b are positive integers.

(a) Continuing Exercise 3.7, express the integer-point transforms of the rational cones

$$\mathcal{K}_1 = \{\lambda_1(0, 1) + \lambda_2(a, b) : \lambda_1, \lambda_2 \geq 0\}$$
$$\mathcal{K}_2 = \{\lambda_1(1, 0) + \lambda_2(a, b) : \lambda_1 > 0, \lambda_2 \geq 0\}$$

in terms of Carlitz polynomials. Here a and b are relatively prime positive integers.

(b) Verify that $\mathbb{R}^2_{\geq 0}$ is the disjoint union of $\mathcal{K}_1$ and $\mathcal{K}_2$, and use this fact to prove the **Carlitz reciprocity law**

$$(v - 1)\, c(u, v; a, b) + (u - 1)\, c(v, u; b, a) = u^{a-1} v^{b-1} - 1.$$

(c) Apply the operators $u\, \partial u$ twice and $v\, \partial v$ once to Carlitz's reciprocity law to deduce (once more) Corollary 8.5.

Open Problems

8.21. Find new relations between various Dedekind sums.

8.22. It is known [31] that the Fourier–Dedekind sums are efficiently computable. Find a fast algorithm that can be implemented in practice.

8.23. For a fixed integer modulus c, find all integer solutions $a, b \in \mathbb{Z}$ to $s(a, c) = s(b, c)$. (See Exercise 8.4 for the case that c is prime.)

8.24. What is
$$\{s(a,b) : a, b \in \mathbb{Z},\ b > 0\},$$

the set of all rational numbers that occur as Dedekind sums? (It is known that this set is dense in $\mathbb{R}$ [17, 136] and that every rational number between 0 and 1 occurs as the fractional part of a Dedekind sum [120]. On the other hand, we know that $12\, s(a,b)$ is never an integer unless $s(a,b) = 0$ [129].)

Chapter 9
Zonotopes

"And what is the use of a book," thought Alice, "without pictures or conversations?"

Lewis Carroll (*Alice in Wonderland*)

We have seen that the discrete volume of a general integral polytope may be quite difficult to compute. It is therefore useful to have an infinite class of integral polytopes whose discrete volume is more tractable, and yet they are robust enough to be "closer" in complexity to generic integral polytopes. One initial class of more tractable polytopes is that of parallelepipeds, and as we will see in Lemma 9.2, the Ehrhart polynomial of a d-dimensional half-open integer parallelepiped $\mathcal{P}$ is equal to $\mathrm{vol}(\mathcal{P})\, t^d$. In this chapter, we generalize parallelepipeds to projections of cubes.

9.1 Definitions and Examples

In order to extend the notion of a parallelepiped, we begin by defining the **Minkowski sum** of the polytopes $\mathcal{P}_1, \mathcal{P}_2, \ldots, \mathcal{P}_n \subset \mathbb{R}^d$ as

$$\mathcal{P}_1 + \mathcal{P}_2 + \cdots + \mathcal{P}_n := \{\mathbf{x}_1 + \mathbf{x}_2 + \cdots + \mathbf{x}_n : \mathbf{x}_j \in \mathcal{P}_j\}.$$

For example, if $\mathcal{P}_1$ is the rectangle $[0,2] \times [0,1] \subset \mathbb{R}^2$ and $\mathcal{P}_2$ is the line segment $\left[\binom{0}{0}, \binom{2}{3}\right] \subset \mathbb{R}^2$, then their Minkowski sum $\mathcal{P}_1 + \mathcal{P}_2$ is the hexagon whose vertices are $\binom{0}{0}, \binom{2}{0}, \binom{0}{1}, \binom{4}{3}, \binom{2}{4}$, and $\binom{4}{4}$, as depicted in Figure 9.1. Parallelepipeds are special instances of Minkowski sums, namely those of line segments whose direction vectors are linearly independent, plus a point.

We will also make use of the following handy construct: given a polynomial $p(z_1, z_2, \ldots, z_d)$ in d variables, the **Newton polytope** $\mathcal{N}(p(z_1, z_2, \ldots, z_d))$ of $p(z_1, z_2, \ldots, z_d)$ is the convex hull of all exponent vectors appearing in the

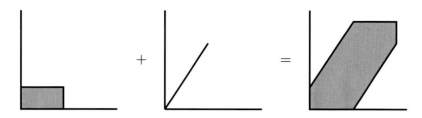

Fig. 9.1 The Minkowski sum $([0,2] \times [0,1]) + \left[\binom{0}{0}, \binom{2}{3} \right]$.

nonzero terms of $p(z_1, z_2, \ldots, z_d)$. For example, the hexagon in Figure 9.1 can be written as

$$\mathcal{N}\left(1 + 3z_1^2 - z_2 - 5z_1^4 z_2^3 + 34z_1^2 z_2^4 + z_1^4 z_2^4\right).$$

It turns out (Exercise 9.1) that the constructions of Newton polytopes and Minkowski sums are intimately related. Namely, if $p(z_1, z_2, \ldots, z_d)$ and $q(z_1, z_2, \ldots, z_d)$ are polynomials, then

$$\mathcal{N}\big(p(z_1, z_2, \ldots, z_d)\, q(z_1, z_2, \ldots, z_d)\big)$$
$$= \mathcal{N}\big(p(z_1, z_2, \ldots, z_d)\big) + \mathcal{N}\big(q(z_1, z_2, \ldots, z_d)\big). \tag{9.1}$$

Suppose that we are now given n line segments in $\mathbb{R}^d$, such that each line segment has one endpoint at the origin and the other endpoint is located at the vector $\mathbf{u}_j \in \mathbb{R}^d$, for $j = 1, \ldots, n$. Then by definition, the Minkowski sum of these n segments is

$$\mathcal{Z}(\mathbf{u}_1, \mathbf{u}_2, \ldots, \mathbf{u}_n) := \{\mathbf{x}_1 + \mathbf{x}_2 + \cdots + \mathbf{x}_n : \mathbf{x}_j = \lambda_j \mathbf{u}_j \text{ with } \lambda_j \in [0, 1]\}.$$

Fig. 9.2 The zonotope $\mathcal{Z}\left(\binom{0}{4}, \binom{3}{3}, \binom{4}{1}\right)$—a hexagon.

Let's rewrite the definition above in matrix form:

$$\mathcal{Z}(\mathbf{u}_1, \mathbf{u}_2, \ldots, \mathbf{u}_n) = \{\lambda_1 \mathbf{u}_1 + \lambda_2 \mathbf{u}_2 + \cdots + \lambda_n \mathbf{u}_n : 0 \le \lambda_j \le 1\}$$

$$= \left\{ (\mathbf{u}_1 \ \mathbf{u}_2 \ \cdots \ \mathbf{u}_n) \begin{pmatrix} \lambda_1 \\ \lambda_2 \\ \vdots \\ \lambda_n \end{pmatrix} : 0 \le \lambda_j \le 1 \right\}$$

$$= \mathbf{A} [0, 1]^n,$$

where $\mathbf{A}$ is the $(d \times n)$ matrix whose j^{th} column is $\mathbf{u}_j$. Just a bit more generally, a **zonotope** is defined to be any translate of $\mathbf{A} [0, 1]^n$, i.e.,

$$\mathbf{A} [0, 1]^n + \mathbf{b},$$

for some vector $\mathbf{b} \in \mathbb{R}^d$. So we now have two equivalent definitions of a zonotope—the first is as a Minkowski sum of line segments, while the second is as a projection of the unit cube $[0, 1]^n$. Figures 9.2–9.4 show example zonotopes.

Fig. 9.3 The rhombic dodecahedron is a zonotope.

It is sometimes useful to translate a zonotope so that the origin becomes its new center of mass. To this end, we now dilate the matrix $\mathbf{A}$ by a factor of 2 and translate the resulting image so that its new center of mass is at the origin. Precisely,

$$2 \mathbf{A} [0, 1]^n - (\mathbf{u}_1 + \cdots + \mathbf{u}_n)$$
$$= \{2\mathbf{u}_1 \lambda_1 + \cdots + 2\mathbf{u}_n \lambda_n - (\mathbf{u}_1 + \cdots + \mathbf{u}_n) : 0 \le \lambda_j \le 1\}$$
$$= \{\mathbf{u}_1 (2\lambda_1 - 1) + \cdots + \mathbf{u}_n (2\lambda_n - 1) : 0 \le \lambda_j \le 1\}$$
$$= \mathbf{A} [-1, 1]^n, \tag{9.2}$$

where the last step holds because $-1 \le 2\lambda_j - 1 \le 1$ when $0 \le \lambda_j \le 1$. In other words, (9.2) is a linear image of the larger cube $[-1, 1]^n$, centered at the origin. We thus define

$$\mathcal{Z}(\pm\mathbf{u}_1, \pm\mathbf{u}_2, \ldots, \pm\mathbf{u}_n) := \mathbf{A}\,[-1,1]^n,$$

and we see that $\mathcal{Z}(\pm\mathbf{u}_1, \pm\mathbf{u}_2, \ldots, \pm\mathbf{u}_n)$ is indeed a zonotope, by definition. We say that a polytope $\mathcal{P}$ is **symmetric about the origin** when it has the property that $\mathbf{x} \in \mathcal{P}$ if and only if $-\mathbf{x} \in \mathcal{P}$. In general, a polytope $\mathcal{P}$ is called **centrally symmetric** if we can translate $\mathcal{P}$ by some vector $\mathbf{b}$ such that $\mathcal{P}+\mathbf{b}$ is symmetric about the origin. The zonotope $\mathcal{Z}(\pm\mathbf{u}_1, \pm\mathbf{u}_2, \ldots, \pm\mathbf{u}_n)$ is an example of a polytope that is symmetric about the origin. We go through the full argument here, because it justifies our choice of a symmetric representation: Choose $\mathbf{x} \in \mathcal{Z}(\pm\mathbf{u}_1, \pm\mathbf{u}_2, \ldots, \pm\mathbf{u}_n)$, so that $\mathbf{x} = \mathbf{A}\,\mathbf{y}$, with $\mathbf{y} \in [-1,1]^n$. Since $-\mathbf{y} \in [-1,1]^n$ as well,

$$-\mathbf{x} = \mathbf{A}\,(-\mathbf{y}) \in \mathcal{Z}(\pm\mathbf{u}_1, \pm\mathbf{u}_2, \ldots, \pm\mathbf{u}_n)\,.$$

It is, moreover, true that each face of a zonotope is again a zonotope and that therefore, every face of a zonotope is centrally symmetric (Exercise 9.2).

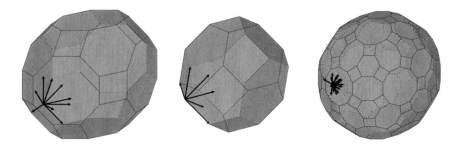

Fig. 9.4 A few more zonotopes.

9.2 Paving a Zonotope

Next we will show that every zonotope can be neatly decomposed into a disjoint union of half-open parallelepipeds. Suppose that $\mathbf{w}_1, \mathbf{w}_2, \ldots, \mathbf{w}_m \in \mathbb{R}^d$ are linearly independent, and let $\sigma_1, \sigma_2, \ldots, \sigma_m \in \{\pm 1\}$. Then we define

$$\Pi_{\mathbf{w}_1, \mathbf{w}_2, \ldots, \mathbf{w}_m}^{\sigma_1, \sigma_2, \ldots, \sigma_m} := \left\{ \lambda_1\mathbf{w}_1 + \lambda_2\mathbf{w}_2 + \cdots + \lambda_m\mathbf{w}_m : \begin{array}{l} 0 \le \lambda_j < 1 \text{ if } \sigma_j = -1 \\ 0 < \lambda_j \le 1 \text{ if } \sigma_j = 1 \end{array} \right\}.$$

In plain English, $\Pi_{\mathbf{w}_1, \mathbf{w}_2, \ldots, \mathbf{w}_m}^{\sigma_1, \sigma_2, \ldots, \sigma_m}$ is a half-open parallelepiped generated by $\mathbf{w}_1, \mathbf{w}_2, \ldots, \mathbf{w}_m$, and the signs $\sigma_1, \sigma_2, \ldots, \sigma_m$ keep track of those facets of the parallelepiped that are included or excluded from the closure of the parallelepiped.

Lemma 9.1. *The zonotope $\mathcal{Z}(\mathbf{u}_1, \mathbf{u}_2, \ldots \mathbf{u}_n)$ can be written as a disjoint union of translates of $\Pi_{\mathbf{w}_1,\mathbf{w}_2,\ldots,\mathbf{w}_m}^{\sigma_1,\sigma_2,\ldots,\sigma_m}$, where $\{\mathbf{w}_1, \mathbf{w}_2, \ldots, \mathbf{w}_m\}$ ranges over all linearly independent subsets of $\{\mathbf{u}_1, \mathbf{u}_2, \ldots \mathbf{u}_n\}$, each equipped with an appropriate choice of signs $\sigma_1, \sigma_2, \ldots, \sigma_m$.*

Figure 9.5 illustrates the decomposition of a zonotope as suggested by Lemma 9.1.

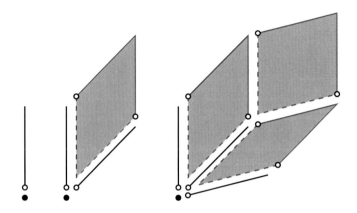

Fig. 9.5 A zonotopal decomposition of $\mathcal{Z}\left(\binom{0}{4}, \binom{3}{3}, \binom{4}{1}\right)$.

Proof of Lemma 9.1. We proceed by induction on n. If $n = 1$, $\mathcal{Z}(\mathbf{u}_1)$ is a line segment and $\mathbf{0} \cup (\mathbf{0}, \mathbf{u}_1]$ is a desired decomposition.

For general $n > 1$, we have by induction the decomposition

$$\mathcal{Z}(\mathbf{u}_1, \mathbf{u}_2, \ldots, \mathbf{u}_{n-1}) = \Pi_1 \cup \Pi_2 \cup \cdots \cup \Pi_k$$

into half-open parallelepipeds of the form given in the statement of Lemma 9.1. Now we define the hyperplane $H := \{\mathbf{x} \in \mathbb{R}^d : \mathbf{x} \cdot \mathbf{u}_n = 0\}$ and let $\pi : \mathbb{R}^d \to H$ denote the orthogonal projection onto H. Then $\pi(\mathbf{u}_1), \pi(\mathbf{u}_2), \ldots, \pi(\mathbf{u}_{n-1})$ are line segments or points, and thus $\mathcal{Z}(\pi(\mathbf{u}_1), \pi(\mathbf{u}_2), \ldots, \pi(\mathbf{u}_{n-1}))$ is a zonotope living in H. Once more by induction, we can decompose

$$\mathcal{Z}(\pi(\mathbf{u}_1), \pi(\mathbf{u}_2), \ldots, \pi(\mathbf{u}_{n-1})) = \Phi_1 \cup \Phi_2 \cup \cdots \cup \Phi_m$$

into half-open parallelepipeds of the form given in Lemma 9.1. Each Φ_j is a half-open parallelepiped generated by some of the vectors $\pi(\mathbf{u}_1), \pi(\mathbf{u}_2), \ldots, \pi(\mathbf{u}_{n-1})$; let $\widetilde{\Phi}_j$ denote the corresponding parallelepiped generated by their unprojected counterparts. Then (Exercise 9.5) the desired disjoint union of $\mathcal{Z}(\mathbf{u}_1, \mathbf{u}_2, \ldots, \mathbf{u}_n)$ is given by $\Pi_1 \cup \Pi_2 \cup \cdots \cup \Pi_k \cup \mathcal{P}_1 \cup \mathcal{P}_2 \cup \cdots \cup \mathcal{P}_m$, where

$$\mathcal{P}_j := \widetilde{\Phi}_j \times (\mathbf{0}, \mathbf{u}_n]. \qquad\qquad \Box$$

This decomposition lemma is useful, e.g., to compute the Ehrhart polynomial of a zonotope. To this end, we now first work out the Ehrhart polynomial of a half-open parallelepiped, which is itself a (particularly simple) zonotope.

Lemma 9.2. *Suppose* $\mathbf{w}_1, \mathbf{w}_2, \dots, \mathbf{w}_d \in \mathbb{Z}^d$ *are linearly independent, and let*

$$\Pi := \{\lambda_1 \mathbf{w}_1 + \lambda_2 \mathbf{w}_2 + \cdots + \lambda_d \mathbf{w}_d : 0 \leq \lambda_1, \lambda_2, \dots, \lambda_d < 1\}.$$

Then

$$\# \left(\Pi \cap \mathbb{Z}^d\right) = \operatorname{vol}\Pi = |\det(\mathbf{w}_1, \dots, \mathbf{w}_d)|,$$

and for every positive integer t,

$$\# \left(t\,\Pi \cap \mathbb{Z}^d\right) = (\operatorname{vol}\Pi)\, t^d.$$

In other words, for the half-open parallelepiped Π, the discrete volume $\# \left(t\,\Pi \cap \mathbb{Z}^d\right)$ coincides with the continuous volume $(\operatorname{vol}\Pi)\, t^d$.

Proof. Because Π is half open, we can tile the tth dilate $t\,\Pi$ by t^d translates of Π, and hence

$$L_\Pi(t) = \# \left(t\,\Pi \cap \mathbb{Z}^d\right) = \# \left(\Pi \cap \mathbb{Z}^d\right) t^d.$$

On the other hand, by the results of Chapter 3, $L_\Pi(t)$ is a polynomial with leading coefficient $\operatorname{vol}\Pi = |\det(\mathbf{w}_1, \dots, \mathbf{w}_d)|$. Since we have equality of these polynomials for all positive integers t, it follows that

$$\# \left(\Pi \cap \mathbb{Z}^d\right) = \operatorname{vol}\Pi. \qquad\qquad \Box$$

Our proof shows that Lemma 9.2 remains true if we switch the $\leq$ and $<$ inequalities for some of the λ_j in the definition of Π, a fact that we will use freely, e.g., in the following tool to compute Ehrhart polynomials of zonotopes, which follows from a combination of Lemmas 9.1 and 9.2.

Corollary 9.3. *Decompose the zonotope* $\mathcal{Z} \subset \mathbb{R}^d$ *into half-open parallelepipeds according to Lemma 9.1. Then the coefficient* c_k *of the Ehrhart polynomial* $L_{\mathcal{Z}}(t) = c_d\, t^d + c_{d-1}\, t^{d-1} + \cdots + c_0$ *equals the sum of the (relative) volumes of the* k-*dimensional parallelepipeds in the decomposition of* $\mathcal{Z}$.

9.3 The Permutahedron

To illustrate that Corollary 9.3 is useful for computations, we now study a famous polytope, the **permutahedron**

$$\mathcal{P}_d := \operatorname{conv}\{(\pi(1) - 1,\, \pi(2) - 1, \dots,\, \pi(d) - 1) : \pi \in S_d\},$$

that is, the convex hull of $(0, 1, \ldots, d-1)$ and all points formed by permuting its entries. It is not hard to show (see Exercise 9.9) that $\mathcal{P}_d$ is $(d-1)$-dimensional and has $d!$ vertices. Figures 9.6 and 9.7 show P_3 and P_4 (the latter projected into $\mathbb{R}^3$).

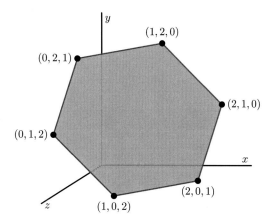

Fig. 9.6 The permutahedron P_3.

The reason that permutahedra make an appearance in this chapter is the following result.

Theorem 9.4.

$$\mathcal{P}_d = [\mathbf{e}_1, \mathbf{e}_2] + [\mathbf{e}_1, \mathbf{e}_3] + \cdots + [\mathbf{e}_{d-1}, \mathbf{e}_d],$$

in words: the permutahedron $\mathcal{P}_d$ is the Minkowski sum of the line segments between each pair of unit vectors in $\mathbb{R}^d$.

Proof. We make use of a matrix that already made its debut in Section 3.7: the permutahedron $\mathcal{P}_d$ is the Newton polytope of the polynomial

$$\det \begin{pmatrix} x_1^{d-1} & x_1^{d-2} & \cdots & x_1 & 1 \\ x_2^{d-1} & x_2^{d-2} & \cdots & x_2 & 1 \\ \vdots & \vdots & & \vdots & \vdots \\ x_d^{d-1} & x_d^{d-2} & \cdots & x_d & 1 \end{pmatrix}.$$

One can see the vertices of $\mathcal{P}_d$ appearing in the exponent vectors by computing this determinant by cofactor expansion. Now we use Exercise 3.21:

$$\det \begin{pmatrix} x_1^{d-1} & x_1^{d-2} & \cdots & x_1 & 1 \\ x_2^{d-1} & x_2^{d-2} & \cdots & x_2 & 1 \\ \vdots & \vdots & & \vdots & \vdots \\ x_d^{d-1} & x_d^{d-2} & \cdots & x_d & 1 \end{pmatrix} = \prod_{1 \le j < k \le d} (x_j - x_k),$$

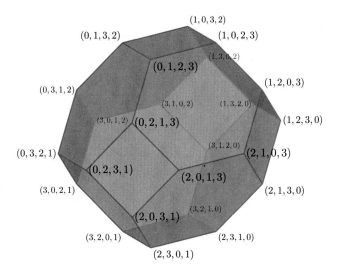

Fig. 9.7 The permutahedron P_4 (projected into $\mathbb{R}^3$).

and so by (9.1),

$$\mathcal{P}_d = \mathcal{N}(x_1 - x_2) + \mathcal{N}(x_1 - x_3) + \cdots + \mathcal{N}(x_{d-1} - x_d).$$

The right-hand side is the Minkowski sum of the line segments $[\mathbf{e}_1, \mathbf{e}_2]$, $[\mathbf{e}_1, \mathbf{e}_3]$, $\ldots$, $[\mathbf{e}_{d-1}, \mathbf{e}_d]$. □

Now we will apply Corollary 9.3 to the special zonotope $\mathcal{P}_d$. A **forest** is a graph that does not contain any closed paths.

Theorem 9.5. *The coefficient c_k of the Ehrhart polynomial*

$$L_{\mathcal{P}_d}(t) = c_{d-1}\, t^{d-1} + c_{d-2}\, t^{d-2} + \cdots + c_0$$

of the permutahedron $\mathcal{P}_d$ equals the number of labeled forests on d nodes with k edges.

For example, we can compute $L_{\mathcal{P}_3}(t) = 3t^2 + 3t + 1$ by looking at the labeled forests in Figure 9.8.

An important ingredient to the proof of Theorem 9.5 is the following correspondence: to a subset $S \subseteq \{\mathbf{e}_1 + \mathbf{e}_2, \mathbf{e}_1 + \mathbf{e}_3, \ldots, \mathbf{e}_{d-1} + \mathbf{e}_d\}$ we associate the graph G_S with node set $[d]$ and edge set

$$\{jk : \mathbf{e}_j + \mathbf{e}_k \in S\}.$$

We invite the reader to prove the following lemma (Exercise 9.13):

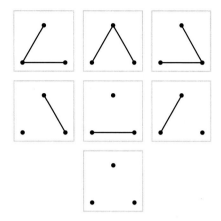

Fig. 9.8 The $3 + 3 + 1$ labeled forests on three nodes.

Lemma 9.6. *A subset $S \subseteq \{e_1 + e_2, e_1 + e_3, \dots, e_{d-1} + e_d\}$ is linearly independent if and only if G_S is a forest.*

Proof of Theorem 9.5. Let I consist of all nonempty linearly independent subsets of

$$\{e_1 + e_2, e_1 + e_3, \dots, e_{d-1} + e_d\}.$$

To each such subset S we will associate the half-open parallelepiped

$$\sum_{e_j + e_k \in S} (\mathbf{0}, e_j + e_k]$$

(written as a Minkowski sum). It has relative volume 1 (Exercise 9.14). These half-open parallelepipeds give rise to the disjoint union

$$\mathcal{P}_d = \mathcal{Z}(e_1 + e_2, e_1 + e_3, \dots, e_{d-1} + e_d)$$

$$= \mathbf{0} \cup \bigcup_{S \in I} \left(\sum_{e_j + e_k \in S} (\mathbf{0}, e_j + e_k] \right), \tag{9.3}$$

as the reader should verify in Exercise 9.15. Corollary 9.3 now says that the coefficient c_k of the Ehrhart polynomial

$$L_{\mathcal{P}_d}(t) = c_{d-1} t^{d-1} + c_{d-2} t^{d-2} + \cdots + c_0$$

equals the sum of the relative volumes of the k-dimensional parallelepipeds in our zonotopal decomposition of $\mathcal{P}_d$ given in (9.3). As mentioned above, each of these volumes is 1, and Lemma 9.6 now gives the statement of Theorem 9.5.

$\square$

The leading coefficient of the Ehrhart polynomial of $\mathcal{P}_d$ is, of course, its relative volume, about which we can say slightly more. A **tree** is a connected forest.

Corollary 9.7. *The (relative) volume of the permutahedron $\mathcal{P}_d$ equals the number of labeled trees on d nodes.*

It turns out (Exercise 9.16) that there are precisely d^{d-2} labeled trees on d nodes, so the relative volume of $\mathcal{P}_d$ equals d^{d-2}.

Proof. By Theorem 9.5, the leading coefficient of $\mathcal{P}_d$ equals the number of labeled forests on d nodes with $d-1$ edges. But (as the reader should show in Exercise 9.17) every such forest is connected.   □

9.4 The Ehrhart Polynomial of a Zonotope

Next we will generalize Lemma 9.2 to refine the formula of Corollary 9.3 for Ehrhart polynomials of zonotopes. Lemma 9.2 computed the (continuous and discrete) volume of a half-open parallelepiped Π spanned by d linearly independent vectors in $\mathbb{Z}^d$; this volume equals the number of integer points in Π. We need a version of this lemma in which Π is spanned by n linearly independent vectors in $\mathbb{Z}^d$ where $n < d$. It turns out that the (continuous and discrete) volume of Π is still given by the number of integer points in Π, but this number is not quite as easily computed as in Lemma 9.2.

Lemma 9.8. *Suppose $\mathbf{w}_1, \mathbf{w}_2, \ldots, \mathbf{w}_n \in \mathbb{Z}^d$ are linearly independent, let*

$$\Pi := \{\lambda_1 \mathbf{w}_1 + \lambda_2 \mathbf{w}_2 + \cdots + \lambda_n \mathbf{w}_n : 0 \le \lambda_1, \lambda_2, \ldots, \lambda_n < 1\},$$

and let V be the greatest common divisor of all $n \times n$ minors of the matrix formed by the column vectors $\mathbf{w}_1, \mathbf{w}_2, \ldots, \mathbf{w}_n$. Then the relative volume of Π equals V. Furthermore,

$$\#\left(\Pi \cap \mathbb{Z}^d\right) = V,$$

and for every positive integer t,

$$\#\left(t\,\Pi \cap \mathbb{Z}^d\right) = V\,t^n.$$

As with Lemma 9.2, the statement of Lemma 9.8 remains true if we switch the $\le$ and $<$ inequalities for some of the λ_j in the definition of Π.

Proof. We will make use of the *Smith normal form* of an integer matrix (or more generally, a matrix with entries in an integral domain): more precisely, one proves in linear algebra that for every full-rank matrix $\mathbf{A} \in \mathbb{Z}^{m \times n}$, say for which $m \ge n$, there exist invertible matrices $\mathbf{S} \in \mathbb{Z}^{m \times m}$ and $\mathbf{T} \in \mathbb{Z}^{n \times n}$ such that

$$\mathbf{S\,A\,T} = \begin{pmatrix} d_1 & 0 & \cdots & & & 0 \\ 0 & d_2 & & & & \\ \vdots & & \ddots & & & \vdots \\ & & & d_{n-1} & 0 & \\ 0 & & \cdots & 0 & d_n & \\ 0 & & \cdots & & & 0 \\ \vdots & & & & & \vdots \\ 0 & & \cdots & & & 0 \end{pmatrix}$$

with diagonal integer entries whose product $d_1 d_2 \cdots d_n$ equals the greatest common divisor of all $n \times n$ minors of $\mathbf{A}$.

Geometrically, our matrices $\mathbf{S}$ and $\mathbf{T}$ transform Π into the half-open rectangular parallelepiped

$$\widetilde{\Pi} := [0, d_1) \times [0, d_2) \times \cdots \times [0, d_n) \subset \mathbb{R}^d$$

in such a way that the integer lattice $\mathbb{Z}^d$ is preserved; in particular, the relative volumes of Π and $\widetilde{\Pi}$ are equal, namely $d_1 d_2 \cdots d_n = V$. Furthermore,

$$\# \left(\Pi \cap \mathbb{Z}^d \right) = \# \left(\widetilde{\Pi} \cap \mathbb{Z}^d \right) = d_1 d_2 \cdots d_n = V.$$

The remainder of our proof follows that of Lemma 9.2: since Π is half open, we can tile the tth dilate $t\,\Pi$ by t^n translates of Π, and so

$$\# \left(t\,\Pi \cap \mathbb{Z}^d \right) = \# \left(\Pi \cap \mathbb{Z}^d \right) t^n. \qquad \square$$

Lemma 9.8 is the crucial ingredient for the following refinement of Corollary 9.3.

Theorem 9.9. *Let $\mathcal{Z} := \mathcal{Z}(\mathbf{u}_1, \ldots, \mathbf{u}_n)$ be a zonotope generated by the integer vectors $\mathbf{u}_1, \ldots, \mathbf{u}_n$. Then the Ehrhart polynomial of $\mathcal{Z}$ is given by*

$$L_{\mathcal{Z}}(t) = \sum_S m(S)\, t^{|S|},$$

where S ranges over all linearly independent subsets of $\{\mathbf{u}_1, \ldots, \mathbf{u}_n\}$, and $m(S)$ is the greatest common divisor of all minors of size $|S|$ of the matrix whose columns are the elements of S.

For example, returning to the zonotope $\mathcal{Z}\left(\binom{0}{4}, \binom{3}{3}, \binom{4}{1}\right)$ featured in Figure 9.5, we compute

$$L_{\mathcal{Z}\left(\binom{0}{4}, \binom{3}{3}, \binom{4}{1}\right)}(t) = \left|\det\begin{pmatrix} 0 & 3 \\ 4 & 3 \end{pmatrix}\right| t^2 + \left|\det\begin{pmatrix} 0 & 4 \\ 4 & 1 \end{pmatrix}\right| t^2 + \left|\det\begin{pmatrix} 3 & 4 \\ 3 & 1 \end{pmatrix}\right| t^2$$

$$+ \gcd(0, 4)\, t + \gcd(3, 3)\, t + \gcd(4, 1)\, t + 1$$

$$= 37\, t^2 + 8\, t + 1.$$

We remark that the numbers $m(S)$ appearing in Theorem 9.9 have geometric meaning: if $V(S)$ and $L(S)$ denote respectively the vector space and lattice generated by S, then $m(S)$ equals the cardinality of the group $\left(V(S) \cap \mathbb{Z}^d\right)/L(S)$.

Proof of Theorem 9.9. Decompose $\mathcal{Z}$ as described in Lemma 9.1. Corollary 9.3 says that the coefficient c_k of the Ehrhart polynomial

$$L_{\mathcal{Z}}(t) = c_d\,t^d + c_{d-1}\,t^{d-1} + \cdots + c_0$$

equals the sum of the relative volumes of the k-dimensional parallelepipeds in this decomposition. Lemma 9.8 implies that these relative volumes are the greatest common divisors of the minors of the matrices formed by the generators of these parallelepipeds. $\square$

Notes

1. The word *zonotope* appears to have originated from the fact that for each line segment $\mathbf{u}_j$ that comes into the Minkowski-sum definition of a zonotope $\mathcal{Z}$, there corresponds a "zone," composed of all of the facets of $\mathcal{Z}$ that contain parallel translates of $\mathbf{u}_j$. This zone separates the zonotope into two isometric pieces, a "northern" hemisphere and a "southern" hemisphere, a property that is sometimes useful in their study.

2. The combinatorial study of zonotopes, starting with Lemma 9.1, was brought to the forefront (if not initiated) by Peter McMullen and Geoffrey Shephard [170, 173, 219].

3. The permutahedron seems to have been first studied by Pieter Hendrik Schoute [212] in 1911. It tends to play a central role whenever one tries to "geometrize" a situation involving the symmetric group. The permutahedron is a *simple* zonotope and as such quite a rare animal. It is not clear from the literature who first discovered Theorem 9.5, but we suspect it was Richard Stanley [230, Exercises 4.63 and 4.64]. There has been a recent flurry of research activity on *generalized permutahedra* (initiated by Alexander Postnikov [196]), which share many of the fascinating geometric and arithmetic properties of permutahedra.

4. Corollary 9.3 and Theorem 9.9 are due to Stanley [228] (see also [226, Example 3.1] and [230, Exercise 4.31]), and we follow his proof closely in this chapter. The formula for the leading coefficient of the Ehrhart polynomial in Theorem 9.9, i.e., the volume of the zonotope, goes back to Shephard [219, equation (57)].

5. Zonotopes have played a central role in the recently established theory of *arithmetic matroids* and *arithmetic Tutte polynomials*, and Corollary 9.3 was restated in this language by Michele D'Adderio and Luca Moci [89, Theorem 3.2]. Ehrhart polynomials of zonotopes and their arithmetic Tutte siblings are in general harder to compute than what this chapter might convey; some examples for computable formulas (for arithmetic Tutte polynomials for the classical root systems) were given by Federico Ardila, Federico Castillo, and Michael Henley [10].

Exercises

9.1. ♣ Prove (9.1): if $p(z_1, z_2, \ldots, z_d)$ and $q(z_1, z_2, \ldots, z_d)$ are polynomials, then

$$\mathcal{N}\big(p(z_1, z_2, \ldots, z_d)\, q(z_1, z_2, \ldots, z_d)\big)$$
$$= \mathcal{N}\big(p(z_1, z_2, \ldots, z_d)\big) + \mathcal{N}\big(q(z_1, z_2, \ldots, z_d)\big).$$

9.2. Prove that every face of a zonotope is a zonotope, and conclude that every face of a zonotope is centrally symmetric.

9.3. Let $\mathcal{P}$ be a d-dimensional integral parallelepiped in $\mathbb{R}^n$. Prove that if there exists only one integer point $\mathbf{x}$ in the interior of $\mathcal{P}$, then $\mathbf{x}$ must be the center of mass of $\mathcal{P}$.

9.4. Let $\mathcal{P}$ be a d-dimensional integral parallelepiped in $\mathbb{R}^n$. Prove that the convex hull of all the integer points in the interior of $\mathcal{P}$ is a centrally symmetric polytope.

9.5. ♣ Complete the proof of Lemma 9.1 by showing (using the notation from the proof) that $\mathcal{Z}(\mathbf{u}_1, \mathbf{u}_2, \ldots, \mathbf{u}_n)$ equals the disjoint union of $\Pi_1, \Pi_2, \ldots, \Pi_k$ and $\mathcal{P}_1, \mathcal{P}_2, \ldots, \mathcal{P}_m$.

9.6. (a) Show that for all real vectors $\mathbf{u}_1, \ldots, \mathbf{u}_n, \mathbf{w} \in \mathbb{R}^d$,

$$\mathcal{Z}(\mathbf{u}_1, \ldots, \mathbf{u}_n, -\mathbf{w}) = \mathcal{Z}(\mathbf{u}_1, \ldots, \mathbf{u}_n, \mathbf{w}) - \mathbf{w},$$

where the latter difference is defined as a translation of the zonotope $\mathcal{Z}(\mathbf{u}_1, \ldots, \mathbf{u}_n, \mathbf{w})$ by the vector $-\mathbf{w}$.
(b) Show that given a zonotope $\mathcal{Z}$, we may always choose all of the generators of a translate of $\mathcal{Z}$ in some half-space containing the origin on its boundary.

9.7. Prove that the following statements are equivalent for a polytope $\mathcal{P}$:

(a) $\mathcal{P}$ is a zonotope;
(b) every 2-dimensional face of $\mathcal{P}$ is a zonotope;
(c) every 2-dimensional face of $\mathcal{P}$ is centrally symmetric.

9.8. Given a zonotope $\mathcal{Z}(\pm\mathbf{u}_1, \pm\mathbf{u}_2, \ldots, \pm\mathbf{u}_n) \subset \mathbb{R}^d$, consider the hyperplane arrangement $\mathcal{H}$ consisting of the n hyperplanes in $\mathbb{R}^d$ through the origin with normal vectors $\mathbf{u}_1, \mathbf{u}_2, \ldots, \mathbf{u}_n$. Show that there is a one-to-one correspondence between the vertices of $\mathcal{Z}(\pm\mathbf{u}_1, \pm\mathbf{u}_2, \ldots, \pm\mathbf{u}_n)$ and the regions of $\mathcal{H}$ (i.e., the maximal connected components of $\mathbb{R}^d \setminus \bigcup \mathcal{H}$). Can you give an analogous correspondence for the other faces of $\mathcal{Z}(\pm\mathbf{u}_1, \pm\mathbf{u}_2, \ldots, \pm\mathbf{u}_n)$?

9.9. ♣ Show that the dimension of the permutahedron

$$\mathcal{P}_d = \text{conv} \{(\pi(1) - 1, \pi(2) - 1, \ldots, \pi(d) - 1) : \pi \in S_d\}$$

is $d - 1$ and that P_d has $d!$ vertices.

9.10. Show that the permutahedron $\mathcal{P}_d$ is the image of the Birkhoff–von Neumann polytope $\mathcal{B}_d$ from Chapter 6 under a suitable linear map.

9.11. Give a hyperplane description of the permutahedron $\mathcal{P}_d$.

9.12. According to Exercise 9.9, the permutahedron $\mathcal{P}_d$ lies in a hyperplane $H \subset \mathbb{R}^d$. Show that $\mathcal{P}_d$ tiles H. (*Hint:* begin by drawing the case $d = 3$. For the general case, the viewpoint of Exercise 9.8 might be useful.)

9.13. Prove Lemma 9.6: A subset $S \subseteq \{\mathbf{e}_1 + \mathbf{e}_2, \mathbf{e}_1 + \mathbf{e}_3, \ldots, \mathbf{e}_{d-1} + \mathbf{e}_d\}$ is linearly independent if and only if G_S is a forest.

9.14. Let S be a linearly independent subset of

$$\{\mathbf{e}_1 + \mathbf{e}_2, \mathbf{e}_1 + \mathbf{e}_3, \ldots, \mathbf{e}_{d-1} + \mathbf{e}_d\}.$$

Show that the half-open parallelepiped

$$\sum_{\mathbf{e}_j + \mathbf{e}_k \in S} (\mathbf{0}, \mathbf{e}_j + \mathbf{e}_k]$$

(written as a Minkowski sum) has relative volume 1.

9.15. Prove (9.3):

$$\mathcal{P}_d = \mathbf{0} \cup \bigcup_{S \in I} \left(\sum_{\mathbf{e}_j + \mathbf{e}_k \in S} (\mathbf{0}, \mathbf{e}_j + \mathbf{e}_k] \right)$$

as a disjoint union.

9.16. Prove that there are precisely d^{d-2} labeled trees on d nodes.

9.17. Show that every forest on d nodes with $d - 1$ edges is connected, i.e., it is a tree.

9.18. Given a graph G with node set $[d]$, we define the **graphical zonotope** $\mathcal{Z}_G$ as the Minkowski sum of the line segments $[\mathbf{e}_j, \mathbf{e}_k]$ for all edges jk of G. (Thus the permutahedron $\mathcal{P}_d$ is the graphical zonotope of the complete graph on d nodes.) Prove that the volume of $\mathcal{Z}_G$ equals the number of spanning trees of G. Give an interpretation of the Ehrhart coefficients of $\mathcal{Z}_G$, in analogy with Theorem 9.5.[1]

9.19. Let G be a graph with node set $[d]$, and let $\deg(j)$ denote the number of edges incident to j (the **degree** of the node j). The vector

$$\deg(G) := (\deg(1), \deg(2), \ldots, \deg(d))$$

is the **labeled degree sequence** of G. Let $\mathcal{Z}_d$ be the zonotope generated by the vectors $\mathbf{e}_j + \mathbf{e}_k$ for $1 \leq j < k \leq d$.

(a) Show that every labeled degree sequence of a graph with d nodes is an integer point in $\mathcal{Z}_d$.
(b) If you know the *Erdős–Gallai theorem* (and if you don't, look it up), prove that every integer point in $\mathcal{Z}_d$ whose coordinate sum is even is a labeled degree sequence of a graph with d nodes.

9.20. Let K be a compact, convex set in $\mathbb{R}^d$, symmetric about the origin, whose volume is greater than 2^d. Prove that K must contain a nonzero integer point. (This result is known as *Minkowski's fundamental theorem*, and it lies at the core of the geometry of numbers, as well as algebraic number theory. Its applications have shown how interesting and useful symmetric bodies can be.)

Open Problems

9.21. Classify the polynomials in $\mathbb{Z}[t]$ that are Ehrhart polynomials of integral zonotopes. (This is hard: if the matrix defining a zonotope is unimodular, then its Ehrhart coefficients are the entries of the f-vector of the underlying matroid—see Note 5; the classification of f-vectors of matroids is an old and difficult problem.)

9.22. The **unlabeled degree sequence** (often simply called **degree sequence**) of a graph G is the vector $\deg(G)$ (defined in Exercise 9.19) redefined such that its entries are in decreasing order. How many distinct degree sequences are there for graphs on n nodes? (For known values for the first

---

[1] An intimately related polytope, which is in fact older than the graphical zonotope associated with G, is the **acyclotope**, the convex hull of net degree vectors of acyclic orientations of G. For example, the acyclotope of the complete graph on d nodes is the shifted permutahedron $\operatorname{conv} \{(\pi(1), \pi(2), \ldots, \pi(d)) : \pi \in S_d\}$. It turns out that the acyclotope associated with G is the Minkowski sum of the line segments $[\mathbf{e}_j - \mathbf{e}_k, \mathbf{e}_k - \mathbf{e}_j]$ for all edges jk of G.

few n, see [1, Sequence A004251]. The analogous question for *labeled* degree sequences was answered in [228]; that proof begins with Exercise 9.19.)

9.23. Let $K \subset \mathbb{R}^d$ be a d-dimensional convex body with the origin as its barycenter. If K contains only the origin as an interior lattice point, then $\text{vol}(K) \leq \frac{(d+1)^d}{d!}$, where equality holds if and only if K is unimodularly equivalent to $(d+1)\Delta$, where Δ is the d-dimensional standard simplex from Section 2.3. (This is a conjecture of Ehrhart from 1964. Ehrhart himself proved this upper bound for all d-dimensional simplices, and also completely in dimension 2, but it remains open in general. The interested reader may consult [185] about the current state of Ehrhart's conjecture and its relationship to the Ricci curvature of Fano manifolds.)

Chapter 10
h-Polynomials and h^*-Polynomials

Life is the twofold internal movement of composition and decomposition at once general and continuous.

Henri de Blainville (1777–1850)

In Chapters 2 and 3, we developed the Ehrhart polynomial and Ehrhart series of an integral polytope $\mathcal{P}$ and realized that the arithmetic information encoded in an Ehrhart polynomial is equivalent to the information encoded in its Ehrhart series. More precisely, when the Ehrhart series is written as a rational function, we introduced the name h^*-*polynomial* for its numerator:

$$\mathrm{Ehr}_{\mathcal{P}}(z) = 1 + \sum_{t \geq 1} L_{\mathcal{P}}(t)\, z^t = \frac{h_{\mathcal{P}}^*(z)}{(1 - z)^{\dim(\mathcal{P})+1}}.$$

Our goal in this chapter is to prove several decomposition formulas for $h_{\mathcal{P}}^*(z)$ based on triangulations of $\mathcal{P}$. As we will see, these decompositions will involve both arithmetic data from the simplices of the triangulation and combinatorial data from the face structure of the triangulation.

10.1 Simplicial Polytopes and (Unimodular) Triangulations

In Chapter 5—more precisely, in Theorem 5.1—we derived relations among the face numbers f_k of a simple polytope. The twin sisters of these *Dehn–Sommerville relations* for **simplicial** polytopes—all of whose nontrivial faces are simplices—were the subject of Exercise 5.9. It turns out they can be stated in a compact way by introducing the h-**polynomial** of the d-polytope $\mathcal{P}$:

$$h_{\mathcal{P}}(z) := \sum_{k=-1}^{d-1} f_k \, z^{k+1} \, (1-z)^{d-1-k} \, ,$$

where we set $f_{-1} := 1$. The following is a nifty restatement of Exercise 5.9, as the reader should check (Exercise 10.1).

Theorem 10.1 (Dehn–Sommerville relations). *If $\mathcal{P}$ is a simplicial d-polytope, then $h_{\mathcal{P}}$ is a palindromic polynomial.*

We need this simplicial version of the Dehn–Sommerville relations because it naturally connects to triangulations. The h-polynomial of $\mathcal{P}$ encodes combinatorial information about the faces in the boundary of $\mathcal{P}$, and so the natural setting in the world of triangulations will consist of a triangulation of the *boundary* of a polytope. By analogy with the face numbers of a polytope, given a triangulation T of the boundary $\partial\mathcal{P}$ of a given d-dimensional polytope $\mathcal{P}$, we define f_k to be the number of k-simplices in T; they will be encoded, as above, in the h-polynomial

$$h_T(z) := \sum_{k=-1}^{d-1} f_k \, z^{k+1} \, (1-z)^{d-1-k} \, ,$$

where again we set $f_{-1} := 1$. The typical scenario we will encounter is that we are given a triangulation T of the polytope $\mathcal{P}$ and then consider the **induced triangulation**

$$\{\Delta \in T : \Delta \subset \partial\mathcal{P}\}$$

of the boundary of $\mathcal{P}$.

Theorem 10.2 (Dehn–Sommerville relations for boundary triangulations). *Given a regular triangulation of the polytope $\mathcal{P}$, the h-polynomial of the induced triangulation of $\partial\mathcal{P}$ is palindromic.*

Proof. Given a regular triangulation T of $\mathcal{P} \subseteq \mathbb{R}^d$, let $\mathcal{Q} \subseteq \mathbb{R}^{d+1}$ be the corresponding lifted polytope as in (3.1). Choose a point $\mathbf{v} \in \mathcal{P}^\circ$ and lift it to $(\mathbf{v}, h+2) \in \mathbb{R}^{d+1}$, where h is the maximal height among the vertices of $\mathcal{Q}$. Let $\mathcal{R}$ be the convex hull of $(\mathbf{v}, h+2)$ and the vertices of the lower hull of $\mathcal{Q}$, and let

$$S := \mathcal{R} \cap \left\{ \mathbf{x} \in \mathbb{R}^{d+1} : x_{d+1} = h+1 \right\}.$$

Figure 10.1 shows an instance of this setup. (The polytope S is a **vertex figure** of $\mathcal{Q}$ at $\mathbf{v}$.) Exercise 10.2 says that S is a simplicial polytope whose face numbers f_k equal the face numbers of the triangulation of $\partial\mathcal{P}$ induced by T. Thus the h-polynomial of this triangulation of $\partial\mathcal{P}$ equals $h_S(z)$, which is palindromic by Theorem 10.1. $\qquad\square$

When we are given a triangulation T of a polytope (rather than of its boundary), we adjust our definition of the accompanying h-polynomial to

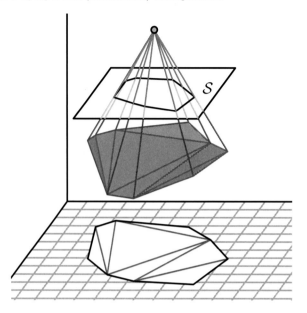

Fig. 10.1 The geometry of our proof of Theorem 10.2.

$$h_T(z) := \sum_{k=-1}^{d} f_k \, z^{k+1} \, (1-z)^{d-k} \, .$$

In general, there is no analogue of Theorem 10.2 for this h-polynomial, though an (important) exception is given in Exercise 10.3.

There is a second reason for us to introduce the h-polynomial of a triangulation. We call a triangulation T **unimodular** if each simplex

$$\Delta = \operatorname{conv}\{\mathbf{v}_0, \mathbf{v}_1, \ldots, \mathbf{v}_k\} \in T$$

has the property that the vectors $\mathbf{v}_1 - \mathbf{v}_0$, $\mathbf{v}_2 - \mathbf{v}_0$, $\ldots$, $\mathbf{v}_k - \mathbf{v}_0$ form a lattice basis of $\operatorname{span}(\Delta) \cap \mathbb{Z}^d$. In this case, we also call the simplex Δ **unimodular**. One example of a unimodular simplex is the standard k-simplex Δ of Section 2.3. We recall that in this case,

$$\operatorname{Ehr}_\Delta(z) = \frac{1}{(1-z)^{k+1}}, \tag{10.1}$$

and we invite the reader to prove that the same Ehrhart series comes with *every* unimodular k-simplex (Exercise 10.4).

Here is the main result of this section.

Theorem 10.3. *If $\mathcal{P}$ is an integral d-polytope that admits a unimodular triangulation T, then*

$$\mathrm{Ehr}_{\mathcal{P}}(z) = \frac{h_T(z)}{(1-z)^{d+1}}.$$

In words, the h^*-polynomial of $\mathcal{P}$ is given by the h-polynomial of the triangulation T.

Fig. 10.2 A triangulation of a hexagon and the corresponding decomposition (10.2).

Proof. We begin by writing $\mathcal{P}$ as the union of all open simplices in T, pictured in an example in Figure 10.2:

$$\mathcal{P} = \bigcup_{\Delta \in T} \Delta^{\circ} \qquad (10.2)$$

(here we are using Exercise 5.3). Since this union is disjoint, we can compute

$$\mathrm{Ehr}_{\mathcal{P}}(z) = 1 + \sum_{\Delta \in T} \mathrm{Ehr}_{\Delta^{\circ}}(z) = 1 + \sum_{\Delta \in T} \left(\frac{z}{1-z} \right)^{\dim \Delta + 1}.$$

Here the last equality follows from the Ehrhart–Macdonald reciprocity theorem (Theorem 4.4) applied to the Ehrhart series (10.1) of a unimodular simplex. Since the above summands depend only on the dimension of each simplex, we can rewrite

$$\mathrm{Ehr}_{\mathcal{P}}(z) = 1 + \sum_{k=0}^{d} f_k \left(\frac{z}{1-z} \right)^{k+1} = \sum_{k=-1}^{d} f_k \left(\frac{z}{1-z} \right)^{k+1}$$

$$= \frac{\sum_{k=-1}^{d} f_k \, z^{k+1} (1-z)^{d-k}}{(1-z)^{d+1}} = \frac{h_T(z)}{(1-z)^{d+1}}. \qquad \square$$

Theorem 10.3 says something remarkable: the arithmetic of a polytope (its discrete volume) is completely determined by the combinatorics of the unimodular triangulation (its face structure). Unfortunately, not all integral polytopes admit unimodular triangulations—in fact, most do not (see Exercise 10.5). Our next goal is to find an analogue of Theorem 10.3 that holds for *every* integral polytope.

10.2 Fundamental Parallelepipeds Open Up, with an h-Twist

Our philosophy in constructing a generalization of Theorem 10.3 rests in revisiting the cone over a given polytope from Chapter 3, using a decomposition of $\mathcal{P}$ into open simplices.

Given an integral d-polytope $\mathcal{P}$, fix a (not necessarily unimodular) triangulation T. As in (10.2), we can write $\mathcal{P}$ as the disjoint union of the open simplices in T; from this point of view, the following definition should look natural. Given a simplex $\Delta \in T$, let $\Pi(\Delta)$ be the fundamental parallelepiped of $\mathrm{cone}(\Delta)$ (as defined in Section 3.3), and let

$$B_\Delta(z) := \sigma_{\Pi(\Delta)^\circ}(1, 1, \ldots, 1, z) .$$

Thus $B_\Delta(z)$ is an "open" variant of $h_\Delta^*(z) = \sigma_{\Pi(\Delta)}(1, 1, \ldots, 1, z)$, an equation we used several times in Chapters 3 and 4.

Looking back at our proof of Theorem 10.3, it makes moral sense to include $\varnothing$ in the collection of faces of a triangulation of a given polytope, with the convention $\dim(\varnothing) := -1$. (This explains in hindsight our convention $f_{-1} := 1$.) We will assume for the rest of this chapter that every triangulation T includes the "empty simplex" $\varnothing$. Along the same lines, we define $B_\varnothing(z) := 1$.

We need one more concept to be able to state our generalization of Theorem 10.3. Given a simplex $\Delta \in T$, let

$$\mathrm{link}_T(\Delta) := \{\Omega \in T : \Omega \cap \Delta = \varnothing,\ \Omega \subseteq \Phi \text{ for some } \Phi \in T \text{ with } \Delta \subseteq \Phi\} ,$$

the **link** of Δ. In words, every simplex in $\mathrm{link}_T(\Delta)$ is disjoint from Δ, yet it is the face of a simplex in T that also contains Δ as a face; see Figure 10.3 for two examples. When the triangulation T is clear from the context, we

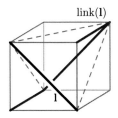

link($\mathbf{v}$) link($\mathbf{l}$)

Fig. 10.3 The links of a vertex and an edge in a triangulation of the 3-cube.

suppress the subscript and simply write $\mathrm{link}(\Delta)$. We remark that by definition, $\mathrm{link}(\varnothing) = T$. In general, $\mathrm{link}(\Delta)$ consists of a collection of simplices in T, the largest dimension of which is $d - \dim(\Delta) - 1$ (Exercise 10.8), and so it is reasonable to define

$$h_{\text{link}(\Delta)}(z) := \sum_{k=-1}^{d-\dim(\Delta)-1} f_k \, z^{k+1} \, (1-z)^{d-\dim(\Delta)-1-k} \,,$$

where f_k denotes the number of k-simplices in $\text{link}(\Delta)$. The h-polynomial of $\text{link}(\Delta)$ encodes combinatorial data coming from the simplices in T that contain Δ. This statement can be made more precise, as we invite the reader to prove in Exercise 10.9:

$$h_{\text{link}(\Delta)}(z) = (1-z)^{d-\dim(\Delta)} \sum_{\Phi \supseteq \Delta} \left(\frac{z}{1-z} \right)^{\dim(\Phi)-\dim(\Delta)} , \qquad (10.3)$$

where the sum is over all simplices $\Phi \in T$ containing Δ.

Theorem 10.4 (Betke–McMullen decomposition of h^*). *Fix a triangulation T of the integer d-polytope $\mathcal{P}$. Then*

$$\text{Ehr}_{\mathcal{P}}(z) = \frac{\sum_{\Delta \in T} h_{\text{link}(\Delta)}(z) \, B_\Delta(z)}{(1-z)^{d+1}} .$$

Before proving Theorem 10.4, we explain its relation to Theorem 10.3. If a simplex $\Delta \in T$ is unimodular, then the corresponding fundamental parallelepiped $\Pi(\Delta)$ of $\text{cone}(\Delta)$ contains the origin as its sole integer lattice point, and thus $B_\Delta(z) = 0$, unless $\Delta = \varnothing$. Thus if the triangulation T is unimodular, then the sum giving the h^*-polynomial in Theorem 10.4 collapses to $h_{\text{link}(\varnothing)}(z) \, B_\varnothing(z) = h_T(z)$, and so Theorem 10.3 follows as a special case of Theorem 10.4.

Proof of Theorem 10.4. We begin, as in our proof of Theorem 10.3, by writing $\mathcal{P}$ as the disjoint union of all open nonempty simplices in T as in (10.2), and thus, using Ehrhart–Macdonald reciprocity (Theorem 4.4),

$$\text{Ehr}_{\mathcal{P}}(z) = 1 + \sum_{\Delta \in T \setminus \{\varnothing\}} \text{Ehr}_{\Delta^\circ}(z) = 1 + \sum_{\Delta \in T \setminus \{\varnothing\}} (-1)^{\dim(\Delta)+1} \frac{h_\Delta^* \left(\frac{1}{z} \right)}{\left(1 - \frac{1}{z} \right)^{\dim \Delta + 1}}$$

$$= 1 + \frac{\sum_{\Delta \in T \setminus \{\varnothing\}} z^{\dim(\Delta)+1} (1-z)^{d-\dim(\Delta)} h_\Delta^* \left(\frac{1}{z} \right)}{(1-z)^{d+1}} . \qquad (10.4)$$

Here $h_\Delta^*(z)$ denotes the h^*-polynomial of the simplex Δ. Now we use Exercise 10.10:

$$\sigma_{\Pi(\Delta)}(\mathbf{z}) = 1 + \sum_{\substack{\Omega \subseteq \Delta \\ \Omega \neq \varnothing}} \sigma_{\Pi(\Omega)^\circ}(\mathbf{z}) \,,$$

where the sum is over all nonempty faces of Δ. This identity specializes, if we choose $\mathbf{z} = (1, 1, \ldots, 1, z)$, to

$$h^*_\Delta(z) = \sum_{\Omega \subseteq \Delta} B_\Omega(z) \,,$$

where now the sum is over all faces of Δ, including $\varnothing$ (recall that $B_\varnothing(z) = 1$). Substituting this back into (10.4) yields

$$\text{Ehr}_\mathcal{P}(z) = 1 + \frac{\sum_{\Delta \in T \setminus \{\varnothing\}} z^{\dim(\Delta)+1}(1-z)^{d-\dim(\Delta)} \sum_{\Omega \subseteq \Delta} B_\Omega(\tfrac{1}{z})}{(1-z)^{d+1}}$$

$$= \frac{\sum_{\Delta \in T} z^{\dim(\Delta)+1}(1-z)^{d-\dim(\Delta)} \sum_{\Omega \subseteq \Delta} B_\Omega(\tfrac{1}{z})}{(1-z)^{d+1}} \,. \qquad (10.5)$$

Recall that the polynomial $B_\Delta(z)$ encodes data about the lattice points in the interior of the fundamental parallelepiped of cone(Δ). The symmetry of this parallelepiped gets translated into the palindromicity of the associated polynomial (Exercise 10.7):

$$B_\Delta(z) = z^{\dim(\Delta)+1} B_\Delta\left(\tfrac{1}{z}\right) . \qquad (10.6)$$

This allows us to rewrite (10.5) further:

$$h^*_\mathcal{P}(z) = \sum_{\Delta \in T} z^{\dim(\Delta)+1}(1-z)^{d-\dim(\Delta)} \sum_{\Omega \subseteq \Delta} B_\Omega\left(\frac{1}{z}\right)$$

$$= \sum_{\Delta \in T} z^{\dim(\Delta)+1}(1-z)^{d-\dim(\Delta)} \sum_{\Omega \subseteq \Delta} z^{-\dim(\Omega)-1} B_\Omega(z)$$

$$= \sum_{\Omega \in T} \sum_{\Delta \supseteq \Omega} z^{\dim(\Delta)-\dim(\Omega)}(1-z)^{d-\dim(\Delta)} B_\Omega(z)$$

$$= \sum_{\Omega \in T} (1-z)^{d-\dim(\Omega)} B_\Omega(z) \sum_{\Delta \supseteq \Omega} \left(\frac{z}{1-z}\right)^{\dim(\Delta)-\dim(\Omega)} .$$

Theorem 10.4 follows now with (10.3). $\qquad \square$

10.3 Palindromic Decompositions of h^*-Polynomials

Our next goal is to refine Theorem 10.4 in the case that T comes from a *boundary* triangulation of $\mathcal{P}$, for which we can exploit Theorem 10.2, or rather its analogue for links, which is the subject of Exercise 10.12. To keep our exposition accessible, we will discuss only the case that $\mathcal{P}$ contains an interior lattice point; the identities in this section and the next have analogues without this condition, which we will address in the notes at the end of the chapter.

Theorem 10.5. *Suppose $\mathcal{P}$ is an integral d-polytope that contains an interior lattice point. Then there exist unique polynomials $a(z)$ and $b(z)$ such that*

$$h_{\mathcal{P}}^*(z) = a(z) + z\, b(z)\,,$$

$a(z) = z^d\, a(\frac{1}{z})$, and $b(z) = z^{d-1}\, b(\frac{1}{z})$.

The identities for $a(z)$ and $b(z)$ say that $a(z)$ and $b(z)$ are palindromic polynomials; the degree of $a(z)$ is necessarily d (because the constant coefficient of $a(z)$ is $h_{\mathcal{P}}^*(0) = 1$), while the degree of $b(z)$ is $d-1$ or less. In fact, $b(z)$ can be the zero polynomial—this happens if and only if $\mathcal{P}$ is the translate of a reflexive polytope, by Theorem 4.6.

Proof. After a harmless lattice translation, we may assume that $\mathbf{0} \in \mathcal{P}^\circ$. Fix a regular boundary triangulation T of $\mathcal{P}$ and let

$$T_0 := T \cup \{\mathrm{conv}(\Delta,\mathbf{0}) : \Delta \in T\}\,.$$

Thus T_0 is a triangulation of $\mathcal{P}$ whose simplices come in two flavors: those on the boundary of $\mathcal{P}$ and those of the form $\mathrm{conv}(\Delta,\mathbf{0})$ for some $\Delta \in T$. (We say that T_0 comes from **coning over** the boundary triangulation T.) Figure 10.4 shows an example. Perhaps not surprisingly, the two kinds of simplices in T_0

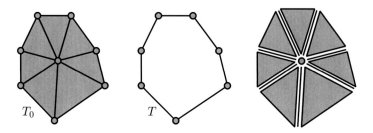

T_0 T

Fig. 10.4 A triangulation T_0 and its two types of simplices; the middle picture shows the boundary triangulation T.

are related: Exercise 10.11 says that for every nonempty simplex $\Delta \in T$,

$$h_{\mathrm{link}_{T_0}(\Delta)}(z) = h_{\mathrm{link}_{T_0}(\mathrm{conv}(\Delta,\mathbf{0}))}(z) = h_{\mathrm{link}_T(\Delta)}(z)\,. \qquad (10.7)$$

Thus Theorem 10.4 gives in this case

$$\begin{aligned}
h_{\mathcal{P}}^*(z) &= \sum_{\Delta \in T_0} h_{\mathrm{link}_{T_0}(\Delta)}(z)\, B_\Delta(z) \\
&= \sum_{\Delta \in T} h_{\mathrm{link}_T(\Delta)}(z) \left(B_\Delta(z) + B_{\mathrm{conv}(\Delta,\mathbf{0})}(z)\right).
\end{aligned}$$

Now let

$$a(z) := \sum_{\Delta \in T} h_{\mathrm{link}_T(\Delta)}(z)\, B_\Delta(z) \tag{10.8}$$

$$b(z) := \frac{1}{z} \sum_{\Delta \in T} h_{\mathrm{link}_T(\Delta)}(z)\, B_{\mathrm{conv}(\Delta, \mathbf{0})}(z). \tag{10.9}$$

The fact that each $B_{\mathrm{conv}(\Delta, \mathbf{0})}(z)$ has constant term zero ensures that $b(z)$ is a polynomial (of degree at most $d - 1$). Furthermore, by Exercise 10.12 and (10.6),

$$z^d\, a\left(\frac{1}{z}\right) = \sum_{\Delta \in T} z^{d - \dim(\Delta) - 1} h_{\mathrm{link}_T(\Delta)}\left(\frac{1}{z}\right) z^{\dim(\Delta) + 1} B_\Delta\left(\frac{1}{z}\right)$$

$$= \sum_{\Delta \in T} h_{\mathrm{link}_T(\Delta)}(z)\, B_\Delta(z) = a(z),$$

and $b(z) = z^{d-1} b(\frac{1}{z})$ follows analogously. Thus our setup gives rise to the decomposition

$$h_{\mathcal{P}}^*(z) = a(z) + z\, b(z)$$

with palindromic polynomials $a(z)$ and $b(z)$, and Exercise 10.13 says that this decomposition is unique. (Note that this means, in particular, that *every* boundary triangulation yields the same decomposition.) □

10.4 Inequalities for h^*-Polynomials

The proof of Theorem 10.5 has a powerful consequence due to the following fact, whose proof would lead us too far astray (though we outline a self-contained proof in Exercises 10.16–10.18).

Theorem 10.6. *The h-polynomial of a link in a regular triangulation has nonnegative coefficients.*

Looking back how the polynomials $a(z)$ and $b(z)$ were constructed (see (10.8) and (10.9)), we can thus deduce the following corollary:

Corollary 10.7. *The polynomials $a(z)$ and $b(z)$ appearing in Theorem 10.5 have nonnegative coefficients.*

Naturally, the fact that the coefficients of $a(z)$ and $b(z)$ have nonnegative coefficients can be translated into inequalities among the coefficients of the accompanying h^*-polynomials: the following corollary of Corollary 10.7 is easily proved (Exercise 10.14).

Corollary 10.8. *Suppose $\mathcal{P}$ is an integral d-polytope that contains an interior lattice point. Then its h^*-polynomial $h_{\mathcal{P}}^*(x) = h_d^* x^d + h_{d-1}^* x^{d-1} + \cdots + h_0^*$ satisfies*

$$h_0^* + h_1^* + \cdots + h_{j+1}^* \geq h_d^* + h_{d-1}^* + \cdots + h_{d-j}^* \quad for \quad 0 \leq j \leq \lfloor \tfrac{d}{2} \rfloor - 1 \quad (10.10)$$

and

$$h_0^* + h_1^* + \cdots + h_j^* \leq h_d^* + h_{d-1}^* + \cdots + h_{d-j}^* \quad for \quad 0 \leq j \leq d. \quad (10.11)$$

Notes

1. Theorem 10.3 is only one indication of how important (and special) unimodular triangulations are; see [98, Chapter 9] for more. If an integral polytope $\mathcal{P}$ admits a unimodular triangulation, then it is **integrally closed**: for every positive integer k and every integer point $\mathbf{x} \in k\mathcal{P}$, there exist integer points $\mathbf{y}_1, \mathbf{y}_2, \ldots, \mathbf{y}_k \in \mathcal{P}$ such that $\mathbf{y}_1 + \mathbf{y}_2 + \cdots + \mathbf{y}_k = \mathbf{x}$.[1] (See Exercise 10.6.) There are integrally closed polytopes that do not admit a unimodular triangulation. These two notions are part of an interesting hierarchy of integral polytopes, with connections to commutative algebra and algebraic geometry; see, e.g., [73].

2. Theorem 10.4 is due to Ulrich Betke and Peter McMullen, as is Theorem 10.5, published in an influential paper [54] in 1985. Corollary 10.7 *should* perhaps have been in [54] (the nonnegativity of the h-polynomials in Betke–McMullen's formulas was established in the 1970s), but it appeared only—in the guise of Corollary 10.8—in the 1990s, in papers by Takayuki Hibi [133] and Richard Stanley [227]. Neither paper gives the impression that the inequalities of Corollary 10.8 follow immediately from Betke–McMullen's work, though both hold without the condition of the existence of an interior lattice point (here d has to be replaced by the degree of $h_\mathcal{P}^*(z)$ in (10.11)) and in situations more general than the realm of Ehrhart series.

3. Theorem 10.5 and Corollary 10.7 are not the end of the story. Building on work of Sam Payne [188] (which gave a multivariate version of Theorem 10.4), Alan Stapledon [234] generalized Theorem 10.5 and Corollary 10.7 to arbitrary integral polytopes. His theorem says that if $h_\mathcal{P}^*(z)$ has degree s (recall from Chapter 4 that in this case, we say $\mathcal{P}$ has degree s), then there exist unique polynomials $a(z)$ and $b(z)$ with nonnegative coefficients such that

$$\left(1 + z + \cdots + z^{d-s}\right) h_\mathcal{P}^*(z) = a(z) + z^{d+1-s} b(z), \quad (10.12)$$

$a(z) = z^d\, a(\tfrac{1}{z}), b(z) = z^{d-l}\, b(\tfrac{1}{z})$, and, writing $a(z) = a_d z^d + a_{d-1} z^{d-1} + \cdots + a_0$,

---

[1] Integrally closed polytopes are also said to have the **integer decomposition property**. There is a related notion for integral polytopes, namely that of *normality*. For full-dimensional polytopes, the terms *integrally closed* and *normal* are equivalent, but this is not the case when the subgroup $\sum_{\mathbf{x},\,\mathbf{y} \in \mathcal{P} \cap \mathbb{Z}^d} \mathbb{Z}(\mathbf{x} - \mathbf{y})$ of $\mathbb{Z}^d$ is not a direct summand of $\mathbb{Z}^d$. For more about this subtlety (and much more), see [73,88].

$$1 = a_0 \leq a_1 \leq a_j \quad \text{for} \quad 2 \leq j \leq d-1. \tag{10.13}$$

These inequalities imply Corollary 10.8 without the condition that $\mathcal{P}$ contain an interior lattice point (again here d has to be replaced by s in (10.11)); see Exercise 10.15. Stapledon has recently improved this theorem further, giving infinitely many classes of linear inequalities among the h^*-coefficients [232]. This exciting new line of research involves additional techniques from additive number theory. He also introduced a weighted variant of the h^*-polynomial that is *always* palindromic, motivated by motivic integration and the cohomology of certain toric varieties [233]. One can easily recover $h_{\mathcal{P}}^*(x)$ from this weighted h^*-polynomial, but one can also deduce the palindromicity of both $a(x)$ and $b(x)$ as coming from the same source (and this perspective has some serious geometric applications).

Exercises

10.1. ♣ Show that the f-vector and the h-vector of a simplicial d-polytope are related via

$$h_k = \sum_{j=0}^{k}(-1)^{k-j}\binom{d-j}{d-k}f_{j-1} \quad \text{and} \quad f_{k-1} = \sum_{j=0}^{k}\binom{d-j}{k-j}h_j,$$

and conclude that the Dehn–Sommerville relations in Exercise 5.9 are equivalent to Theorem 10.1.

10.2. ♣ As in the setup of our proof of Theorem 10.2, let $\mathcal{Q} \subseteq \mathbb{R}^{d+1}$ be the lifted polytope giving rise to a regular triangulation T of the polytope $\mathcal{P} \subseteq \mathbb{R}^d$, choose $\mathbf{v} \in \mathcal{P}^\circ$, and lift it to $(\mathbf{v}, h+2) \in \mathbb{R}^{d+1}$, where h is the maximal height among the vertices of $\mathcal{Q}$. Let $\mathcal{R}$ be the convex hull of $(\mathbf{v}, h+2)$ and the vertices of the lower hull of $\mathcal{Q}$, and let

$$S := \mathcal{R} \cap \{\mathbf{x} \in \mathbb{R}^{d+1} : x_{d+1} = h+1\}.$$

Prove that S is a simplicial polytope whose face numbers f_k equal the face numbers of the triangulation of $\partial \mathcal{P}$ induced by T.

10.3. Given a triangulation T of the boundary of a d-polytope $\mathcal{P}$ and a point $\mathbf{v} \in \mathcal{P}^\circ$, construct a triangulation K of $\mathcal{P}$ consisting of T with the simplices $\text{conv}(\Delta, \mathbf{v})$ for all $\Delta \in T$ appended; i.e., the new triangulation K comes from coning over T. Prove that

$$h_K(z) = h_T(z).$$

10.4. ♣ Show that if Δ is a unimodular k-simplex, then

$$\mathrm{Ehr}_\Delta(z) = \frac{1}{(1-z)^{k+1}} \, .$$

10.5. Show that every integral polygon admits a unimodular triangulation. Give an example of an integral d-polytope that does not admit a unimodular triangulation for any $d \geq 3$.

10.6. Recall from the notes that an integral polytope $\mathcal{P}$ is **integrally closed** if for every positive integer k and every integer point $\mathbf{x} \in k\mathcal{P}$, there exist integer points $\mathbf{y}_1, \mathbf{y}_2, \ldots, \mathbf{y}_k \in \mathcal{P}$ such that $\mathbf{y}_1 + \mathbf{y}_2 + \cdots + \mathbf{y}_k = \mathbf{x}$. Prove that every integral polytope that admits a unimodular triangulation is integrally closed.

10.7. ♣ Prove (10.6): $B_\Delta(z) = z^{\dim(\Delta)+1} B_\Delta\left(\frac{1}{z}\right)$.

10.8. ♣ Let T be a triangulation of the d-polytope $\mathcal{P}$. Given $\Delta \in T$, show that the largest dimension occurring among the simplices in $\mathrm{link}(\Delta)$ is $d - \dim(\Delta) - 1$.

10.9. ♣ Prove (10.3): given a triangulation T of a d-dimensional polytope, then for every simplex $\Delta \in T$,

$$h_{\mathrm{link}(\Delta)}(z) = (1-z)^{d-\dim(\Delta)} \sum_{\Phi \supseteq \Delta} \left(\frac{z}{1-z}\right)^{\dim(\Phi)-\dim(\Delta)},$$

where the sum is over all simplices $\Phi \in T$ that contain Δ.

10.10. ♣ For an integral simplex Δ, let $\Pi(\Delta)$ denote the fundamental parallelepiped of $\mathrm{cone}(\Delta)$. Show that

$$\sigma_{\Pi(\Delta)}(\mathbf{z}) = 1 + \sum_{\Omega \subseteq \Delta} \sigma_{\Pi(\Omega)^\circ}(\mathbf{z}),$$

where the sum is over all nonempty faces of Δ.

10.11. ♣ Prove (10.7): Fix a regular triangulation T_0 of a polytope $\mathcal{P}$ with $\mathbf{0} \in \mathcal{P}^\circ$ whose vertices are $\mathbf{0}$ and the vertices of $\mathcal{P}$, and let $T := \{\Delta \in T_0 : \Delta \subseteq \partial\mathcal{P}\}$. Then for every nonempty $\Delta \in T$,

$$h_{\mathrm{link}(\Delta)}(z) = h_{\mathrm{link}(\mathrm{conv}(\Delta,\mathbf{0}))}(z) \, .$$

10.12. ♣ Let Δ be a simplex in a regular boundary triangulation of a d-dimensional polytope. Prove that $h_{\mathrm{link}(\Delta)}(z)$ is palindromic. (*Hint:* establish a one-to-one correspondence between $\mathrm{link}(\Delta)$ and the boundary faces of a polytope of dimension $d - \dim(\Delta) - 1$, respecting the face relations.)

10.13. ♣ Let $p(z)$ be a polynomial of degree d. Show that there are unique polynomials $a(z)$ and $b(z)$ such that

$$p(z) = a(z) + z\, b(z),$$

$a(z) = z^d\, a(\frac{1}{z})$, and $b(z) = z^{d-1}\, b(\frac{1}{z})$.

10.14. ♣ Consider the polynomial decomposition $h_{\mathcal{P}}^*(z) = a(z) + z\, b(z)$ of an integral d-polytope that contains an interior lattice point, given in Theorem 10.5. Prove that the fact that $a(z)$ has nonnegative coefficients implies (10.10):

$$h_0^* + h_1^* + \cdots + h_{j+1}^* \geq h_d^* + h_{d-1}^* + \cdots + h_{d-j}^* \quad \text{for} \quad 0 \leq j \leq \lfloor \tfrac{d}{2} \rfloor - 1,$$

and the fact that $b(z)$ has nonnegative coefficients implies (10.11):

$$h_0^* + h_1^* + \cdots + h_j^* \leq h_d^* + h_{d-1}^* + \cdots + h_{d-j}^* \quad \text{for} \quad 0 \leq j \leq d.$$

10.15. Derive the analogue of Corollary 10.8 without the condition that $\mathcal{P}$ contain an interior lattice point, using (10.12) and (10.13).

10.16. A **shelling** of the d-polyhedron $\mathcal{P}$ is a linear ordering of the facets $\mathcal{F}_1, \mathcal{F}_2, \ldots, \mathcal{F}_m$ of $\mathcal{P}$ such that the following recursive conditions hold:

(1) $\mathcal{F}_1$ has a shelling.
(2) For every $j > 1$, $(\mathcal{F}_1 \cup \mathcal{F}_2 \cup \cdots \cup \mathcal{F}_{j-1}) \cap \mathcal{F}_j$ is of dimension $d - 2$ and

$$(\mathcal{F}_1 \cup \mathcal{F}_2 \cup \cdots \cup \mathcal{F}_{j-1}) \cap \mathcal{F}_j = \mathcal{G}_1 \cup \mathcal{G}_2 \cup \cdots \cup \mathcal{G}_k,$$

where $\mathcal{G}_1 \cup \mathcal{G}_2 \cup \cdots \cup \mathcal{G}_k \cup \cdots \cup \mathcal{G}_n$ is a shelling of $\mathcal{F}_j$.[2]

Fig. 10.5 Constructing a line shelling of a pentagon.

Given a d-polytope $\mathcal{P} \subset \mathbb{R}^d$, choose a generic vector $\mathbf{v} \in \mathbb{R}^d$ (e.g., you can choose a vector at random). If we think of $\mathcal{P}$ as a planet and we place a hot-air balloon on one of the facets of $\mathcal{P}$, we will slowly see more and more facets, one at a time, as the balloon rises in the direction of $\mathbf{v}$ (see Figure 10.5). Here

---

[2] Note that this condition implies that $(\mathcal{F}_1 \cup \mathcal{F}_2 \cup \cdots \cup \mathcal{F}_{j-1}) \cap \mathcal{F}_j$ is connected for $d \geq 3$.

seeing means the following: we say that a facet $\mathcal{F}$ is **visible** from the point $\mathbf{x} \notin \mathcal{P}$ if the line segment between $\mathbf{x}$ and any point $\mathbf{y} \in \mathcal{F}$ intersects $\mathcal{P}$ only in $\mathbf{y}$.

Our balloon ride gives rise (no pun intended) to an ordering $\mathcal{F}_1, \mathcal{F}_2, \ldots, \mathcal{F}_m$ of the facets of $\mathcal{P}$. Namely, $\mathcal{F}_1$ is the facet on which we are beginning our journey, $\mathcal{F}_2$ is the next visible facet, etc., until we are high enough so that no more visible facets can be added. At this point, we "pass through infinity" and let the balloon approach the polytope planet from the opposite side. The next facet in our list is the first one that will disappear as we move toward the polytope, then the next facet that will disappear, etc., until we are landing back on the polytope. (That is, the second half of our list of facets is the reversed list of what we would get had we started our ride at $\mathcal{F}_m$.)

Prove that this ordering $\mathcal{F}_1, \mathcal{F}_2, \ldots, \mathcal{F}_m$ of the facets of $\mathcal{P}$ is a shelling (called a **line shelling**).

10.17. Suppose $\mathcal{P}$ is a simplicial d-polytope with facets $\mathcal{F}_1, \mathcal{F}_2, \ldots, \mathcal{F}_m$, ordered by some line shelling. We collect the vertices of $\mathcal{F}_j$ in the set F_j. Let $\widetilde{F}_j$ be the set of all vertices $\mathbf{v} \in F_j$ such that $F_j \setminus \{\mathbf{v}\}$ is contained in one of $F_1, F_2, \ldots, F_{j-1}$ (which define the facets coming earlier in the shelling order).

(a) Prove that the new faces appearing in the jth step of the construction of the line shelling, i.e., the faces in $\mathcal{F}_j \setminus (\mathcal{F}_1 \cup \mathcal{F}_2 \cup \cdots \cup \mathcal{F}_{j-1})$, are precisely those sets $\operatorname{conv}(G)$ for some $\widetilde{F}_j \subseteq G \subseteq F_j$.

(b) Show that

$$f_{k-1} = \sum_{m=0}^{k} \binom{d-m}{k-m} \widetilde{h}_m,$$

where $\widetilde{h}_m$ denotes the number of sets $\widetilde{F}_j$ of cardinality m.

(c) Conclude that $\widetilde{h}_m = h_m$ and thus that $h_m \geq 0$.

10.18. Given a regular triangulation of the polytope $\mathcal{P}$, prove that the h-polynomial of the induced triangulation of $\partial \mathcal{P}$ has nonnegative coefficients, as does the h-polynomial of the link of every simplex in this triangulation.

10.19. Show that the construction in Exercise 10.17 gives the following characterization of the h-polynomial of a *simple* polytope $\mathcal{P}$: fix a generic direction vector $\mathbf{v}$ and orient the graph of $\mathcal{P}$ so that each oriented edge (viewed as a vector) forms an acute angle with $\mathbf{v}$. Then h_k equals the number of vertices of this oriented graph with in-degree k.

Open Problems

10.20. Which integral polytopes admit a unimodular triangulation? (The property of admitting unimodular triangulations is part of an interesting hierarchy of integral polytopes; see [73].)

10.21. Find a complete set of inequalities for the coefficients of all h^*-polynomials of degree ≤ 3, i.e., piecewise linear regions in $\mathbb{R}^3$ all of whose integer points (h_1^*, h_2^*, h_3^*) come from an h^*-polynomial of a 3-dimensional integral polytope. (See also Open Problem 3.43.)

10.22. A linear inequality $a_0 x_0 + a_1 x_1 + \ldots a_d x_d \geq 0$ is **balanced** if $a_0 + a_1 + \cdots + a_d = 0$. Find a complete set of balanced inequalities for the coefficients of every h^*-polynomial of degree ≤ 6. (For degree ≤ 5, see [232].)

10.23. Prove that every integrally closed reflexive polytope has a unimodal h^*-polynomial. More generally, prove that every integrally closed polytope has a unimodal h^*-polynomial. Even more generally, classify integral polytopes with a unimodal h^*-polynomial. (See [12, 63, 74, 210] for possible starting points.)

Chapter 11
The Decomposition of a Polytope into Its Cones

Mathematics compares the most diverse phenomena and discovers the secret analogies that unite them.

Jean Baptiste Joseph Fourier (1768–1830)

In this chapter, we return to integer-point transforms of rational cones and polytopes and connect them in a magical way that was first discovered by Michel Brion. The power of Brion's theorem has been applied to various domains, such as *Barvinok's algorithm* in integer linear programming, and to higher-dimensional *Euler–Maclaurin summation formulas*, which we study in Chapter 12. In a sense, Brion's theorem is the natural extension of the familiar finite geometric series identity $\sum_{m=a}^{b} z^m = \frac{z^{b+1} - z^a}{z-1}$ to higher dimensions.

11.1 The Identity "$\sum_{m \in \mathbb{Z}} z^m = 0$"...
Or "Much Ado About Nothing"

We begin gently by illustrating Brion's theorem in dimension 1. To this end, let's consider the line segment $\mathcal{I} := [20, 34]$. We recall that its integer-point transform lists the lattice points in $\mathcal{I}$ in the form of monomials:

$$\sigma_{\mathcal{I}}(z) = \sum_{m \in \mathcal{I} \cap \mathbb{Z}} z^m = z^{20} + z^{21} + \cdots + z^{34}.$$

Already in this simple example, we are too lazy to list all integers in $\mathcal{I}$ and use $\cdots$ to write the polynomial $\sigma_{\mathcal{I}}$. Is there a more compact way to write $\sigma_{\mathcal{I}}$? The reader might have guessed it even before we asked the question: this integer-point transform equals the rational function

$$\sigma_{\mathcal{I}}(z) = \frac{z^{20} - z^{35}}{1 - z}.$$

This last sentence is not quite correct: the definition of $\sigma_{\mathcal{I}}(z)$ yielded a polynomial in z, whereas the rational function above is not defined at $z = 1$. We can overcome this deficiency by noticing that the limit of this rational function as $z \to 1$ equals the evaluation of the polynomials $\sigma_{\mathcal{I}}(1) = 15$, by L'Hôpital's rule. Notice that the rational-function representation of $\sigma_{\mathcal{I}}$ has the unquestionable advantage of being much more compact than the original polynomial representation. The reader who is not convinced of this advantage should replace the right vertex 34 of $\mathcal{I}$ by 3400.

Now let's rewrite the rational form of the integer-point transform of $\mathcal{I}$ slightly:

$$\sigma_{\mathcal{I}}(z) = \frac{z^{20} - z^{35}}{1 - z} = \frac{z^{20}}{1 - z} + \frac{z^{34}}{1 - \frac{1}{z}}. \tag{11.1}$$

There is a natural geometric interpretation of the two summands on the right-hand side. The first term represents the integer-point transform of the interval $[20, \infty)$:

$$\sigma_{[20,\infty)}(z) = \sum_{m \geq 20} z^m = \frac{z^{20}}{1 - z}.$$

The second term in (11.1) corresponds to the integer-point transform of the interval $(-\infty, 34]$:

$$\sigma_{(-\infty,34]}(z) = \sum_{m \leq 34} z^m = \frac{z^{34}}{1 - \frac{1}{z}}.$$

So (11.1) says that on a rational-function level,

$$\sigma_{[20,\infty)}(z) + \sigma_{(-\infty,34]}(z) = \sigma_{[20,34]}(z). \tag{11.2}$$

Fig. 11.1 Decomposing a line segment into two infinite rays.

This identity, which we illustrate graphically in Figure 11.1, should come as a mild surprise. Two rational functions that represent infinite sequences somehow collapse, when being summed up, to a polynomial with a finite number of terms. We emphasize that (11.2) does not make sense on the level of infinite series; in fact, the two infinite series involved here have disjoint regions of convergence.

Even more magical is the geometry behind this identity: on the right-hand side, we have a polynomial that lists the integer points in a *finite* interval $\mathcal{P}$, while on the left-hand side, each of the rational generating functions represents the integer points in an *infinite* ray that begins at a vertex of $\mathcal{P}$. The two half-lines will be called *vertex cones* below, and indeed, the remainder of this chapter is devoted to proving that an identity similar to (11.2) holds in general dimension.

We recall a definition that was touched on only briefly in Exercises 3.19 and 9.8: A **hyperplane arrangement** $\mathcal{H}$ is a finite collection of hyperplanes. An arrangement $\mathcal{H}$ is **rational** if all its hyperplanes are rational, that is, if each hyperplane in $\mathcal{H}$ is of the form $\left\{ \mathbf{x} \in \mathbb{R}^d : a_1 x_1 + a_2 x_2 + \cdots + a_d x_d = b \right\}$ for some $a_1, a_2, \ldots, a_d, b \in \mathbb{Z}$. An arrangement $\mathcal{H}$ is called a **central** hyperplane arrangement if its hyperplanes meet in (at least) one point.

Our next definition generalizes (finally) the notion of a pointed cone, defined in Chapter 3. A **cone** is the intersection of finitely many half-spaces of the form

$$\left\{ \mathbf{x} \in \mathbb{R}^d : a_1 x_1 + a_2 x_2 + \cdots + a_d x_d \leq b \right\}$$

for which the corresponding hyperplanes

$$\left\{ \mathbf{x} \in \mathbb{R}^d : a_1 x_1 + a_2 x_2 + \cdots + a_d x_d = b \right\}$$

form a central arrangement. This definition extends that of a pointed cone: a cone is pointed if the defining hyperplanes meet in *exactly* one point. Thus nonpointed cones are cones that contain a line. A cone is **rational** if all of its defining hyperplanes are rational. Cones and polytopes are special cases of **polyhedra**, which are convex bodies defined as the intersection of finitely many half-spaces. Our next goal is to prove that every rational cone that contains a line essentially has integer-point transform equal to zero, in a sense that we have to make mathematically precise.

11.2 Technically Speaking...

To understand integer-point transforms of general cones, we need a small aspect of the theory of modules, the technically necessary tools to make things rigorous. To see what kind of problems might arise, consider the situation of Alice who decides to glue together a few pointed cones A_j to form some (not necessarily convex) polyhedral region K, and her friend Bob decides to glue together some other pointed cones B_j, to form the same region K. Why should it be true that the sum of the rational functions that we have associated with the pointed cones A_j equals the sum of the rational functions that we have associated with the pointed cones B_j? We must account for this inherent potential ambiguity and show that it is in fact unambiguous—Alice and Bob will always get the same rational functions.

We recall that a **module** M over a ring R is a generalization of a vector space over a field, where the salient new feature is that the *field of scalars* is replaced by a *ring of scalars*. Consider, for example, the set of all polynomials in d variables with coefficients in $\mathbb{Z}$, under addition. We can think intuitively of this particular module over the ring $\mathbb{Z}$ as a discretized version of a vector space over the field of real numbers.

The only module notions that are important to us in this chapter are that M forms an Abelian group under addition, that we can multiply elements of R by elements of M in a sensible way (e.g., compatible with ring multiplication in R), and that we preserve the definition of a **linear map** ϕ between two modules M and N over the same ring R, namely, ϕ satisfies

$$\phi(m_1 + r\,m_2) = \phi(m_1) + r\,\phi(m_2)$$

for every $m_1, m_2 \in M$ and $r \in R$.

From the point of departure for this chapter, it might not come as a surprise that the two modules we will use are formed by Laurent series and by rational functions. We cannot multiply two Laurent series in a meaningful way (and a reader who has not thought about this should construct an example in which things go wrong), but we *can* multiply a Laurent series by a Laurent *polynomial*, i.e., a finite Laurent series. This motivates the choice of ring with which we will be operating, namely, the ring

$$\mathbb{C}\left[\mathbf{z}^{\pm}\right] := \mathbb{C}\left[z_1^{\pm}, z_2^{\pm}, \ldots, z_d^{\pm}\right]$$

of Laurent polynomials in d variables with coefficients in $\mathbb{C}$. We will consider two modules over $\mathbb{C}\left[\mathbf{z}^{\pm}\right]$: on the one hand, the set CL_d of all (formal) Laurent series of rational cones in $\mathbb{R}^d$, and on the other hand, the set $\mathbb{C}(\mathbf{z}) := \mathbb{C}(z_1, z_2, \ldots, z_d)$ of rational functions in d variables.

In Chapter 3, we saw that certain Laurent series, namely integer-point transforms of rational simplicial cones, naturally evaluate to rational functions, by a natural tiling argument. The following lemma says that we can think of this evaluation as one instance of a linear map between the modules CL_d and $\mathbb{C}(\mathbf{z})$ over $\mathbb{C}\left[\mathbf{z}^{\pm}\right]$.

Lemma 11.1. *There is a unique linear map* $\phi : \mathrm{CL}_d \to \mathbb{C}(\mathbf{z})$ *that maps an integer-point transform (viewed as a Laurent series) of a rational simplicial cone*

$$\{\mathbf{v} + \lambda_1\mathbf{w}_1 + \lambda_2\mathbf{w}_2 + \cdots + \lambda_k\mathbf{w}_k : \lambda_1, \lambda_2, \ldots, \lambda_k \geq 0\} \subseteq \mathbb{R}^d$$

to the rational function

$$\frac{\sigma_\Pi(\mathbf{z})}{(1 - \mathbf{z}^{\mathbf{w}_1})(1 - \mathbf{z}^{\mathbf{w}_2}) \cdots (1 - \mathbf{z}^{\mathbf{w}_k})},$$

where $\sigma_\Pi(\mathbf{z})$ is the integer-point transform of the half-open parallelepiped

$$\Pi := \{\mathbf{v} + \lambda_1 \mathbf{w}_1 + \lambda_2 \mathbf{w}_2 + \cdots + \lambda_k \mathbf{w}_k : 0 \le \lambda_1, \lambda_2, \ldots, \lambda_k < 1\}. \quad (11.3)$$

To understand the significance of this lemma, we note that CL_d is generated by the integer-point transforms of rational simplicial cones, since every rational cone can be triangulated by rational simplicial cones. Thus Lemma 11.1 says that the rational forms of integer-point generating functions of simplicial cones, which we derived in Chapter 3, extend uniquely to rational forms of integer-point generating functions of all rational cones.

Proof of Lemma 11.1. Let

$$\mathcal{K} := \{\mathbf{v} + \lambda_1 \mathbf{w}_1 + \lambda_2 \mathbf{w}_2 + \cdots + \lambda_k \mathbf{w}_k : \lambda_1, \lambda_2, \ldots, \lambda_k \ge 0\}$$

be a simplicial rational k-cone in $\mathbb{R}^d$. For clarity, let's consider $\sigma_{\mathcal{K}}(\mathbf{z})$ as an element in CL_d, and for this Laurent series we proved in Chapter 3 the identity

$$(1 - \mathbf{z}^{\mathbf{w}_1})(1 - \mathbf{z}^{\mathbf{w}_2}) \cdots (1 - \mathbf{z}^{\mathbf{w}_k}) \sigma_{\mathcal{K}}(\mathbf{z}) = \sigma_\Pi(\mathbf{z}), \quad (11.4)$$

where Π is given by (11.3). We remark that (11.4) is an identity in the module CL_d over $\mathbb{C}[\mathbf{z}^\pm]$: the factor $(1 - \mathbf{z}^{\mathbf{w}_1})(1 - \mathbf{z}^{\mathbf{w}_2}) \cdots (1 - \mathbf{z}^{\mathbf{w}_k})$ on the left-hand side and the right-hand side $\sigma_\Pi(\mathbf{z})$ are Laurent polynomials in $\mathbb{C}[\mathbf{z}^\pm]$, whereas $\sigma_{\mathcal{K}}(\mathbf{z})$ is in CL_d.

If $\mathcal{K}$ is now a general rational cone, we can triangulate it into simplicial cone, each of which comes with a version of (11.4). The integer-point transform $\sigma_{\mathcal{K}}(\mathbf{z}) \in \mathrm{CL}_d$ of our general cone $\mathcal{K}$ can naturally be written in an inclusion–exclusion form as a sum (with positive and negative terms) of integer-point transforms of these simplicial cones and their faces, which are also simplicial cones. Applying the same sum to the identities of the form (11.4) for these simplicial cones gives an identity

$$g(\mathbf{z}) \sigma_{\mathcal{K}}(\mathbf{z}) = f(\mathbf{z})$$

for some Laurent monomials $f(\mathbf{z})$ and $g(\mathbf{z})$. This yields our sought-after linear map: we define

$$\phi(\sigma_{\mathcal{K}}(\mathbf{z})) := \frac{f(\mathbf{z})}{g(\mathbf{z})} \in \mathbb{C}(\mathbf{z}).$$

That this map ϕ is linear follows by construction, and that it is unique follows from the uniqueness of the rational-function form of $\sigma_{\mathcal{K}}(\mathbf{z})$ when $\mathcal{K}$ is simplicial. $\square$

The linear map ϕ in Lemma 11.1 allows us to make precise our aforementioned statement that integer-point transforms of rational cones that contain a line are philosophically equal to zero.

Lemma 11.2. *Let* $\phi : \mathrm{CL}_d \to \mathbb{C}(\mathbf{z})$ *be the linear map in Lemma 11.1, and let* $\mathcal{K} \subseteq \mathbb{R}^d$ *be a rational cone that contains a line. Then*

$$\phi\left(\sigma_{\mathcal{K}}(\mathbf{z})\right) = 0 \, .$$

Proof. Let $\mathcal{K} \subseteq \mathbb{R}^d$ be a rational cone that contains a line. This implies that there exists a vector $\mathbf{w} \in \mathbb{Z}^d \setminus \{\mathbf{0}\}$ such that $\mathbf{w} + \mathcal{K} = \mathcal{K}$. Translated into the language of Laurent series, this means that $\mathbf{z}^{\mathbf{w}} \, \sigma_{\mathcal{K}}(\mathbf{z}) = \sigma_{\mathcal{K}}(\mathbf{z})$, and thus, since ϕ is linear,

$$\mathbf{z}^{\mathbf{w}} \, \phi\left(\sigma_{\mathcal{K}}(\mathbf{z})\right) = \phi\left(\sigma_{\mathcal{K}}(\mathbf{z})\right) .$$

But this gives the identity $(1 - \mathbf{z}^{\mathbf{w}}) \, \phi\left(\sigma_{\mathcal{K}}(\mathbf{z})\right) = 0$ in the world $\mathbb{C}(\mathbf{z})$ of rational functions. Since $1 - \mathbf{z}^{\mathbf{w}}$ is not a zero divisor in this world, we conclude that $\phi\left(\sigma_{\mathcal{K}}(\mathbf{z})\right) = 0$. □

11.3 Tangent Cones and Their Rational Generating Functions

The goal of this section, apart from developing the language that allows us to prove Brion's theorem, is to prove a sort of analogue of (11.2) in general dimension.

We now attach a cone to each face $\mathcal{F}$ of $\mathcal{P}$, namely its **tangent cone**, defined by

$$\mathcal{K}_{\mathcal{F}} := \left\{ \mathbf{x} + \lambda\left(\mathbf{y} - \mathbf{x}\right) : \mathbf{x} \in \mathcal{F}, \, \mathbf{y} \in \mathcal{P}, \, \lambda \in \mathbb{R}_{\geq 0} \right\}.$$

We note that $\mathcal{K}_{\mathcal{P}} = \mathrm{span}\,\mathcal{P}$. For a vertex $\mathbf{v}$ of $\mathcal{P}$, the tangent cone $\mathcal{K}_{\mathbf{v}}$ is often called a **vertex cone**; it is pointed, and we show an example in Figure 11.2. For a k-face $\mathcal{F}$ of $\mathcal{P}$ with $k > 0$, the tangent cone $\mathcal{K}_{\mathcal{F}}$ is not pointed. For example, the tangent cone of an edge of a 3-polytope is a wedge.

Lemma 11.3. *For every face $\mathcal{F}$ of $\mathcal{P}$, we have* $\mathrm{span}\,\mathcal{F} \subseteq \mathcal{K}_{\mathcal{F}}$.

Proof. As $\mathbf{x}$ and $\mathbf{y}$ vary over all points of $\mathcal{F}$, the points $\mathbf{x} + \lambda\left(\mathbf{y} - \mathbf{x}\right)$ vary over $\mathrm{span}\,\mathcal{F}$. □

We note that this lemma implies that $\mathcal{K}_{\mathcal{F}}$ contains a line unless $\mathcal{F}$ is a vertex. More precisely, if $\mathcal{K}_{\mathcal{F}}$ is not pointed, it contains the affine space $\mathrm{span}\,\mathcal{F}$, which is called the **apex** of the tangent cone. (A pointed cone has a point as apex. We also remark that $\mathcal{K}_{\mathcal{F}}$ is the smallest convex cone with apex $\mathrm{span}\,\mathcal{F}$ that contains $\mathcal{P}$.) Consequently, Lemma 11.2 implies the following:

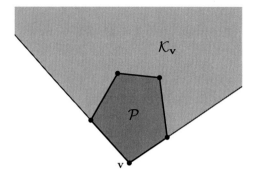

Fig. 11.2 A vertex cone.

Corollary 11.4. *Let $\phi : \mathrm{CL}_d \to \mathbb{C}(\mathbf{z})$ be the linear map in Lemma 11.1, and let $\mathcal{F}$ be a face of $\mathcal{P}$ that is not a vertex. Then*

$$\phi\left(\sigma_{\mathcal{K}_{\mathcal{F}}}(\mathbf{z})\right) = 0\,.$$

11.4 Brion's Theorem

The following theorem is a classical identity of convex geometry named after Charles Julien Brianchon (1783–1864)[1] and Jørgen Pedersen Gram (1850–1916).[2] It holds for every convex polytope. However, its proof for *simplices* is considerably simpler than that for the general case. We need only the Brianchon–Gram identity for simplices, so we restrict ourselves to this special case. (One could prove the general case along similar lines as below; however, we would need some additional machinery not covered in this book.) The **indicator function** 1_S of a set $S \subseteq \mathbb{R}^d$ is defined by

$$1_S(\mathbf{x}) := \begin{cases} 1 & \text{if } \mathbf{x} \in S, \\ 0 & \text{if } \mathbf{x} \notin S. \end{cases}$$

Theorem 11.5 (Brianchon–Gram identity for simplices). *Let Δ be a d-simplex. Then*

$$1_\Delta(\mathbf{x}) = \sum_{\mathcal{F} \subseteq \Delta} (-1)^{\dim \mathcal{F}} 1_{\mathcal{K}_{\mathcal{F}}}(\mathbf{x})\,,$$

where the sum is taken over all nonempty faces $\mathcal{F}$ of Δ.

Proof. We distinguish between two disjoint cases: whether or not $\mathbf{x}$ is in the simplex.

---

[1] For more information about Brianchon, see
http://www-history.mcs.st-andrews.ac.uk/Biographies/Brianchon.html.

[2] For more information about Gram, see
http://www-history.mcs.st-andrews.ac.uk/Biographies/Gram.html.

Case 1: $\mathbf{x} \in \Delta$. Then $\mathbf{x} \in \mathcal{K}_{\mathcal{F}}$ for each $\mathcal{F}$ of Δ, and the identity becomes

$$1 = \sum_{\mathcal{F} \subseteq \Delta} (-1)^{\dim \mathcal{F}} = \sum_{k=0}^{\dim \Delta} (-1)^k f_k \, .$$

This is the Euler relation for simplices, proved in Exercise 5.6.

Case 2: $\mathbf{x} \notin \Delta$. Then there is a unique minimal face $\mathcal{F}$ of Δ (minimal with respect to dimension) such that $\mathbf{x} \in \mathcal{K}_{\mathcal{F}}$ and $\mathbf{x} \in \mathcal{K}_{\mathcal{G}}$ for all faces $\mathcal{G} \subseteq \Delta$ that contain $\mathcal{F}$ (Exercise 11.4). The identity to be proved is now

$$0 = \sum_{\mathcal{G} \supseteq \mathcal{F}} (-1)^{\dim \mathcal{G}} \, . \tag{11.5}$$

The validity of this identity again follows from the logic of Exercise 5.6; the proof of (11.5) is the subject of Exercise 11.6. □

Corollary 11.6 (Brion's theorem for simplices). *Suppose Δ is a rational simplex. Then we have the following identity of rational functions:*

$$\sigma_\Delta(\mathbf{z}) = \sum_{\mathbf{v} \text{ a vertex of } \Delta} \sigma_{\mathcal{K}_{\mathbf{v}}}(\mathbf{z}) \, .$$

Proof. We translate the Brianchon–Gram theorem into the language of integer-point transforms: we sum both sides of the identity in Theorem 11.5 for all $\mathbf{m} \in \mathbb{Z}^d$,

$$\sum_{\mathbf{m} \in \mathbb{Z}^d} 1_\Delta(\mathbf{m}) \, \mathbf{z}^{\mathbf{m}} = \sum_{\mathbf{m} \in \mathbb{Z}^d} \sum_{\mathcal{F} \subseteq \Delta} (-1)^{\dim \mathcal{F}} 1_{\mathcal{K}_{\mathcal{F}}}(\mathbf{m}) \, \mathbf{z}^{\mathbf{m}},$$

which is equivalent to

$$\sigma_\Delta(\mathbf{z}) = \sum_{\mathcal{F} \subseteq \Delta} (-1)^{\dim \mathcal{F}} \sigma_{\mathcal{K}_{\mathcal{F}}}(\mathbf{z}) \, .$$

Now we apply the linear map $\phi : \mathrm{CL}_d \to \mathbb{C}(\mathbf{z})$ from Lemma 11.1 to this identity:

$$\phi(\sigma_\Delta(\mathbf{z})) = \sum_{\mathcal{F} \subseteq \Delta} (-1)^{\dim \mathcal{F}} \phi(\sigma_{\mathcal{K}_{\mathcal{F}}}(\mathbf{z})) \, .$$

The left-hand side is the Laurent polynomial $\sigma_\Delta(\mathbf{z})$. On the right-hand side, we can see the rational-function versions (which we may still denote by $\sigma_{\mathcal{K}_{\mathbf{v}}}(\mathbf{z})$) of the integer-point transforms of the (simplicial) vertex cones $\mathcal{K}_{\mathbf{v}}$, and the remaining terms, namely $\phi(\sigma_{\mathcal{K}_{\mathcal{F}}}(\mathbf{z}))$ for nonvertices $\mathcal{F}$, disappear by Corollary 11.4. □

Now we extend Corollary 11.6 to every convex rational polytope:

Theorem 11.7 (Brion's theorem). *Suppose $\mathcal{P}$ is a rational convex polytope. Then we have the following identity of rational functions:*

$$\sigma_{\mathcal{P}}(\mathbf{z}) = \sum_{\mathbf{v} \text{ a vertex of } \mathcal{P}} \sigma_{\mathcal{K}_{\mathbf{v}}}(\mathbf{z}). \tag{11.6}$$

Proof. We use the same irrational trick as in the proofs of Theorems 3.12 and 4.3. Namely, we begin by triangulating $\mathcal{P}$ into the simplices $\Delta_1, \Delta_2, \ldots, \Delta_m$ (using no new vertices). Consider the hyperplane arrangement

$$\mathcal{H} := \{\operatorname{span} \mathcal{F} : \mathcal{F} \text{ is a facet of } \Delta_1, \Delta_2, \ldots, \text{ or } \Delta_m\}.$$

We will now shift the hyperplanes in $\mathcal{H}$, obtaining a new hyperplane arrangement $\mathcal{H}^{\text{shift}}$. Those hyperplanes of $\mathcal{H}$ that defined $\mathcal{P}$ now define, after shifting, a new polytope that we will call $\mathcal{P}^{\text{shift}}$. Exercise 11.8 ensures that we can shift $\mathcal{H}$ in such a way that:

- no hyperplane in $\mathcal{H}^{\text{shift}}$ contains a lattice point;
- $\mathcal{H}^{\text{shift}}$ yields a triangulation of $\mathcal{P}^{\text{shift}}$;
- the lattice points contained in a vertex cone of $\mathcal{P}$ are precisely the lattice points contained in the corresponding vertex cone of $\mathcal{P}^{\text{shift}}$.

This setup implies that

- the lattice points in $\mathcal{P}$ are precisely the lattice points in $\mathcal{P}^{\text{shift}}$;
- the lattice points in a vertex cone of $\mathcal{P}^{\text{shift}}$ can be written as a *disjoint* union of lattice points in vertex cones of simplices of the triangulation that $\mathcal{H}^{\text{shift}}$ induces on $\mathcal{P}^{\text{shift}}$.

The latter two conditions, in turn, mean that Brion's identity (11.6) follows from Brion's theorem for simplices: the integer-point transforms on both sides of the identity can be written as a sum of integer-point transforms of simplices and their vertex cones. $\qquad\square$

11.5 Brion Implies Ehrhart

We conclude this chapter by showing that Ehrhart's theorem (Theorem 3.23) for rational polytopes (which includes the integral case, Theorem 3.8) follows from Brion's theorem (Theorem 11.7) in a relatively straightforward manner.

Second Proof of Theorem 3.23. As in our first proof of Ehrhart's theorem, it suffices to prove Theorem 3.23 for simplices, because we can triangulate any polytope (using only the vertices). So suppose Δ is a rational d-simplex whose vertices have coordinates with denominator p. Our goal is to show that for a fixed $0 \leq r < p$, the function $L_{\Delta}(r + pt)$ is a polynomial in t; this means that L_{Δ} is a quasipolynomial with period dividing p.

First, if $r = 0$, then $L_\Delta(pt) = L_{p\Delta}(t)$, which is a polynomial by Ehrhart's theorem (Theorem 3.8), because $p\Delta$ is an integer simplex.

Now we assume $r > 0$. By Theorem 11.7,

$$L_\Delta(r + pt) = \sum_{\mathbf{m} \in (r+pt)\Delta \cap \mathbb{Z}^d} 1$$

$$= \lim_{\mathbf{z} \to \mathbf{1}} \sigma_{(r+pt)\Delta}(\mathbf{z}) \tag{11.7}$$

$$= \lim_{\mathbf{z} \to \mathbf{1}} \sum_{\mathbf{v} \text{ vertex of } \Delta} \sigma_{(r+pt)\mathcal{K}_\mathbf{v}}(\mathbf{z}) .$$

We used the limit computation for the integer-point transform $\sigma_{(r+pt)\Delta}$ rather than the evaluation $\sigma_{(r+pt)\Delta}(\mathbf{1})$, because that evaluation would have yielded singularities in the rational generating functions of the vertex cones. Note that the vertex cones $\mathcal{K}_\mathbf{v}$ are all simplicial, because Δ is a simplex. So suppose

$$\mathcal{K}_\mathbf{v} = \{\mathbf{v} + \lambda_1 \mathbf{w}_1 + \lambda_2 \mathbf{w}_2 + \cdots + \lambda_d \mathbf{w}_d : \lambda_1, \lambda_2, \ldots, \lambda_d \geq 0\} ;$$

then

$$
\begin{aligned}
(r + pt)\mathcal{K}_\mathbf{v} &= \{(r + pt)\mathbf{v} + \lambda_1 \mathbf{w}_1 + \lambda_2 \mathbf{w}_2 + \cdots + \lambda_d \mathbf{w}_d : \lambda_1, \lambda_2, \ldots, \lambda_d \geq 0\} \\
&= tp\mathbf{v} + \{r\mathbf{v} + \lambda_1 \mathbf{w}_1 + \lambda_2 \mathbf{w}_2 + \cdots + \lambda_d \mathbf{w}_d : \lambda_1, \lambda_2, \ldots, \lambda_d \geq 0\} \\
&= tp\mathbf{v} + r\mathcal{K}_\mathbf{v} .
\end{aligned}
$$

What is important to note here is that $p\mathbf{v}$ is an integer vector. In particular, we can safely write

$$\sigma_{(r+pt)\mathcal{K}_\mathbf{v}}(\mathbf{z}) = \mathbf{z}^{tp\mathbf{v}} \sigma_{r\mathcal{K}_\mathbf{v}}(\mathbf{z})$$

(we say *safely* because $tp\mathbf{v} \in \mathbb{Z}^d$, so $\mathbf{z}^{tp\mathbf{v}}$ is indeed a monomial). Now we can rewrite (11.7) as

$$L_\Delta(r + pt) = \lim_{\mathbf{z} \to \mathbf{1}} \sum_{\mathbf{v} \text{ vertex of } \Delta} \mathbf{z}^{tp\mathbf{v}} \sigma_{r\mathcal{K}_\mathbf{v}}(\mathbf{z}) . \tag{11.8}$$

The exact form of the rational functions $\sigma_{r\mathcal{K}_\mathbf{v}}(\mathbf{z})$ is not important for the purposes of the proof, except for the fact that they do not depend on t. We know that the sum of the generating functions of all vertex cones is a Laurent polynomial in $\mathbf{z}$; that is, the singularities of the rational functions cancel. To compute $L_\Delta(r + pt)$ from (11.8), we combine all of the rational functions on the right-hand side over one denominator and use L'Hôpital's rule to compute the limit of this one huge rational function. The variable t appears only in the simple monomials $\mathbf{z}^{tp\mathbf{v}}$, so the effect of L'Hôpital's rule is that we obtain linear factors of t every time we differentiate the numerator of this rational function. At the end, we evaluate the remaining rational function at $\mathbf{z} = \mathbf{1}$. The result is a polynomial in t. $\qquad \square$

Notes

1. Theorem 11.5 (in its general form for convex polytopes) has an interesting history. In 1837, Charles Brianchon proved a version of this theorem involving volumes of polytopes in $\mathbb{R}^3$ [65]. In 1874, Jørgen Gram gave a proof of the same result [121]; apparently, he was unaware of Brianchon's paper. In 1927, Duncan Sommerville published a proof for general d [223], which was corrected in the 1960s by Victor Klee [150], Branko Grünbaum [126, Section 14.1], and many others.

2. Michel Brion discovered Theorem 11.7 in 1988 [66]. His proof involved the Baum–Fulton–Quart Riemann–Roch formula for equivariant K-theory of toric varieties. A more elementary proof of Theorem 11.7 was found by Masa-Nori Ishida a few years later [142]. Our approach in this chapter follows [40].

3. Around the time Theorem 11.7 was conceived, Jim Lawrence [161] and Alexander Varchenko [249] discovered a companion theorem—it also gives the integer-point transform of a rational polytope as a sum of rational functions stemming from cones based at the vertices of the polytopes, but slightly different from the vertex cones. Both the Brion and the Lawrence–Varchenko theorems can be neatly understood using the theory of *valuations*, which we do not use here (see, for example, [160]).

4. As we have already remarked earlier, Brion's theorem led to an efficient algorithm by Alexander Barvinok to compute Ehrhart quasipolynomials [24]. More precisely, Barvinok proved that in fixed dimension, one can efficiently[3] compute the Ehrhart series $\sum_{t \geq 0} L_{\mathcal{P}}(t) \, z^t$ as a short sum of rational functions.[4] Brion's theorem essentially reduces the problem to computing the integer-point transforms of the rational tangent cones of the polytope. Barvinok's ingenious idea was to use a *signed decomposition* of a rational cone to compute its integer-point transform: the cone is written as a sum and difference of *unimodular* cones, which we will encounter in Section 12.4 and which have a trivial integer-point transform. Finding a signed decomposition involves triangulations, Minkowski's theorem on lattice points in convex bodies (see, for example, [82, 178, 186, 220]), and the LLL algorithm, which finds a short vector in a lattice [163]. At any rate, Barvinok proved that one can find a signed decomposition quickly, which is the main step toward computing the Ehrhart series of the polytope. Barvinok's algorithm has been implemented in the software packages `barvinok` [250] and `LattE` [94, 96, 154]. Barvinok's algorithm is described in detail in [22].

---

[3] *Efficiently* here means that for every dimension, there exists a polynomial that gives an upper bound on the running time of the algorithm when evaluated at the logarithm of the input data of the polytope (e.g., its vertices).

[4] *Short* means that the set of data needed to output this sum of rational functions is also of polynomial size in the logarithm of the input data of the polytope.

Exercises

11.1. Prove that the cones in $\mathbb{R}^d$ are precisely the sets of the form

$$\mathbf{v} + \left\{\mathbf{x} \in \mathbb{R}^d : \mathbf{A}\,\mathbf{x} \leq \mathbf{0}\right\}$$

for some $\mathbf{v} \in \mathbb{R}^d$ and $\mathbf{A} \in \mathbb{R}^{m \times d}$.

11.2. Recall that the **orthogonal complement** $\mathbf{A}^\perp$ of an affine space $\mathbf{A}$ is defined by

$$\mathbf{A}^\perp := \left\{\mathbf{x} \in \mathbb{R}^d : \mathbf{x} \cdot \mathbf{v} = 0 \text{ for all } \mathbf{v} \in \mathbf{A}\right\}.$$

Prove that for every face $\mathcal{F}$ of a polytope, $(\operatorname{span}\mathcal{F})^\perp \cap \mathcal{K}_\mathcal{F}$ is a pointed cone. (*Hint:* Show that if H is a defining hyperplane for $\mathcal{F}$, then $H \cap (\operatorname{span}\mathcal{F})^\perp$ is a hyperplane in the vector space $(\operatorname{span}\mathcal{F})^\perp$.)

11.3. Let $\mathcal{P} \subset \mathbb{R}^d$ be a full-dimensional polytope. A face $\mathcal{F}$ of $\mathcal{P}$ is **visible** from the point $\mathbf{x} \in \mathbb{R}^d$ if the line segment between $\mathbf{x}$ and any point $\mathbf{y} \in \mathcal{F}$ intersects $\mathcal{P}$ only in $\mathbf{y}$. (We used this notion already in Exercise 10.16.) Show that $\mathcal{F}$ is visible from $\mathbf{x}$ if and only if $\mathbf{x} \notin \mathcal{K}_\mathcal{F}$.

11.4. ♣ Suppose Δ is a simplex and $\mathbf{x} \notin \Delta$. Prove that there is a unique minimal face $\mathcal{F} \subseteq \Delta$ (minimal with respect to dimension) such that the corresponding tangent cone $\mathcal{K}_\mathcal{F}$ contains $\mathbf{x}$. Show that $\mathbf{x} \in \mathcal{K}_\mathcal{G}$ for all faces $\mathcal{G} \subseteq \Delta$ that contain $\mathcal{F}$, and $\mathbf{x} \notin \mathcal{K}_\mathcal{G}$ for all other faces $\mathcal{G}$.

11.5. Show that Exercise 11.4 fails to be true if Δ is a quadrilateral (for example). Show that the Brianchon–Gram identity holds for your quadrilateral.

11.6. ♣ Prove (11.5): for a face $\mathcal{F}$ of a simplex Δ,

$$\sum_{\mathcal{G} \supseteq \mathcal{F}} (-1)^{\dim \mathcal{G}} = 0,$$

where the sum is taken over all faces of Δ that contain $\mathcal{F}$.

11.7. Give a direct proof of Brion's theorem for the 1-dimensional case.

11.8. ♣ Provide the details of the irrational-shift argument in the proof of Theorem 11.7: Given a rational polytope $\mathcal{P}$, triangulate it into the simplices $\Delta_1, \Delta_2, \ldots, \Delta_m$ (using no new vertices). Consider the hyperplane arrangement

$$\mathcal{H} := \{\operatorname{span}\mathcal{F} : \mathcal{F} \text{ is a facet of } \Delta_1, \Delta_2, \ldots, \Delta_m\}.$$

We will now shift the hyperplanes in $\mathcal{H}$, obtaining a new hyperplane arrangement $\mathcal{H}^{\text{shift}}$. Those hyperplanes of $\mathcal{H}$ that defined $\mathcal{P}$ now define, after shifting, a new polytope that we will call $\mathcal{P}^{\text{shift}}$. Prove that we can shift $\mathcal{H}$ in such a way that:

- no hyperplane in $\mathcal{H}^{\text{shift}}$ contains any lattice point;
- $\mathcal{H}^{\text{shift}}$ yields a triangulation of $\mathcal{P}^{\text{shift}}$;
- the lattice points contained in a vertex cone of $\mathcal{P}$ are precisely the lattice points contained in the corresponding vertex cone of $\mathcal{P}^{\text{shift}}$.

11.9. ♣ Prove the following "open polytope" analogue for Brion's theorem: If $\mathcal{P}$ is a rational convex polytope, then we have the identity of rational functions

$$\sigma_{\mathcal{P}^\circ}(\mathbf{z}) = \sum_{\mathbf{v} \text{ a vertex of } \mathcal{P}} \sigma_{\mathcal{K}_{\mathbf{v}}^\circ}(\mathbf{z}).$$

11.10. Prove the following extension of Ehrhart's theorem (Theorem 3.23): Suppose $\mathcal{P} \subset \mathbb{R}^d$ is a rational convex polytope and q is a polynomial in d variables. Then

$$L_{\mathcal{P}}^q(t) := \sum_{\mathbf{m} \in t\mathcal{P} \cap \mathbb{Z}^d} q(\mathbf{m})$$

is a quasipolynomial in t. (*Hint:* Modify the proof in Section 11.5 by introducing a differential operator.)

Chapter 12
Euler–Maclaurin Summation in $\mathbb{R}^d$

All means (even continuous) sanctify the discrete end.

Doron Zeilberger

Thus far we have often been concerned with the difference between the discrete volume of a polytope $\mathcal{P}$ and its continuous volume. In other words, the quantity

$$\sum_{\mathbf{m} \in \mathcal{P} \cap \mathbb{Z}^d} 1 - \int_{\mathcal{P}} d\mathbf{y}\,, \tag{12.1}$$

which is by definition $L_{\mathcal{P}}(1) - \operatorname{vol}(\mathcal{P})$, has been on our minds for a long time and has arisen naturally in many different contexts. An important extension is the difference between the discrete integer-point transform and its continuous sibling:

$$\sum_{\mathbf{m} \in \mathcal{P} \cap \mathbb{Z}^d} e^{\mathbf{m} \cdot \mathbf{x}} - \int_{\mathcal{P}} e^{\mathbf{y} \cdot \mathbf{x}} d\mathbf{y}\,, \tag{12.2}$$

where we have replaced the variable $\mathbf{z}$ that we have commonly used in generating functions by the exponential variable $(z_1, z_2, \ldots, z_d) = (e^{x_1}, e^{x_2}, \ldots, e^{x_d})$. Note that on setting $\mathbf{x} = 0$ in (12.2), we get the former quantity (12.1). Relations between the two quantities $\sum_{\mathbf{m} \in \mathcal{P} \cap \mathbb{Z}^d} e^{\mathbf{m} \cdot \mathbf{x}}$ and $\int_{\mathcal{P}} e^{\mathbf{y} \cdot \mathbf{x}} d\mathbf{y}$ are known as *Euler–Maclaurin summation* formulas for polytopes. The "behind-the-scenes" operators that are responsible for affording us with such connections are the differential operators known as *Todd operators*, whose definition utilizes the Bernoulli numbers in a surprising way.

12.1 Todd Operators and Bernoulli Numbers

Recall the Bernoulli numbers B_k from Section 2.4, defined by the generating function

$$\frac{z}{e^z - 1} = \sum_{k \geq 0} \frac{B_k}{k!} z^k.$$

We now introduce a differential operator via essentially the same generating function, namely

$$\text{Todd}_h := \sum_{k \geq 0} (-1)^k \frac{B_k}{k!} \left(\frac{d}{dh} \right)^k. \tag{12.3}$$

This **Todd operator** is often abbreviated as

$$\text{Todd}_h = \frac{\frac{d}{dh}}{1 - e^{-\frac{d}{dh}}},$$

but we should keep in mind that this is only a shorthand notation for the infinite series (12.3). We first show that the exponential function is an eigenfunction of the Todd operator.

Lemma 12.1. *For $z \in \mathbb{C} \setminus \{0\}$ with $|z| < 2\pi$,*

$$\text{Todd}_h\, e^{zh} = \frac{z\, e^{zh}}{1 - e^{-z}}.$$

Proof.

$$\text{Todd}_h\, e^{zh} = \sum_{k \geq 0} (-1)^k \frac{B_k}{k!} \left(\frac{d}{dh} \right)^k e^{zh}$$

$$= \sum_{k \geq 0} (-1)^k \frac{B_k}{k!} z^k e^{zh}$$

$$= e^{zh} \sum_{k \geq 0} (-z)^k \frac{B_k}{k!}$$

$$= e^{zh} \frac{-z}{e^{-z} - 1}.$$

The condition $|z| < 2\pi$ is needed in the last step, by Exercise 2.14.   □

The Todd operator is a discretizing operator, in the sense that it transforms a continuous integral into a discrete sum, as the following theorem shows.

Theorem 12.2 (Euler–Maclaurin in dimension 1). *For all $a < b \in \mathbb{Z}$ and $z \in \mathbb{C}$ with $|z| < 2\pi$,*

$$\text{Todd}_{h_1} \text{Todd}_{h_2} \int_{a-h_2}^{b+h_1} e^{zx} dx \bigg|_{h_1=h_2=0} = \sum_{k=a}^{b} e^{kz}.$$

Proof. Case 1: $z = 0$. Then $e^{zx} = 1$, and so

$$\text{Todd}_{h_1} \text{Todd}_{h_2} \int_{a-h_2}^{b+h_1} e^{zx} dx \bigg|_{h_1=h_2=0}$$

$$= \text{Todd}_{h_1} \text{Todd}_{h_2} \int_{a-h_2}^{b+h_1} dx \bigg|_{h_1=h_2=0}$$

$$= b - a + \text{Todd}_{h_1} h_1 + \text{Todd}_{h_2} h_2 |_{h_1=h_2=0}$$

$$= b - a + h_1 + \frac{1}{2} + h_2 + \frac{1}{2} \bigg|_{h_1=h_2=0}$$

$$= b - a + 1$$

by Exercise 12.1. Since $\sum_{k=a}^{b} e^{k \cdot 0} = b - a + 1$, we have verified the theorem in this case.

Case 2: $z \neq 0$. Then

$$\text{Todd}_{h_1} \text{Todd}_{h_2} \int_{a-h_2}^{b+h_1} e^{zx} dx = \text{Todd}_{h_1} \text{Todd}_{h_2} \frac{1}{z} \left(e^{z(b+h_1)} - e^{z(a-h_2)} \right)$$

$$= \frac{1}{z} \left(\text{Todd}_{h_1} e^{zb+zh_1} - \text{Todd}_{h_2} e^{za-zh_2} \right)$$

$$= \frac{e^{zb}}{z} \text{Todd}_{h_1} e^{zh_1} - \frac{e^{za}}{z} \text{Todd}_{h_2} e^{-zh_2}$$

$$= \frac{e^{zb}}{z} \frac{z e^{zh_1}}{1 - e^{-z}} - \frac{e^{za}}{z} \frac{-z e^{-zh_2}}{1 - e^z},$$

where the last step follows from Lemma 12.1. Hence

$$\text{Todd}_{h_1} \text{Todd}_{h_2} \int_{a-h_2}^{b+h_1} e^{zx} dx \bigg|_{h_1=h_2=0} = e^{zb} \frac{1}{1 - e^{-z}} + e^{za} \frac{1}{1 - e^z}$$

$$= \frac{e^{z(b+1)} - e^{za}}{e^z - 1}$$

$$= \sum_{k=a}^{b} e^{kz}. \qquad \square$$

We will need a multivariate version of the Todd operator later, so we define for $\mathbf{h} = (h_1, h_2, \ldots, h_m)$,

$$\mathrm{Todd_h} := \prod_{j=1}^{m} \left(\frac{\frac{\partial}{\partial h_j}}{1 - \exp\left(-\frac{\partial}{\partial h_j}\right)} \right),$$

keeping in mind that this is a product over infinite series of the form (12.3).

12.2 A Continuous Version of Brion's Theorem

In the following two sections, we develop the tools that, once fused with the Todd operator, will enable us to extend Euler–Maclaurin summation to higher dimensions. We first give an integral analogue of Theorem 11.7 for simple rational polytopes. We begin by translating Brion's integer-point transforms

$$\sigma_{\mathcal{P}}(\mathbf{z}) = \sum_{\mathbf{v} \text{ a vertex of } \mathcal{P}} \sigma_{\mathcal{K}_{\mathbf{v}}}(\mathbf{z})$$

into an exponential form:

$$\sigma_{\mathcal{P}}(\exp \mathbf{z}) = \sum_{\mathbf{v} \text{ a vertex of } \mathcal{P}} \sigma_{\mathcal{K}_{\mathbf{v}}}(\exp \mathbf{z}),$$

where we used the notation $\exp \mathbf{z} = (e^{z_1}, e^{z_2}, \ldots, e^{z_d})$. For the continuous analogue of Brion's theorem, we replace the sum on the left-hand side,

$$\sigma_{\mathcal{P}}(\exp \mathbf{z}) = \sum_{\mathbf{m} \in \mathcal{P} \cap \mathbb{Z}^d} (\exp \mathbf{z})^{\mathbf{m}} = \sum_{\mathbf{m} \in \mathcal{P} \cap \mathbb{Z}^d} \exp(\mathbf{m} \cdot \mathbf{z}),$$

by an integral.

Theorem 12.3 (Brion's theorem: continuous form). *Suppose $\mathcal{P}$ is a simple rational convex d-polytope. For each vertex cone $\mathcal{K}_{\mathbf{v}}$ of $\mathcal{P}$, fix a set of generators $\mathbf{w}_1(\mathbf{v}), \mathbf{w}_2(\mathbf{v}), \ldots, \mathbf{w}_d(\mathbf{v}) \in \mathbb{Z}^d$. Then*

$$\int_{\mathcal{P}} \exp(\mathbf{x} \cdot \mathbf{z}) \, d\mathbf{x} = (-1)^d \sum_{\mathbf{v} \text{ a vertex of } \mathcal{P}} \frac{\exp(\mathbf{v} \cdot \mathbf{z}) \, |\det(\mathbf{w}_1(\mathbf{v}), \ldots, \mathbf{w}_d(\mathbf{v}))|}{\prod_{k=1}^{d} (\mathbf{w}_k(\mathbf{v}) \cdot \mathbf{z})}$$

for all $\mathbf{z}$ such that the denominators on the right-hand side do not vanish.

Proof. We begin with the assumption that $\mathcal{P}$ is an *integral* polytope; we will see in the process of the proof that this assumption can be relaxed. Let's write out the exponential form of Brion's theorem (Theorem 11.7), using the assumption that the vertex cones are simplicial (because $\mathcal{P}$ is simple). By Theorem 3.5,

$$\sum_{\mathbf{m} \in \mathcal{P} \cap \mathbb{Z}^d} \exp(\mathbf{m} \cdot \mathbf{z}) = \sum_{\mathbf{v} \text{ a vertex of } \mathcal{P}} \frac{\exp(\mathbf{v} \cdot \mathbf{z}) \, \sigma_{\Pi_{\mathbf{v}}}(\exp \mathbf{z})}{\prod_{k=1}^{d} (1 - \exp(\mathbf{w}_k(\mathbf{v}) \cdot \mathbf{z}))}, \quad (12.4)$$

where

$$\Pi_{\mathbf{v}} = \{\lambda_1 \mathbf{w}_1(\mathbf{v}) + \lambda_2 \mathbf{w}_2(\mathbf{v}) + \cdots + \lambda_d \mathbf{w}_d(\mathbf{v}) : 0 \le \lambda_1, \lambda_2, \ldots, \lambda_d < 1\}$$

is the fundamental parallelepiped of the vertex cone $\mathcal{K}_{\mathbf{v}}$. We would like to rewrite (12.4) with the lattice $\mathbb{Z}^d$ replaced by the refined lattice $\left(\frac{1}{n}\mathbb{Z}\right)^d$, because then, the left-hand side of (12.4) will give rise to the sought-after integral by letting n approach infinity. The right-hand side of (12.4) changes accordingly; now every integral point has to be scaled down by $\frac{1}{n}$:

$$\sum_{\mathbf{m} \in \mathcal{P} \cap \left(\frac{1}{n}\mathbb{Z}\right)^d} \exp(\mathbf{m} \cdot \mathbf{z}) = \sum_{\mathbf{v} \text{ a vertex of } \mathcal{P}} \frac{\exp(\mathbf{v} \cdot \mathbf{z}) \sum_{\mathbf{m} \in \Pi_{\mathbf{v}} \cap \mathbb{Z}^d} \exp\left(\frac{\mathbf{m}}{n} \cdot \mathbf{z}\right)}{\prod_{k=1}^d \left(1 - \exp\left(\frac{\mathbf{w}_k(\mathbf{v})}{n} \cdot \mathbf{z}\right)\right)}.$$

$$(12.5)$$

The proof of this identity is in essence the same as that of Theorem 3.5; we leave it as Exercise 12.2. Now our sought-after integral is

$$\int_{\mathcal{P}} \exp(\mathbf{x} \cdot \mathbf{z}) \, d\mathbf{x} = \lim_{n \to \infty} \frac{1}{n^d} \sum_{\mathbf{m} \in \mathcal{P} \cap \left(\frac{1}{n}\mathbb{Z}\right)^d} \exp(\mathbf{m} \cdot \mathbf{z})$$

$$= \lim_{n \to \infty} \frac{1}{n^d} \sum_{\mathbf{v} \text{ a vertex of } \mathcal{P}} \frac{\exp(\mathbf{v} \cdot \mathbf{z}) \sum_{\mathbf{m} \in \Pi_{\mathbf{v}} \cap \mathbb{Z}^d} \exp\left(\frac{\mathbf{m}}{n} \cdot \mathbf{z}\right)}{\prod_{k=1}^d \left(1 - \exp\left(\frac{\mathbf{w}_k(\mathbf{v})}{n} \cdot \mathbf{z}\right)\right)}.$$

$$(12.6)$$

At this point, we can see that our assumption that $\mathcal{P}$ has integral vertices can be relaxed to the rational case, since we may compute the limit only for n's that are multiples of the denominator of $\mathcal{P}$. The numerators of the terms on the right-hand side have a simple limit:

$$\lim_{n \to \infty} \exp(\mathbf{v} \cdot \mathbf{z}) \sum_{\mathbf{m} \in \Pi_{\mathbf{v}} \cap \mathbb{Z}^d} \exp\left(\frac{\mathbf{m}}{n} \cdot \mathbf{z}\right) = \exp(\mathbf{v} \cdot \mathbf{z}) \sum_{\mathbf{m} \in \Pi_{\mathbf{v}} \cap \mathbb{Z}^d} 1$$

$$= \exp(\mathbf{v} \cdot \mathbf{z}) \left|\det\left(\mathbf{w}_1(\mathbf{v}), \ldots, \mathbf{w}_d(\mathbf{v})\right)\right|,$$

where the last identity follows from Lemma 9.2. Hence (12.6) simplifies to

$$\int_{\mathcal{P}} \exp(\mathbf{x} \cdot \mathbf{z}) \, d\mathbf{x} = \sum_{\mathbf{v} \text{ a vertex of } \mathcal{P}} \frac{\exp(\mathbf{v} \cdot \mathbf{z}) \left|\det\left(\mathbf{w}_1(\mathbf{v}), \ldots, \mathbf{w}_d(\mathbf{v})\right)\right|}{\prod_{k=1}^d \lim_{n \to \infty} n \left(1 - \exp\left(\frac{\mathbf{w}_k(\mathbf{v})}{n} \cdot \mathbf{z}\right)\right)}.$$

Finally, using L'Hôpital's rule,

$$\lim_{n \to \infty} n \left(1 - \exp\left(\frac{\mathbf{w}_k(\mathbf{v})}{n} \cdot \mathbf{z}\right)\right) = -\mathbf{w}_k(\mathbf{v}) \cdot \mathbf{z},$$

and the theorem follows. $\square$

It turns out (Exercise 12.6) that for each vertex cone $\mathcal{K}_\mathbf{v}$,

$$\int_{\mathcal{K}_\mathbf{v}} \exp(\mathbf{x} \cdot \mathbf{z}) \, d\mathbf{x} = (-1)^d \, \frac{\exp(\mathbf{v} \cdot \mathbf{z}) \, |\det(\mathbf{w}_1(\mathbf{v}), \ldots, \mathbf{w}_d(\mathbf{v}))|}{\prod_{k=1}^d (\mathbf{w}_k(\mathbf{v}) \cdot \mathbf{z})}, \qquad (12.7)$$

and Theorem 12.3 shows that the Fourier–Laplace transform of $\mathcal{P}$ equals the sum of the Fourier–Laplace transforms of the vertex cones. In other words,

$$\int_{\mathcal{P}} \exp(\mathbf{x} \cdot \mathbf{z}) \, d\mathbf{x} = \sum_{\mathbf{v} \text{ a vertex of } \mathcal{P}} \int_{\mathcal{K}_\mathbf{v}} \exp(\mathbf{x} \cdot \mathbf{z}) \, d\mathbf{x}.$$

We also remark that $|\det(\mathbf{w}_1(\mathbf{v}), \ldots, \mathbf{w}_d(\mathbf{v}))|$ has a geometric meaning: it is the volume of the fundamental parallelepiped of the vertex cone $\mathcal{K}_\mathbf{v}$.

The curious reader might wonder what happens to the statement of Theorem 12.3 if we scale each of the generators $\mathbf{w}_k(\mathbf{v})$ by a different factor. It is immediate (Exercise 12.7) that the right-hand side of Theorem 12.3 remains invariant.

There is an important difference between the vertex cone generating functions (integrals) that appear in the continuous version of Brion's theorem (Theorem 12.3) and the vertex cone generating functions (sums) that appear in the discrete Brion theorem (Theorem 11.7). To see the difference, consider the following example:

Let K_0 be the first quadrant in $\mathbb{R}^2$, having generators $(1, 0)$ and $(0, 1)$. Let K_1 be the cone defined as the nonnegative real span of $(1, 0)$ and $(1, k)$. For $k = 2^{100}$, say, we see that for all practical purposes, K_1 is very close to K_0 in its geometry, in the sense that their angles are almost the same for computational purposes, and thus their continuous Brion generating functions are almost the same, computationally.

However, $\sigma_{K_0}(z)$ is quite far from $\sigma_{K_1}(z)$, since the latter has 2^{100} terms in its numerator, while the former has only 1 as its trivial numerator. Thus, tangent cones that are "arbitrarily close" geometrically may simultaneously be "arbitrarily far" from each other in the discrete sense dictated by the integer points in their fundamental domains.

12.3 Polytopes Have Their Moments

The most common notion for moments of a set $\mathcal{P} \subset \mathbb{R}^d$ is the collection of **monomial moments** defined by the real numbers

$$\mu_\mathbf{a} := \int_{\mathcal{P}} \mathbf{y}^\mathbf{a} \, d\mathbf{y} = \int_{\mathcal{P}} y_1^{a_1} y_2^{a_2} \cdots y_d^{a_d} \, d\mathbf{y}, \qquad (12.8)$$

for a fixed vector $\mathbf{a} = (a_1, a_2, \ldots, a_d) \in \mathbb{C}^d$. For example, when $\mathbf{a} = \mathbf{0} = (0, 0, \ldots, 0)$, we get $\mu_\mathbf{0} = \text{vol} \, \mathcal{P}$. As an application of moments, consider the problem of finding the center of mass of $\mathcal{P}$, which is defined by

$$\frac{1}{\text{vol}\,\mathcal{P}}\left(\int_{\mathcal{P}} y_1\,d\mathbf{y}\,,\,\int_{\mathcal{P}} y_2\,d\mathbf{y}\,,\,\ldots,\,\int_{\mathcal{P}} y_d\,d\mathbf{y}\right).$$

This integral is equal to

$$\frac{1}{\mu_0}\left(\mu_{(1,0,0,\ldots,0)},\mu_{(0,1,0,\ldots,0)},\cdots,\mu_{(0,\ldots,0,1)}\right).$$

Similarly, one can define the variance of $\mathcal{P}$ and other statistical data attached to $\mathcal{P}$ and use moments to compute them.

Our next task is to relate the monomial moments $\mu_{\mathbf{a}}$ to the formula of Theorem 12.3. One way to do this is to differentiate the continuous Brion identity in Theorem 12.3 repeatedly with respect to each z_j, and in fact, interchanging these derivatives with the integral is valid because we are integrating over a compact region $\mathcal{P}$. Thus, for a simple rational polytope $\mathcal{P}$,

$$\int_{\mathcal{P}} x_1^{a_1} x_2^{a_2} \cdots x_d^{a_d} \exp(\mathbf{x}\cdot\mathbf{z})\,d\mathbf{x}$$

$$= \int_{\mathcal{P}} \left(\frac{\partial}{\partial z_1}\right)^{a_1}\cdots\left(\frac{\partial}{\partial z_d}\right)^{a_d} \exp(\mathbf{x}\cdot\mathbf{z})\,d\mathbf{x}$$

$$= (-1)^d \sum_{\mathbf{v}\text{ a vertex of }\mathcal{P}} \left(\frac{\partial}{\partial z_1}\right)^{a_1}\cdots\left(\frac{\partial}{\partial z_d}\right)^{a_d} \frac{\exp\left(\mathbf{v}\cdot\mathbf{z}\right)\left|\det\left(\mathbf{w}_1(\mathbf{v}),\ldots,\mathbf{w}_d(\mathbf{v})\right)\right|}{\prod_{k=1}^{d}\left(\mathbf{w}_k(\mathbf{v})\cdot\mathbf{z}\right)}.$$

If we take the limit as each $z_j \to 0$ of both sides of the latter identity, we have a formula (albeit a messy one) for the monomial moments:

$$(-1)^d \mu_{\mathbf{a}} =$$

$$\lim_{\mathbf{z}\to 0} \sum_{\mathbf{v}\text{ a vertex of }\mathcal{P}} \left(\frac{\partial}{\partial z_1}\right)^{a_1}\cdots\left(\frac{\partial}{\partial z_d}\right)^{a_d} \frac{\exp\left(\mathbf{v}\cdot\mathbf{z}\right)\left|\det\left(\mathbf{w}_1(\mathbf{v}),\ldots,\mathbf{w}_d(\mathbf{v})\right)\right|}{\prod_{k=1}^{d}\left(\mathbf{w}_k(\mathbf{v})\cdot\mathbf{z}\right)}.$$

This formula for the monomial moments may be used effectively for relatively small integer values of the exponent $\mathbf{a} = (a_1, a_2, \ldots, a_d)$. Going a step further, we can use Theorem 12.3 to obtain information about a different set of moments, which are sometimes called *axial moments*. Along the way, we stumble upon the following amazing formula for the continuous volume of a polytope.

Theorem 12.4. *Suppose $\mathcal{P}$ is a simple rational convex d-polytope. For each vertex cone $\mathcal{K}_{\mathbf{v}}$ of $\mathcal{P}$, fix a set of generators $\mathbf{w}_1(\mathbf{v}), \mathbf{w}_2(\mathbf{v}), \ldots, \mathbf{w}_d(\mathbf{v}) \in \mathbb{Z}^d$. Then*

$$\text{vol}\,\mathcal{P} = \frac{(-1)^d}{d!} \sum_{\mathbf{v}\text{ a vertex of }\mathcal{P}} \frac{(\mathbf{v}\cdot\mathbf{z})^d\left|\det\left(\mathbf{w}_1(\mathbf{v}),\ldots,\mathbf{w}_d(\mathbf{v})\right)\right|}{\prod_{k=1}^{d}\left(\mathbf{w}_k(\mathbf{v})\cdot\mathbf{z}\right)}$$

for all $\mathbf{z}$ such that the denominators on the right-hand side do not vanish. More generally, for every integer $j \geq 0$,

$$\int_{\mathcal{P}} (\mathbf{x} \cdot \mathbf{z})^j \, d\mathbf{x} = \frac{(-1)^d j!}{(j+d)!} \sum_{\mathbf{v} \text{ a vertex of } \mathcal{P}} \frac{(\mathbf{v} \cdot \mathbf{z})^{j+d} \left| \det (\mathbf{w}_1(\mathbf{v}), \ldots, \mathbf{w}_d(\mathbf{v})) \right|}{\prod_{k=1}^d (\mathbf{w}_k(\mathbf{v}) \cdot \mathbf{z})}.$$

Proof. We replace the variable $\mathbf{z}$ in the identity of Theorem 12.3 by $s\mathbf{z}$, where s is a scalar:

$$\int_{\mathcal{P}} \exp (\mathbf{x} \cdot (s\mathbf{z})) \, d\mathbf{x} = (-1)^d \sum_{\mathbf{v} \text{ a vertex of } \mathcal{P}} \frac{\exp (\mathbf{v} \cdot (s\mathbf{z})) \left| \det (\mathbf{w}_1(\mathbf{v}), \ldots, \mathbf{w}_d(\mathbf{v})) \right|}{\prod_{k=1}^d (\mathbf{w}_k(\mathbf{v}) \cdot (s\mathbf{z}))},$$

which can be rewritten as

$$\int_{\mathcal{P}} \exp (s (\mathbf{x} \cdot \mathbf{z})) \, d\mathbf{x} = (-1)^d \sum_{\mathbf{v} \text{ a vertex of } \mathcal{P}} \frac{\exp (s (\mathbf{v} \cdot \mathbf{z})) \left| \det (\mathbf{w}_1(\mathbf{v}), \ldots, \mathbf{w}_d(\mathbf{v})) \right|}{s^d \prod_{k=1}^d (\mathbf{w}_k(\mathbf{v}) \cdot (\mathbf{z}))}.$$

The general statement of the theorem follows now by first expanding the exponential functions as Taylor series in s, and then comparing coefficients on both sides:

$$\sum_{j \geq 0} \int_{\mathcal{P}} (\mathbf{x} \cdot \mathbf{z})^j \, d\mathbf{x} \, \frac{s^j}{j!}$$

$$= (-1)^d \sum_{\mathbf{v} \text{ a vertex of } \mathcal{P}} \sum_{j \geq 0} (\mathbf{v} \cdot \mathbf{z})^j \, \frac{s^{j-d}}{j!} \, \frac{\left| \det (\mathbf{w}_1(\mathbf{v}), \ldots, \mathbf{w}_d(\mathbf{v})) \right|}{\prod_{k=1}^d (\mathbf{w}_k(\mathbf{v}) \cdot (\mathbf{z}))}$$

$$= \sum_{j \geq -d} (-1)^d \sum_{\mathbf{v} \text{ a vertex of } \mathcal{P}} \frac{(\mathbf{v} \cdot \mathbf{z})^{j+d} \left| \det (\mathbf{w}_1(\mathbf{v}), \ldots, \mathbf{w}_d(\mathbf{v})) \right|}{\prod_{k=1}^d (\mathbf{w}_k(\mathbf{v}) \cdot (\mathbf{z}))} \, \frac{s^j}{(j+d)!}. \qquad \square$$

The proof of this theorem reveals yet more identities between rational functions. Namely, the coefficients of the negative powers of s in the last line of the proof have to be zero. This immediately yields the following curious set of d identities for simple d-polytopes:

Corollary 12.5. *Suppose $\mathcal{P}$ is a simple rational convex d-polytope. For each vertex cone $\mathcal{K}_{\mathbf{v}}$ of $\mathcal{P}$, fix a set of generators $\mathbf{w}_1(\mathbf{v}), \mathbf{w}_2(\mathbf{v}), \ldots, \mathbf{w}_d(\mathbf{v}) \in \mathbb{Z}^d$. Then for each $0 \leq j \leq d - 1$,*

$$\sum_{\mathbf{v} \text{ a vertex of } \mathcal{P}} \frac{(\mathbf{v} \cdot \mathbf{z})^j \left| \det (\mathbf{w}_1(\mathbf{v}), \ldots, \mathbf{w}_d(\mathbf{v})) \right|}{\prod_{k=1}^d (\mathbf{w}_k(\mathbf{v}) \cdot (\mathbf{z}))} = 0. \qquad \square$$

12.4 Computing the Discrete Continuously

In this section, we apply the Todd operator to a perturbation of the continuous volume. Namely, consider a simple full-dimensional polytope $\mathcal{P}$, which we may write as

$$\mathcal{P} = \left\{ \mathbf{x} \in \mathbb{R}^d : \mathbf{A}\,\mathbf{x} \le \mathbf{b} \right\}.$$

Then we define the perturbed polytope

$$\mathcal{P}(\mathbf{h}) := \left\{ \mathbf{x} \in \mathbb{R}^d : \mathbf{A}\,\mathbf{x} \le \mathbf{b} + \mathbf{h} \right\}$$

for a small vector $\mathbf{h} \in \mathbb{R}^m$ (we will quantify the word *small* in a moment). A famous theorem due to Askold Khovanskiĭ and Aleksandr Pukhlikov says that the integer-point count in $\mathcal{P}$ can be obtained by applying the Todd operator to $\operatorname{vol}(\mathcal{P}(\mathbf{h}))$. Here we prove the theorem for a certain class of polytopes, which we need to define first.

We call a rational pointed d-cone **unimodular** if its generators are a basis of $\mathbb{Z}^d$. An integral polytope is **unimodular** if each of its vertex cones is unimodular.[1]

Theorem 12.6 (Khovanskiĭ–Pukhlikov theorem). *For a unimodular d-polytope $\mathcal{P}$,*

$$\#\left(\mathcal{P} \cap \mathbb{Z}^d\right) = \operatorname{Todd}_{\mathbf{h}} \operatorname{vol}\left(\mathcal{P}(\mathbf{h})\right)\big|_{\mathbf{h}=0}.$$

More generally,

$$\sigma_{\mathcal{P}}(\exp \mathbf{z}) = \operatorname{Todd}_{\mathbf{h}} \int_{\mathcal{P}(\mathbf{h})} \exp(\mathbf{x} \cdot \mathbf{z})\, d\mathbf{x} \bigg|_{\mathbf{h}=0}.$$

Proof. We use Theorem 12.3, the continuous version of Brion's theorem; note that if $\mathcal{P}$ is unimodular, then $\mathcal{P}$ is automatically simple. For each vertex cone $\mathcal{K}_{\mathbf{v}}$ of $\mathcal{P}$, denote its generators by $\mathbf{w}_1(\mathbf{v}), \mathbf{w}_2(\mathbf{v}), \ldots, \mathbf{w}_d(\mathbf{v}) \in \mathbb{Z}^d$. Then Theorem 12.3 states that

$$\int_{\mathcal{P}} \exp(\mathbf{x} \cdot \mathbf{z})\, d\mathbf{x} = (-1)^d \sum_{\mathbf{v} \text{ a vertex of } \mathcal{P}} \frac{\exp(\mathbf{v} \cdot \mathbf{z}) \left| \det(\mathbf{w}_1(\mathbf{v}), \ldots, \mathbf{w}_d(\mathbf{v})) \right|}{\prod_{k=1}^d (\mathbf{w}_k(\mathbf{v}) \cdot \mathbf{z})}$$

$$= (-1)^d \sum_{\mathbf{v} \text{ a vertex of } \mathcal{P}} \frac{\exp(\mathbf{v} \cdot \mathbf{z})}{\prod_{k=1}^d (\mathbf{w}_k(\mathbf{v}) \cdot \mathbf{z})}, \qquad (12.9)$$

where the last identity follows from Exercise 12.3. A similar formula holds for $\mathcal{P}(\mathbf{h})$, except that we have to account for the shift of the vertices. The vector $\mathbf{h}$ shifts the facet-defining hyperplanes. This shift of the facets induces a shift of the vertices; let's say that the vertex $\mathbf{v}$ gets moved along each edge direction $\mathbf{w}_k$ (the vectors that generate the vertex cone $\mathcal{K}_{\mathbf{v}}$) by $h_k(\mathbf{v})$, so that

---

[1] Unimodular polytopes go by two additional names, namely **smooth** and **Delzant**.

$\mathcal{P}(\mathbf{h})$ has now the vertex $\mathbf{v} - \sum_{k=1}^{d} h_k(\mathbf{v}) \mathbf{w}_k(\mathbf{v})$. If $\mathbf{h}$ is small enough, $\mathcal{P}(\mathbf{h})$ will still be simple,[2] and we can apply Theorem 12.3 to $\mathcal{P}(\mathbf{h})$:

$$\int_{\mathcal{P}(\mathbf{h})} \exp(\mathbf{x} \cdot \mathbf{z}) \, d\mathbf{x} = (-1)^d \sum_{\mathbf{v} \text{ a vertex of } \mathcal{P}} \frac{\exp\left(\left(\mathbf{v} - \sum_{k=1}^{d} h_k(\mathbf{v}) \mathbf{w}_k(\mathbf{v})\right) \cdot \mathbf{z}\right)}{\prod_{k=1}^{d} \left(\mathbf{w}_k(\mathbf{v}) \cdot \mathbf{z}\right)}$$

$$= (-1)^d \sum_{\mathbf{v} \text{ a vertex of } \mathcal{P}} \frac{\exp\left(\mathbf{v} \cdot \mathbf{z} - \sum_{k=1}^{d} h_k(\mathbf{v}) \mathbf{w}_k(\mathbf{v}) \cdot \mathbf{z}\right)}{\prod_{k=1}^{d} \left(\mathbf{w}_k(\mathbf{v}) \cdot \mathbf{z}\right)}$$

$$= (-1)^d \sum_{\mathbf{v} \text{ a vertex of } \mathcal{P}} \frac{\exp(\mathbf{v} \cdot \mathbf{z}) \prod_{k=1}^{d} \exp\left(-h_k(\mathbf{v}) \mathbf{w}_k(\mathbf{v}) \cdot \mathbf{z}\right)}{\prod_{k=1}^{d} \left(\mathbf{w}_k(\mathbf{v}) \cdot \mathbf{z}\right)}.$$

Strictly speaking, this formula holds only for $\mathbf{h} \in \mathbb{Q}^m$, so that the vertices of $\mathcal{P}(\mathbf{h})$ are rational. Since we will eventually set $\mathbf{h} = 0$, this is a harmless restriction. Now we apply the Todd operator:

$$\text{Todd}_{\mathbf{h}} \int_{\mathcal{P}(\mathbf{h})} \exp(\mathbf{x} \cdot \mathbf{z}) \, d\mathbf{x} \bigg|_{\mathbf{h}=0}$$

$$= (-1)^d \sum_{\mathbf{v} \text{ vertex of } \mathcal{P}} \text{Todd}_{\mathbf{h}} \frac{\exp(\mathbf{v} \cdot \mathbf{z}) \prod_{k=1}^{d} \exp\left(-h_k(\mathbf{v}) \mathbf{w}_k(\mathbf{v}) \cdot \mathbf{z}\right)}{\prod_{k=1}^{d} \left(\mathbf{w}_k(\mathbf{v}) \cdot \mathbf{z}\right)} \bigg|_{\mathbf{h}=0}$$

$$= (-1)^d \sum_{\mathbf{v} \text{ vertex of } \mathcal{P}} \frac{\exp(\mathbf{v} \cdot \mathbf{z})}{\prod_{k=1}^{d} \left(\mathbf{w}_k(\mathbf{v}) \cdot \mathbf{z}\right)}$$

$$\times \prod_{k=1}^{d} \text{Todd}_{h_k(\mathbf{v})} \exp\left(-h_k(\mathbf{v}) \mathbf{w}_k(\mathbf{v}) \cdot \mathbf{z}\right) \bigg|_{h_k(\mathbf{v})=0}.$$

By a multivariate version of Lemma 12.1,

$$\text{Todd}_{\mathbf{h}} \int_{\mathcal{P}(\mathbf{h})} \exp(\mathbf{x} \cdot \mathbf{z}) \, d\mathbf{x} \bigg|_{\mathbf{h}=0}$$

$$= (-1)^d \sum_{\mathbf{v} \text{ vertex of } \mathcal{P}} \frac{\exp\left(\mathbf{v} \cdot \mathbf{z}\right)}{\prod_{k=1}^{d} \left(\mathbf{w}_k(\mathbf{v}) \cdot \mathbf{z}\right)} \prod_{k=1}^{d} \frac{-\mathbf{w}_k(\mathbf{v}) \cdot \mathbf{z}}{1 - \exp(\mathbf{w}_k(\mathbf{v}) \cdot \mathbf{z})}$$

$$= \sum_{\mathbf{v} \text{ vertex of } \mathcal{P}} \exp\left(\mathbf{v} \cdot \mathbf{z}\right) \prod_{k=1}^{d} \frac{1}{1 - \exp(\mathbf{w}_k(\mathbf{v}) \cdot \mathbf{z})}.$$

However, Brion's theorem (Theorem 11.7), together with the fact that $\mathcal{P}$ is unimodular, says that the right-hand side of this last formula is precisely the integer-point transform of $\mathcal{P}$ (see also (12.9)), and thus

---

[2] The cautious reader may consult [258, p. 66] to confirm this fact.

$$\text{Todd}_{\mathbf{h}} \int_{\mathcal{P}(\mathbf{h})} \exp(\mathbf{x} \cdot \mathbf{z}) \, d\mathbf{x} \bigg|_{\mathbf{h}=0} = \sigma_{\mathcal{P}}(\exp \mathbf{z}).$$

Finally, setting $\mathbf{z} = 0$ gives

$$\text{Todd}_{\mathbf{h}} \int_{\mathcal{P}(\mathbf{h})} d\mathbf{x} \bigg|_{\mathbf{h}=0} = \sum_{\mathbf{m} \in \mathcal{P} \cap \mathbb{Z}^d} 1,$$

as claimed.   □

We note that $\int_{\mathcal{P}(\mathbf{h})} \exp(\mathbf{x} \cdot \mathbf{z}) \, d\mathbf{x}$ is, by definition, the continuous Fourier–Laplace transform of $\mathcal{P}(\mathbf{h})$. Upon being acted on by the discretizing operator $\text{Todd}_{\mathbf{h}}$, the integral $\int_{\mathcal{P}(\mathbf{h})} \exp(\mathbf{x} \cdot \mathbf{z}) \, d\mathbf{x}$ gives us the discrete integer-point transform $\sigma_{\mathcal{P}}(\mathbf{z})$.

Notes

1. The classical Euler–Maclaurin formula states that

$$\sum_{k=1}^{n} f(k) = \int_0^n f(x) \, dx + \frac{f(0) + f(n)}{2} + \sum_{m=1}^{p} \frac{B_{2m}}{(2m)!} \left[f^{(2m-1)}(x) \right]_0^n$$
$$+ \frac{1}{(2p+1)!} \int_0^n B_{2p+1}(\{x\}) \, f^{(2p+1)}(x) \, dx,$$

where $B_k(x)$ denotes the kth Bernoulli polynomial. It was discovered independently by Leonhard Euler and Colin Maclaurin (1698–1746).[3] This formula provides an explicit error term, whereas Theorem 12.2 provides a summation formula with no error term.

2. The Todd operator was introduced by Friedrich Hirzebruch in the 1950s [138], following a more complicated definition by John A. Todd [243, 244] some twenty years earlier. The Khovanskiĭ–Pukhlikov theorem (Theorem 12.6) can be interpreted as a combinatorial analogue of the algebrogeometric Hirzebruch–Riemann–Roch theorem, in which the Todd operator plays a prominent role.

3. Theorem 12.3, the continuous form of Brion's theorem, was generalized by Alexander Barvinok to *every* polytope [19]. In fact, [19] contains a certain extension of Brion's theorem to irrational polytopes as well. The decomposition formula for moments of a polytope in Theorem 12.4 is due to Michel Brion and Michèle Vergne [67].

---

[3] For more information about Maclaurin, see
http://www-history.mcs.st-andrews.ac.uk/Biographies/Maclaurin.html.

4. A natural question arises regarding moments of a given polytope, namely the following inverse problem: how many axial moments of a polytope $\mathcal{P}$ do we need to know in order to reconstruct the vertices of $\mathcal{P}$? A solution was given in [122] recently, and it turns out to be at most $O(dN)$, where $d = \dim(\mathcal{P})$ and N is the number of vertices of P. Practical implementations of this algorithm, which find applications in computer graphics and computer vision, are still being developed. A nice resource for related moment problems is the book [158].

5. Theorem 12.6 was first proved in 1992 by Askold Khovanskiĭ and Aleksandr Pukhlikov [147]. The proof we give here is essentially theirs. Their paper [147] also draws parallels between toric varieties and lattice polytopes. Subsequently, many attempts to provide formulas for Ehrhart quasipolynomials—some based on Theorem 12.6—have provided fertile ground for deeper connections and future work; a long but by no means complete list of references is $[15, 16, 50, 67, 79, 85, 86, 106, 127, 145, 146, 159, 168, 182, 193, 240]$.

Exercises

12.1. ♣ Show that $\mathrm{Todd}_h\, h = h + \frac{1}{2}$. More generally, prove that

$$\mathrm{Todd}_h\, h^k = B_k(h+1)$$

for $k \geq 1$, where $B_k(x)$ denotes the kth Bernoulli polynomial. Thus, the Todd operators take the usual monomial basis to the basis of Bernoulli polynomials.

12.2. ♣ Prove (12.5): Suppose $\mathcal{P}$ is a simple integral d-polytope. For each vertex cone $\mathcal{K}_\mathbf{v}$ of $\mathcal{P}$, denote its generators by $\mathbf{w}_1(\mathbf{v}), \mathbf{w}_2(\mathbf{v}), \ldots, \mathbf{w}_d(\mathbf{v}) \in \mathbb{Z}^d$ and its fundamental parallelepiped by $\Pi_\mathbf{v}$. Then

$$\sum_{\mathbf{m} \in \mathcal{P} \cap (\frac{1}{n}\mathbb{Z})^d} \exp(\mathbf{m} \cdot \mathbf{z}) = \sum_{\mathbf{v}\ \text{a vertex of}\ \mathcal{P}} \frac{\exp(\mathbf{v} \cdot \mathbf{z}) \sum_{\mathbf{m} \in \Pi_\mathbf{v} \cap \mathbb{Z}^d} \exp\left(\frac{\mathbf{m}}{n} \cdot \mathbf{z}\right)}{\prod_{k=1}^{d}\left(1 - \exp\left(\frac{\mathbf{w}_k(\mathbf{v})}{n} \cdot \mathbf{z}\right)\right)}.$$

12.3. ♣ Given a unimodular cone

$$\mathcal{K} = \{\mathbf{v} + \lambda_1 \mathbf{w}_1 + \lambda_2 \mathbf{w}_2 + \cdots + \lambda_d \mathbf{w}_d : \lambda_1, \lambda_2, \ldots, \lambda_d \geq 0\},$$

where $\mathbf{v}, \mathbf{w}_1, \mathbf{w}_2, \ldots, \mathbf{w}_d \in \mathbb{Z}^d$ such that $\mathbf{w}_1, \mathbf{w}_2, \ldots, \mathbf{w}_d$ are a basis for $\mathbb{Z}^d$, show that

$$\sigma_\mathcal{K}(\mathbf{z}) = \frac{\mathbf{z}^\mathbf{v}}{\prod_{k=1}^{d}\left(1 - \mathbf{z}^{\mathbf{w}_k}\right)}$$

and $|\det(\mathbf{w}_1, \ldots, \mathbf{w}_d)| = 1$.

12.4. Compute the integral (Fourier–Laplace transform) given by the formula of Theorem 12.4 for the standard simplex in $\mathbb{R}^d$,

$$\Delta = \left\{ (x_1, x_2 \ldots, x_d) \in \mathbb{R}^d : x_1 + x_2 + \cdots + x_d \leq 1 \text{ and all } x_k \geq 0 \right\}.$$

12.5. Using the previous exercise, compute the center of mass of the standard simplex in $\mathbb{R}^d$.

12.6. ♣ Prove (12.7). That is, for the simplicial cone

$$\mathcal{K} = \left\{ \mathbf{v} + \sum_{k=1}^d \lambda_k \mathbf{w}_k : \lambda_k \geq 0 \right\}$$

with $\mathbf{v}, \mathbf{w}_1, \mathbf{w}_2, \ldots, \mathbf{w}_d \in \mathbb{Q}^d$, show that

$$\int_{\mathcal{K}} \exp(\mathbf{x} \cdot \mathbf{z}) \, d\mathbf{x} = (-1)^d \frac{\exp(\mathbf{v} \cdot \mathbf{z}) \left| \det(\mathbf{w}_1(\mathbf{v}), \ldots, \mathbf{w}_d(\mathbf{v})) \right|}{\prod_{k=1}^d (\mathbf{w}_k(\mathbf{v}) \cdot \mathbf{z})}.$$

12.7. Show that in the statement of Theorem 12.3, the expression

$$\frac{\left| \det(\mathbf{w}_1(\mathbf{v}), \ldots, \mathbf{w}_d(\mathbf{v})) \right|}{\prod_{k=1}^d (\mathbf{w}_k(\mathbf{v}) \cdot \mathbf{z})}$$

remains invariant on scaling each $\mathbf{w}_k(\mathbf{v})$ by an independent positive integer.

12.8. The **binomial transform** of a given sequence $(a_n)_{n=0}^\infty$ is the sequence $b_n := \sum_{k=0}^n (-1)^k \binom{n}{k} a_k$. Prove that the Bernoulli numbers B_n are self-dual with respect to this transform:

$$B_n = \sum_{k=0}^n (-1)^k \binom{n}{k} B_k.$$

12.9. Prove that if $\mathcal{P}$ is an integral d-polytope and $f(x_1, x_2, \ldots, x_d)$ is a polynomial in d variables, then the following expression is a polynomial in t:

$$L_{\mathcal{P}, f}(t) := \sum_{\mathbf{m} \in t\mathcal{P} \cap \mathbb{Z}^d} f(\mathbf{m}).$$

Open Problems

12.10. Find all differentiable eigenfunctions of the Todd operator.

12.11. (a) Classify all polytopes whose discrete and continuous volumes coincide, that is, $L_{\mathcal{P}}(1) = \text{vol}(\mathcal{P})$.
(b) Classify all (not necessarily closed) polytopes for which $L_{\mathcal{P}}(t) = \text{vol}(\mathcal{P}) \, t^d$.

Chapter 13
Solid Angles

Everything you've learned in school as "obvious" becomes less and less obvious as you begin to study the universe. For example, there are no solids in the universe. There's not even a suggestion of a solid. There are no absolute continuums. There are no surfaces. There are no straight lines.

Buckminster Fuller (1895–1983)

The natural generalization of a 2-dimensional angle to higher dimensions is called a *solid angle*. Given a pointed cone $\mathcal{K} \subset \mathbb{R}^d$, the solid angle at its apex is the proportion of space that the cone $\mathcal{K}$ occupies. In slightly different words, if we pick a point $\mathbf{x} \in \mathbb{R}^d$ "at random," then the probability that $\mathbf{x} \in \mathcal{K}$ is precisely the solid angle at the apex of $\mathcal{K}$. Yet another view of solid angles is that they are in fact volumes of spherical polytopes: the region of intersection of a cone with a sphere. There is a theory here that parallels the Ehrhart theory of Chapters 3 and 4, but which has some genuinely new ideas.

13.1 A New Discrete Volume Using Solid Angles

Suppose $\mathcal{P} \subset \mathbb{R}^d$ is a convex rational d-polyhedron. The **solid angle** $\omega_\mathcal{P}(\mathbf{x})$ of a point $\mathbf{x}$ (with respect to $\mathcal{P}$) is a real number equal to the proportion of a small ball centered at $\mathbf{x}$ that is contained in $\mathcal{P}$. That is, we let $B_\epsilon(\mathbf{x})$ denote the ball of radius ϵ centered at $\mathbf{x}$ and define

$$\omega_\mathcal{P}(\mathbf{x}) := \frac{\mathrm{vol}\,(B_\epsilon(\mathbf{x}) \cap \mathcal{P})}{\mathrm{vol}\,B_\epsilon(\mathbf{x})}$$

for all positive ϵ sufficiently small; this notion is depicted in Figure 13.1. We note that when $\mathbf{x} \notin \mathcal{P}$, $\omega_\mathcal{P}(\mathbf{x}) = 0$; when $\mathbf{x} \in \mathcal{P}^\circ$, $\omega_\mathcal{P}(\mathbf{x}) = 1$; when $\mathbf{x} \in \partial\mathcal{P}$,

$0 < \omega_{\mathcal{P}}(\mathbf{x}) < 1$. The **solid angle of a face** $\mathcal{F}$ of $\mathcal{P}$ is defined by choosing a point $\mathbf{x}$ in the relative interior $\mathcal{F}^{\circ}$ and setting $\omega_{\mathcal{P}}(\mathcal{F}) = \omega_{\mathcal{P}}(\mathbf{x})$.

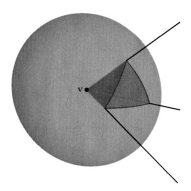

Fig. 13.1 A solid angle.

Example 13.1. We compute the solid angles of the faces belonging to the standard 3-simplex $\Delta = \operatorname{conv}\{(0,0,0),(1,0,0),(0,1,0),(0,0,1)\}$. As we just mentioned, a point interior to Δ has solid angle 1. Every facet has solid angle $\frac{1}{2}$ (and this remains true for every polytope).

The story gets interesting with the edges: here we are computing dihedral angles. The **dihedral angle** of a 1-dimensional edge is defined by the angle between the outward-pointing normal to one of its defining facets and the inward-pointing normal to its other defining facet.

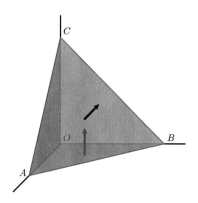

Fig. 13.2 The simplex Δ from Example 13.1.

Each of the edges OA, OB, and OC in Figure 13.2 has the same solid angle $\frac{1}{4}$. Turning to the edge AB, we compute the angle between its defining facets as follows:

$$\frac{1}{2\pi}\cos^{-1}\left(\frac{1}{\sqrt{3}}(-1,-1,-1)\cdot(0,0,-1)\right)=\frac{1}{2\pi}\cos^{-1}\left(\frac{1}{\sqrt{3}}\right).$$

The edges AC and BC have the same solid angle by symmetry.

Finally, we compute the solid angle of the vertices: the origin has solid angle $\frac{1}{8}$, and the other three vertices all have the same solid angle ω. With Corollary 13.9 below (the *Brianchon–Gram relation*), we can compute this angle via

$$0=\sum_{\mathcal{F}\subseteq\mathcal{P}}(-1)^{\dim\mathcal{F}}\omega_{\mathcal{P}}(\mathcal{F})=-1+4\cdot\frac{1}{2}-3\cdot\frac{1}{4}-3\cdot\frac{1}{2\pi}\cos^{-1}\left(\frac{1}{\sqrt{3}}\right)+\frac{1}{8}+3\cdot\omega,$$

which gives $\omega=\frac{1}{2\pi}\cos^{-1}\left(\frac{1}{\sqrt{3}}\right)-\frac{1}{8}$.   □

We now introduce another measure of discrete volume; namely, we let

$$A_{\mathcal{P}}(t):=\sum_{\mathbf{m}\in t\mathcal{P}\cap\mathbb{Z}^{d}}\omega_{t\mathcal{P}}(\mathbf{m}),$$

the sum of the solid angles at all integer points in $t\mathcal{P}$; recalling that $\omega_{\mathcal{P}}(\mathbf{x})=0$ if $\mathbf{x}\notin\mathcal{P}$, we can also write

$$A_{\mathcal{P}}(t)=\sum_{\mathbf{m}\in\mathbb{Z}^{d}}\omega_{t\mathcal{P}}(\mathbf{m}).$$

This new discrete volume measure differs in a substantial way from the Ehrhart counting function $L_{\mathcal{P}}(t)$. Namely, suppose $\mathcal{P}$ is a d-polytope that can be written as the union of the polytopes $\mathcal{P}_1$ and $\mathcal{P}_2$ such that $\dim(\mathcal{P}_1\cap\mathcal{P}_2)<d$, that is, $\mathcal{P}_1$ and $\mathcal{P}_2$ are glued along a lower-dimensional subset. Then at each lattice point $\mathbf{m}\in\mathbb{Z}^{d}$, we have $\omega_{\mathcal{P}_1}(\mathbf{m})+\omega_{\mathcal{P}_2}(\mathbf{m})=\omega_{\mathcal{P}}(\mathbf{m})$, and so the function $A_{\mathcal{P}}$ has an additive property:

$$A_{\mathcal{P}}(t)=A_{\mathcal{P}_1}(t)+A_{\mathcal{P}_2}(t).\tag{13.1}$$

In contrast, the Ehrhart counting functions satisfy

$$L_{\mathcal{P}}(t)=L_{\mathcal{P}_1}(t)+L_{\mathcal{P}_2}(t)-L_{\mathcal{P}_1\cap\mathcal{P}_2}(t).$$

On the other hand, we can transfer computational effort from the Ehrhart counting functions to the solid-angle sum and vice versa, with the use of the following lemma.

Lemma 13.2. *Let $\mathcal{P}$ be a polytope. Then*

$$A_{\mathcal{P}}(t)=\sum_{\mathcal{F}\subseteq\mathcal{P}}\omega_{\mathcal{P}}(\mathcal{F})\,L_{\mathcal{F}^{\circ}}(t).$$

Proof. The dilated polytope $t\mathcal{P}$ is the disjoint union of its relative open faces $t\mathcal{F}^\circ$ (Exercise 5.3), so that we can write

$$A_\mathcal{P}(t) = \sum_{\mathbf{m}\in\mathbb{Z}^d} \omega_{t\mathcal{P}}(\mathbf{m}) = \sum_{\mathcal{F}\subseteq\mathcal{P}} \sum_{\mathbf{m}\in\mathbb{Z}^d} \omega_{t\mathcal{P}}(\mathbf{m}) \, 1_{t\mathcal{F}^\circ}(\mathbf{m}).$$

But $\omega_{t\mathcal{P}}(\mathbf{m})$ is constant on each relatively open face $t\mathcal{F}^\circ$, and we called this constant $\omega_\mathcal{P}(\mathcal{F})$, whence

$$A_\mathcal{P}(t) = \sum_{\mathcal{F}\subseteq\mathcal{P}} \omega_\mathcal{P}(\mathcal{F}) \sum_{\mathbf{m}\in\mathbb{Z}^d} 1_{t\mathcal{F}^\circ}(\mathbf{m}) = \sum_{\mathcal{F}\subseteq\mathcal{P}} \omega_\mathcal{P}(\mathcal{F}) \, L_{\mathcal{F}^\circ}(t). \qquad \square$$

Thus $A_\mathcal{P}(t)$ is a polynomial (respectively quasipolynomial) in t for an integral (respectively rational) polytope $\mathcal{P}$. We claim that Lemma 13.2 is in fact useful in practice. To drive the point home, we illustrate this identity by computing the solid-angle sum over all integer points of Δ in Example 13.1.

Example 13.3. We continue the solid-angle computation for the 3-simplex $\Delta = \mathrm{conv}\,\{(0,0,0),(1,0,0),(0,1,0),(0,0,1)\}$. We recall from Section 2.3 that $L_{\Delta^\circ} = \binom{t-1}{3}$. The facets of Δ are three standard triangles and one triangle that appeared in the context of the Frobenius problem. All four facets have the same interior Ehrhart polynomial $\binom{t-1}{2}$. A similar phenomenon holds for the edges of Δ: all six of them have the same interior Ehrhart polynomial $t-1$. These polynomials add up, by Lemma 13.2 and Example 13.1, to the solid-angle sum

$$A_\Delta(t) = \binom{t-1}{3} + 4\cdot\frac{1}{2}\binom{t-1}{2} + \left(3\cdot\frac{1}{4} + 3\cdot\frac{1}{2\pi}\cos^{-1}\left(\frac{1}{\sqrt{3}}\right)\right)(t-1)$$

$$+ \frac{1}{8} + 3\cdot\left(\frac{1}{2\pi}\cos^{-1}\left(\frac{1}{\sqrt{3}}\right) - \frac{1}{8}\right)$$

$$= \frac{1}{6}t^3 + \left(\frac{3}{2\pi}\cos^{-1}\left(\frac{1}{\sqrt{3}}\right) - \frac{5}{12}\right)t.$$

The magic cancellation of the even powers of this polynomial is not a coincidence, as we will discover in Theorem 13.7. The curious reader may notice that the coefficient of t in this example is not a rational number, in stark contrast with Ehrhart polynomials. $\qquad \square$

The analogue of Ehrhart's theorem (Theorem 3.23) in the world of solid angles is as follows.

Theorem 13.4 (Macdonald's theorem). *Suppose $\mathcal{P}$ is a rational convex d-polytope. Then $A_\mathcal{P}$ is a quasipolynomial of degree d whose leading coefficient is $\mathrm{vol}\,\mathcal{P}$ and whose period divides the denominator of $\mathcal{P}$.*

Proof. The denominator of a face $\mathcal{F} \subset \mathcal{P}$ divides the denominator of $\mathcal{P}$, and hence so does the period of $L_\mathcal{F}$, by Ehrhart's theorem (Theorem 3.23). By

Lemma 13.2, $A_{\mathcal{P}}$ is a quasipolynomial with period dividing the denominator of $\mathcal{P}$. The leading term of $A_{\mathcal{P}}$ equals the leading term of $L_{\mathcal{P}^\circ}$, which is $\operatorname{vol}\mathcal{P}$, by Corollary 3.20 and its extension in Exercise 3.34. □

13.2 Solid-Angle Generating Functions and a Brion-Type Theorem

By analogy with the integer-point transform of a polyhedron $\mathcal{P} \subseteq \mathbb{R}^d$, which lists all lattice points in $\mathcal{P}$, we form the **solid-angle generating function**

$$\alpha_{\mathcal{P}}(\mathbf{z}) := \sum_{\mathbf{m}\in\mathcal{P}\cap\mathbb{Z}^d} \omega_{\mathcal{P}}(\mathbf{m})\,\mathbf{z}^{\mathbf{m}}.$$

Using the same reasoning as in (13.1) for $A_{\mathcal{P}}$, this function satisfies a nice additivity relation. Namely, if the d-polyhedron $\mathcal{P}$ equals $\mathcal{P}_1 \cup \mathcal{P}_2$, where $\dim(\mathcal{P}_1 \cap \mathcal{P}_2) < d$, then

$$\alpha_{\mathcal{P}}(\mathbf{z}) = \alpha_{\mathcal{P}_1}(\mathbf{z}) + \alpha_{\mathcal{P}_2}(\mathbf{z})\,. \tag{13.2}$$

The solid-angle generating function obeys the following reciprocity relation, which parallels both the statement and proof of Theorem 4.3:

Theorem 13.5. *Suppose $\mathcal{K}$ is a rational pointed d-cone with the origin as apex, and $\mathbf{v} \in \mathbb{R}^d$. Then the solid-angle generating function $\alpha_{\mathbf{v}+\mathcal{K}}(\mathbf{z})$ of the pointed d-cone $\mathbf{v} + \mathcal{K}$ is a rational function that satisfies*

$$\alpha_{\mathbf{v}+\mathcal{K}}\left(\frac{1}{\mathbf{z}}\right) = (-1)^d\,\alpha_{-\mathbf{v}+\mathcal{K}}(\mathbf{z})\,.$$

Proof. Because solid angles are additive by (13.2), it suffices to prove this theorem for simplicial cones. The proof for this case proceeds along the same lines as the proof of Theorem 4.2; the main geometric ingredient is Exercise 4.2. We invite the reader to finish the proof (Exercise 13.6). □

The analogue of Brion's theorem in terms of solid angles is as follows.

Theorem 13.6. *Suppose $\mathcal{P}$ is a rational convex polytope. Then we have the following identity of rational functions:*

$$\alpha_{\mathcal{P}}(\mathbf{z}) = \sum_{\mathbf{v}\text{ a vertex of }\mathcal{P}} \alpha_{\mathcal{K}_{\mathbf{v}}}(\mathbf{z})\,.$$

Proof. As in the proof of Theorem 11.7, it suffices to prove Theorem 13.6 for simplices. So let Δ be a rational simplex. We write Δ as the disjoint union of its open faces and use Brion's theorem for open polytopes (Exercise 11.9) on each face. That is, if we denote the vertex cone of $\mathcal{F}$ at vertex $\mathbf{v}$ by $\mathcal{K}_{\mathbf{v}}(\mathcal{F})$, then by a monomial version of Lemma 13.2,

$$\alpha_{\Delta}(\mathbf{z}) = \sum_{\mathcal{F} \subseteq \Delta} \omega_{\Delta}(\mathcal{F}) \, \sigma_{\mathcal{F}^{\circ}}(\mathbf{z})$$

$$= \sum_{\mathbf{v} \text{ a vertex of } \Delta} \omega_{\Delta}(\mathbf{v}) \, \mathbf{z}^{\mathbf{v}} + \sum_{\substack{\mathcal{F} \subseteq \Delta \\ \dim \mathcal{F} > 0}} \omega_{\Delta}(\mathcal{F}) \sum_{\mathbf{v} \text{ a vertex of } \mathcal{F}} \sigma_{\mathcal{K}_{\mathbf{v}}(\mathcal{F})^{\circ}}(\mathbf{z}),$$

where we used Brion's theorem for open polytopes (Exercise 11.9) in the second step. By Exercise 13.7,

$$\sum_{\substack{\mathcal{F} \subseteq \Delta \\ \dim \mathcal{F} > 0}} \omega_{\Delta}(\mathcal{F}) \sum_{\mathbf{v} \text{ a vertex of } \mathcal{F}} \sigma_{\mathcal{K}_{\mathbf{v}}(\mathcal{F})^{\circ}}(\mathbf{z}) = \sum_{\mathbf{v} \text{ a vertex of } \Delta} \sum_{\substack{\mathcal{F} \subseteq \mathcal{K}_{\mathbf{v}} \\ \dim \mathcal{F} > 0}} \omega_{\mathcal{K}_{\mathbf{v}}}(\mathcal{F}) \, \sigma_{\mathcal{F}^{\circ}}(\mathbf{z}),$$

and so

$$\alpha_{\Delta}(\mathbf{z}) = \sum_{\mathbf{v} \text{ a vertex of } \Delta} \omega_{\Delta}(\mathbf{v}) \, \mathbf{z}^{\mathbf{v}} + \sum_{\mathbf{v} \text{ a vertex of } \Delta} \sum_{\substack{\mathcal{F} \subseteq \mathcal{K}_{\mathbf{v}} \\ \dim \mathcal{F} > 0}} \omega_{\mathcal{K}_{\mathbf{v}}}(\mathcal{F}) \, \sigma_{\mathcal{F}^{\circ}}(\mathbf{z})$$

$$= \sum_{\mathbf{v} \text{ a vertex of } \Delta} \sum_{\mathcal{F} \subseteq \mathcal{K}_{\mathbf{v}}} \omega_{\mathcal{K}_{\mathbf{v}}}(\mathcal{F}) \, \sigma_{\mathcal{F}^{\circ}}(\mathbf{z})$$

$$= \sum_{\mathbf{v} \text{ a vertex of } \Delta} \alpha_{\mathcal{K}_{\mathbf{v}}}(\mathbf{z}) . \qquad \square$$

13.3 Solid-Angle Reciprocity and the Brianchon–Gram Relations

With the help of Theorems 13.5 and 13.6, we can now prove the solid-angle analogue of Ehrhart–Macdonald reciprocity (Theorem 4.1):

Theorem 13.7 (Macdonald's reciprocity theorem). *Suppose $\mathcal{P}$ is a rational convex polytope. Then the quasipolynomial $A_{\mathcal{P}}$ satisfies*

$$A_{\mathcal{P}}(-t) = (-1)^{\dim \mathcal{P}} A_{\mathcal{P}}(t) .$$

Proof. We give the proof for an *integral* polytope $\mathcal{P}$ and invite the reader to generalize it to the rational case. The solid-angle counting function of $\mathcal{P}$ can be computed through the solid-angle generating function of $t\mathcal{P}$:

$$A_{\mathcal{P}}(t) = \alpha_{t\mathcal{P}}(1, 1, \dots, 1) = \lim_{\mathbf{z} \to \mathbf{1}} \alpha_{t\mathcal{P}}(\mathbf{z}) .$$

By Theorem 13.6,

$$A_{\mathcal{P}}(t) = \lim_{\mathbf{z} \to \mathbf{1}} \sum_{\mathbf{v} \text{ a vertex of } \mathcal{P}} \alpha_{t\mathcal{K}_{\mathbf{v}}}(\mathbf{z}),$$

where $\mathcal{K}_{\mathbf{v}}$ is the tangent cone of $\mathcal{P}$ at the vertex $\mathbf{v}$. We write $\mathcal{K}_{\mathbf{v}} = \mathbf{v} + \mathcal{K}(\mathbf{v})$, where $\mathcal{K}(\mathbf{v}) := \mathcal{K}_{\mathbf{v}} - \mathbf{v}$ is a rational cone with the origin as its apex. Then $t\mathcal{K}_{\mathbf{v}} = t\mathbf{v} + \mathcal{K}(\mathbf{v})$, because a cone whose apex is the origin does not change under dilation. Hence we obtain, with the help of Exercise 13.5,

$$A_{\mathcal{P}}(t) = \lim_{\mathbf{z} \to \mathbf{1}} \sum_{\mathbf{v} \text{ a vertex of } \mathcal{P}} \alpha_{t\mathbf{v}+\mathcal{K}(\mathbf{v})}(\mathbf{z}) = \lim_{\mathbf{z} \to \mathbf{1}} \sum_{\mathbf{v} \text{ a vertex of } \mathcal{P}} \mathbf{z}^{t\mathbf{v}} \alpha_{\mathcal{K}(\mathbf{v})}(\mathbf{z}).$$

The rational functions $\alpha_{\mathcal{K}(\mathbf{v})}(\mathbf{z})$ on the right-hand side do not depend on t. If we think of the sum over all vertices as one big rational function, to which we apply L'Hôpital's rule to compute the limit as $\mathbf{z} \to \mathbf{1}$, this gives an alternative proof that $A_{\mathcal{P}}(t)$ is a polynomial, in line with our proof for the polynomiality of $L_{\mathcal{P}}(t)$ in Section 11.5. At the same time, this means that we can view the identity

$$A_{\mathcal{P}}(t) = \lim_{\mathbf{z} \to \mathbf{1}} \sum_{\mathbf{v} \text{ a vertex of } \mathcal{P}} \mathbf{z}^{t\mathbf{v}} \alpha_{\mathcal{K}(\mathbf{v})}(\mathbf{z})$$

in a purely algebraic fashion: on the left-hand side, we have a polynomial that makes sense for every complex t, and on the right-hand side, we have a rational function of $\mathbf{z}$, whose limit we compute, for example, by L'Hôpital's rule. So the right-hand side, as a function of t, makes sense for every integer t. Hence we have the algebraic relation, for integral t,

$$A_{\mathcal{P}}(-t) = \lim_{\mathbf{z} \to \mathbf{1}} \sum_{\mathbf{v} \text{ a vertex of } \mathcal{P}} \mathbf{z}^{-t\mathbf{v}} \alpha_{\mathcal{K}(\mathbf{v})}(\mathbf{z}).$$

But now by Theorem 13.5, $\alpha_{\mathcal{K}(\mathbf{v})}(\mathbf{z}) = (-1)^d \alpha_{\mathcal{K}(\mathbf{v})}\left(\frac{1}{\mathbf{z}}\right)$, and so

$$A_{\mathcal{P}}(-t) = \lim_{\mathbf{z} \to \mathbf{1}} \sum_{\mathbf{v} \text{ a vertex of } \mathcal{P}} \mathbf{z}^{-t\mathbf{v}} (-1)^d \alpha_{\mathcal{K}(\mathbf{v})}\left(\frac{1}{\mathbf{z}}\right)$$

$$= (-1)^d \lim_{\mathbf{z} \to \mathbf{1}} \sum_{\mathbf{v} \text{ a vertex of } \mathcal{P}} \left(\frac{1}{\mathbf{z}}\right)^{t\mathbf{v}} \alpha_{\mathcal{K}(\mathbf{v})}\left(\frac{1}{\mathbf{z}}\right)$$

$$= (-1)^d \lim_{\mathbf{z} \to \mathbf{1}} \sum_{\mathbf{v} \text{ a vertex of } \mathcal{P}} \alpha_{t\mathbf{v}+\mathcal{K}(\mathbf{v})}\left(\frac{1}{\mathbf{z}}\right)$$

$$= (-1)^d \lim_{\mathbf{z} \to \mathbf{1}} \alpha_{t\mathcal{P}}\left(\frac{1}{\mathbf{z}}\right)$$

$$= (-1)^d A_{\mathcal{P}}(t).$$

In the third step, we used Exercise 13.5 again.

This proves Theorem 13.7 for integral polytopes. The proof for *rational* polytopes follows along the same lines; one deals with rational vertices in the same manner as in our second proof of Ehrhart's theorem in Section 11.5. We invite the reader to finish the details in Exercise 13.8.   □

We remark that throughout the proof, we cannot simply take the limit inside the finite sum over the vertices of $\mathcal{P}$, since $\mathbf{z} = \mathbf{1}$ is a pole of each rational function $\alpha_{\mathcal{K}(\mathbf{v})}$. It is precisely the magic of Brion's theorem that makes these poles cancel each other, to yield $A_{\mathcal{P}}(t)$.

If $\mathcal{P}$ is an *integral* polytope, then $A_{\mathcal{P}}$ is a polynomial, and Theorem 13.7 tells us that $A_{\mathcal{P}}$ is always even or odd:

$$A_{\mathcal{P}}(t) = c_d\, t^d + c_{d-2} t^{d-2} + \cdots + c_0\,.$$

We can say more.

Theorem 13.8. *Suppose $\mathcal{P}$ is a rational convex polytope. Then $A_{\mathcal{P}}(0) = 0$.*

This is a meaningful zero. We note that the constant term of $A_{\mathcal{P}}$ is given by

$$A_{\mathcal{P}}(0) = \sum_{\mathcal{F} \subseteq \mathcal{P}} \omega_{\mathcal{P}}(\mathcal{F})\, L_{\mathcal{F}^\circ}(0) = \sum_{\mathcal{F} \subseteq \mathcal{P}} \omega_{\mathcal{P}}(\mathcal{F})\, (-1)^{\dim \mathcal{F}},$$

by Lemma 13.2 and Ehrhart–Macdonald reciprocity (Theorem 4.1). Hence Theorem 13.8 implies a classical and useful geometric identity:

Corollary 13.9 (Brianchon–Gram relation). *For a rational convex polytope $\mathcal{P}$,*

$$\sum_{\mathcal{F} \subseteq \mathcal{P}} (-1)^{\dim \mathcal{F}} \omega_{\mathcal{P}}(\mathcal{F}) = 0\,.$$

Example 13.10. Consider a triangle $\mathcal{T}$ in $\mathbb{R}^2$ with vertices $\mathbf{v}_1, \mathbf{v}_2, \mathbf{v}_3$ and edges E_1, E_2, E_3. The Brianchon–Gram relation tells us that for this triangle,

$$\omega_{\mathcal{T}}(\mathbf{v}_1) + \omega_{\mathcal{T}}(\mathbf{v}_2) + \omega_{\mathcal{T}}(\mathbf{v}_3) - (\omega_{\mathcal{T}}(E_1) + \omega_{\mathcal{T}}(E_2) + \omega_{\mathcal{T}}(E_3)) + \omega_{\mathcal{T}}(\mathcal{T}) = 0\,.$$

Since the solid angles of the edges are all $\frac{1}{2}$ and $\omega_{\mathcal{T}}(\mathcal{T}) = 1$, we recover our friendly high-school identity *the sum of the angles in a triangle is 180 degrees*:

$$\omega_{\mathcal{T}}(\mathbf{v}_1) + \omega_{\mathcal{T}}(\mathbf{v}_2) + \omega_{\mathcal{T}}(\mathbf{v}_3) = \frac{1}{2}\,.$$

Thus the Brianchon–Gram relation is the extension of this well-known fact to every dimension and every convex polytope.   □

Proof of Theorem 13.8. It suffices to prove $A_\Delta(0) = 0$ for a rational simplex Δ, since solid angles of a triangulation simply add, by (13.1). Theorem 13.7 gives $A_\Delta(0) = 0$ if $\dim \Delta$ is odd.

So now suppose Δ is a rational d-simplex, where d is even, with vertices $\mathbf{v}_1, \mathbf{v}_2, \ldots, \mathbf{v}_{d+1}$. Let $\mathcal{P}(n)$ be the $(d+1)$-dimensional pyramid that we obtain

Fig. 13.3 The pyramid
$\mathcal{P}(n)$ for a triangle Δ.

by taking the convex hull of $(\mathbf{v}_1, 0), (\mathbf{v}_2, 0), \ldots, (\mathbf{v}_{d+1}, 0)$, and $(0, 0, \ldots, 0, n)$, where n is a positive integer (see Figure 13.3). Note that since $d + 1$ is odd,

$$A_{\mathcal{P}(n)}(0) = \sum_{\mathcal{F}(n) \subseteq \mathcal{P}(n)} (-1)^{\dim \mathcal{F}(n)} \, \omega_{\mathcal{P}(n)}(\mathcal{F}(n)) = 0 \,.$$

We will conclude from this identity that $\sum_{\mathcal{F} \subseteq \Delta} (-1)^{\dim \mathcal{F}} \omega_\Delta(\mathcal{F}) = 0$, which implies that $A_\Delta(0) = 0$. To this end, we consider two types of faces of $\mathcal{P}(n)$:

(a) those that are also faces of Δ;
(b) those that are not contained in Δ.

We begin with the latter: Aside from the vertex $(0, 0, \ldots, 0, n)$, every face $\mathcal{F}(n)$ of $\mathcal{P}(n)$ that is not a face of Δ is the pyramid over a face $\mathcal{G}$ of Δ; let's denote this pyramid by $\mathrm{Pyr}\,(\mathcal{G}, n)$. Further, as n grows, the solid angle of $\mathrm{Pyr}\,(\mathcal{G}, n)$ (in $\mathcal{P}(n)$) approaches the solid angle of $\mathcal{G}$ (in Δ):

$$\lim_{n \to \infty} \omega_{\mathcal{P}(n)}\left(\mathrm{Pyr}\,(\mathcal{G}, n)\right) = \omega_\Delta(\mathcal{G}) \,,$$

since we are forming $\Delta \times [0, \infty)$ in the limit. On the other hand, a face $\mathcal{F}(n) = \mathcal{G}$ of $\mathcal{P}(n)$ that is also a face of Δ obeys the following limit behavior:

$$\lim_{n \to \infty} \omega_{\mathcal{P}(n)}\left(\mathcal{F}(n)\right) = \frac{1}{2}\,\omega_\Delta(\mathcal{G}) \,.$$

The only face of $\mathcal{P}(n)$ that we still have to account for is the vertex $\mathbf{v} := (0, 0, \ldots, 0, n)$. Hence

$$0 = \sum_{\mathcal{F}(n) \subseteq \mathcal{P}(n)} (-1)^{\dim \mathcal{F}(n)} \omega_{\mathcal{P}(n)}(\mathcal{F}(n))$$

$$= \omega_{\mathcal{P}(n)}(\mathbf{v}) + \sum_{\mathcal{G} \subseteq \Delta} (-1)^{\dim \mathcal{G}+1} \omega_{\mathcal{P}(n)}\left(\mathrm{Pyr}\,(\mathcal{G}, n)\right) + \sum_{\mathcal{G} \subseteq \Delta} (-1)^{\dim \mathcal{G}} \omega_{\mathcal{P}(n)}(\mathcal{G}) \,.$$

Now we take the limit as $n \to \infty$ on both sides; note that $\lim_{n \to \infty} \omega_{\mathcal{P}(n)}(\mathbf{v}) = 0$, so that we obtain

$$0 = \sum_{\mathcal{G} \subseteq \Delta} (-1)^{\dim \mathcal{G}+1} \omega_\Delta(\mathcal{G}) + \sum_{\mathcal{G} \subseteq \Delta} (-1)^{\dim \mathcal{G}} \frac{1}{2} \omega_\Delta(\mathcal{G})$$

$$= \frac{1}{2} \sum_{\mathcal{G} \subseteq \Delta} (-1)^{\dim \mathcal{G}+1} \omega_\Delta(\mathcal{G}),$$

and so

$$A_\Delta(0) = \sum_{\mathcal{G} \subseteq \Delta} (-1)^{\dim \mathcal{G}} \omega_\Delta(\mathcal{G}) = 0. \qquad \square$$

The combination of Theorems 13.7 and 13.8 implies that summing solid angles in a polygon is equivalent to computing its area:

Corollary 13.11. *Suppose $\mathcal{P}$ is a 2-dimensional integral polytope with area a. Then $A_\mathcal{P}(t) = a\,t^2$.*

13.4 The Generating Function of Macdonald's Solid-Angle Polynomials

We conclude this chapter with the study of the solid-angle analogue of Ehrhart series. Given an integral polytope $\mathcal{P}$, we define the **solid-angle series** of $\mathcal{P}$ as the generating function of the solid-angle polynomial, encoding the solid-angle sum over all dilates of $\mathcal{P}$ simultaneously:

$$\mathrm{Solid}_\mathcal{P}(z) := \sum_{t \geq 0} A_\mathcal{P}(t)\, z^t.$$

The following theorem is the solid-angle analogue to Theorems 3.12 and 4.4, with the added bonus that we get the palindromicity of the numerator of $\mathrm{Solid}_\mathcal{P}$ for free.

Theorem 13.12. *Suppose $\mathcal{P}$ is an integral d-polytope. Then $\mathrm{Solid}_\mathcal{P}$ is a rational function of the form*

$$\mathrm{Solid}_\mathcal{P}(z) = \frac{a_d z^d + a_{d-1} z^{d-1} + \cdots + a_1 z}{(1-z)^{d+1}}.$$

Furthermore, we have the identity

$$\mathrm{Solid}_\mathcal{P}\left(\frac{1}{z}\right) = (-1)^{d+1} \mathrm{Solid}_\mathcal{P}(z),$$

or equivalently, $a_k = a_{d+1-k}$ for $1 \leq k \leq \frac{d}{2}$.

Proof. The form of the rational function Solid$_\mathcal{P}$ follows, by Lemma 3.9, from the fact that $A_\mathcal{P}$ is a polynomial. The palindromicity of $a_1, a_2, \ldots, a_d$ is equivalent to the relation

$$\text{Solid}_\mathcal{P}\left(\frac{1}{z}\right) = (-1)^{d+1} \text{Solid}_\mathcal{P}(z),$$

which, in turn, follows from Theorem 13.7:

$$\text{Solid}_\mathcal{P}(z) = \sum_{t \geq 0} A_\mathcal{P}(t)\, z^t = \sum_{t \geq 0} (-1)^d A_\mathcal{P}(-t)\, z^t = (-1)^d \sum_{t \leq 0} A_\mathcal{P}(t)\, z^{-t}.$$

Now we use Exercise 4.7:

$$(-1)^d \sum_{t \leq 0} A_\mathcal{P}(t)\, z^{-t} = (-1)^{d+1} \sum_{t \geq 1} A_\mathcal{P}(t)\, z^{-t} = (-1)^{d+1} \text{Solid}_\mathcal{P}\left(\frac{1}{z}\right).$$

In the last step, we used the fact that $A_\mathcal{P}(0) = 0$ (Theorem 13.8). $\square$

Notes

1. I. G. Macdonald inaugurated the systematic study of solid-angle sums in integral polytopes. The fundamental Theorems 13.4, 13.7, and 13.8 can be found in his 1971 paper [166]. The proof of Theorem 13.7 we give here follows [38].

2. The Brianchon–Gram relation (Corollary 13.9) is the solid-angle analogue of the Euler relation for face numbers (Theorem 5.2). The 2-dimensional case discussed in Example 13.10 is ancient; it was most certainly known to Euclid. The 3-dimensional case of Corollary 13.9 was discovered by Charles Julien Brianchon in 1837 and—as far as we know—was independently re-proved by Jørgen Gram in 1874 [121]. It is not clear who first proved the general d-dimensional case of Corollary 13.9. The oldest proofs we could find were from the 1960s, by Branko Grünbaum [126], Micha A. Perles, and Geoffrey C. Shephard [189, 218].

3. Theorem 13.5 is a particular case of a reciprocity relation for *simple lattice-invariant valuations* due to Peter McMullen [172], who also proved a parallel extension of Ehrhart–Macdonald reciprocity to general lattice-invariant valuations. There is a current resurgence of activity on solid angles; see, for example, [45, 76, 103, 115, 204].

Exercises

13.1. Compute $A_{\mathcal{P}}(t)$, where $\mathcal{P}$ is the regular tetrahedron with vertices $(0,0,0)$, $(1,1,0)$, $(1,0,1)$, and $(0,1,1)$ (see Exercise 2.13).

13.2. Compute $A_{\mathcal{P}}(t)$, where $\mathcal{P}$ is the rational triangle with vertices $(0,0)$, $\left(\frac{1}{2},\frac{1}{2}\right)$, and $(1,0)$.

13.3. For a simplex Δ, let $S(\Delta)$ denote the sum of the solid angles at the vertices of Δ.

(a) Prove that $S(\Delta) \le \frac{1}{2}$.
(b) Construct a sequence Δ_n of simplices in a fixed dimension, such that $\lim_{n\to\infty} S(\Delta_n) = \frac{1}{2}$.

13.4. Let $\mathcal{Z}$ be a d-dimensional integral zonotope. Show that $A_{\mathcal{Z}}(t) = \operatorname{vol}(\mathcal{Z})\,t^d$.

13.5. ♣ Let $\mathcal{K}$ be a rational d-cone and let $\mathbf{m} \in \mathbb{Z}^d$. By analogy with Exercise 3.9, show that $\alpha_{\mathbf{m}+\mathcal{K}}(\mathbf{z}) = \mathbf{z}^{\mathbf{m}}\,\alpha_{\mathcal{K}}(\mathbf{z})$.

13.6. ♣ Complete the proof of Theorem 13.5: For a rational pointed d-cone $\mathcal{K}$, $\alpha_{\mathcal{K}}(\mathbf{z})$ is a rational function that satisfies

$$\alpha_{\mathcal{K}}\left(\frac{1}{\mathbf{z}}\right) = (-1)^d\,\alpha_{\mathcal{K}}\left(\mathbf{z}\right).$$

13.7. ♣ Suppose Δ is a rational simplex. Prove that

$$\sum_{\substack{\mathcal{F}\subseteq\Delta \\ \dim\mathcal{F}>0}} \omega_\Delta(\mathcal{F}) \sum_{\mathbf{v}\ \text{a vertex of }\mathcal{F}} \sigma_{\mathcal{K}_{\mathbf{v}}(\mathcal{F})^\circ}(\mathbf{z}) = \sum_{\mathbf{v}\ \text{a vertex of }\Delta} \sum_{\substack{\mathcal{F}\subseteq\mathcal{K}_{\mathbf{v}} \\ \dim\mathcal{F}>0}} \omega_{\mathcal{K}_{\mathbf{v}}}(\mathcal{F})\,\sigma_{\mathcal{F}^\circ}(\mathbf{z}).$$

13.8. ♣ Provide the details of the proof of Theorem 13.7 for *rational* polytopes: Prove that if $\mathcal{P}$ is a rational convex polytope, then the quasipolynomial $A_{\mathcal{P}}$ satisfies

$$A_{\mathcal{P}}(-t) = (-1)^{\dim\mathcal{P}}A_{\mathcal{P}}(t).$$

13.9. Recall from Exercise 3.2 that to a permutation $\pi \in S_d$ on d elements, we can associate the simplex

$$\Delta_\pi := \operatorname{conv}\left\{0, \mathbf{e}_{\pi(1)}, \mathbf{e}_{\pi(1)}+\mathbf{e}_{\pi(2)}, \dots, \mathbf{e}_{\pi(1)}+\mathbf{e}_{\pi(2)}+\cdots+\mathbf{e}_{\pi(d)}\right\}.$$

Prove that for all $\pi \in S_n$, $A_{\Delta_\pi}(t) = \frac{1}{d!}\,t^d$.

13.10. Give a direct proof of Corollary 13.11, e.g., using Pick's theorem (Theorem 2.8).

13.11. State and prove the analogue of Theorem 13.12 for *rational* polytopes.

Open Problems

13.12. Study the roots of solid-angle polynomials.

13.13. Classify all polytopes that have only rational solid angles at their vertices.

13.14. Classify all rational polytopes $\mathcal{P}$ such that $A_{\mathcal{P}}(t) = \text{vol}(\mathcal{P}) \, t^d$.

13.15. Which integral polytopes $\mathcal{P}$ have an integral solid-angle sum? More generally, which integral polytopes $\mathcal{P}$ have solid-angle polynomials $A_{\mathcal{P}}(t) \in \mathbb{Q}[t]$? That is, for which integral polytopes $\mathcal{P}$ are all the coefficients of $A_{\mathcal{P}}(t)$ rational? (For one such class of polytopes, see [123].)

Chapter 14
A Discrete Version of Green's Theorem Using Elliptic Functions

The shortest route between two truths in the real domain passes through the complex domain.

Jacques Salomon Hadamard (1865–1963)

We now allow ourselves the luxury of using basic complex analysis. In particular, we assume that the reader is familiar with contour integration and the residue theorem. We may view the residue theorem as yet another result that intimately connects the continuous and the discrete: it transforms a continuous integral into a discrete sum of residues.

Using the *Weierstraß $\wp$- and ζ-functions*, we show here that Pick's theorem is a discrete version of Green's theorem in the plane. As a bonus, we also obtain an integral formula (Theorem 14.5 below) for the discrepancy between the area enclosed by a general curve C and the number of integer points contained in C.

14.1 The Residue Theorem

We begin this chapter by reviewing a few concepts from complex analysis. Suppose the complex-valued function f has an **isolated singularity** $w \in G$; that is, there is an open set $G \subset \mathbb{C}$ such that f is analytic on $G \setminus \{w\}$. Then f can be expressed locally by the **Laurent series**

$$f(z) = \sum_{n \in \mathbb{Z}} c_n \left(z - w \right)^n,$$

valid for all $z \in G \setminus \{w\}$; here $c_n \in \mathbb{C}$. The coefficient c_{-1} is called the **residue** of f at w; we will denote it by $\mathrm{Res}(z = w)$. The reason to give c_{-1}

a special name can be found in the following theorem. We call a function **meromorphic** if it is analytic in $\mathbb{C}$ with the exception of isolated poles.

Theorem 14.1 (Residue theorem). *Suppose f is meromorphic and C is a positively oriented, piecewise differentiable simple closed curve that does not pass through any pole of f. Then*

$$\int_C f = 2\pi i \sum_w \operatorname{Res}(z = w),$$

where the sum is taken over all singularities w inside C. $\square$

If f is a rational function, Theorem 14.1 gives the same result as a partial fraction expansion of f. We illustrate this philosophy by returning to the elementary beginnings of Chapter 1.

Example 14.2. Recall our constant-term identity for the restricted partition function for $A = \{a_1, a_2, \ldots, a_d\}$ in Chapter 1:

$$p_A(n) = \operatorname{const} \left(\frac{1}{(1 - z^{a_1})(1 - z^{a_2}) \cdots (1 - z^{a_d}) z^n} \right).$$

Computing the constant term of the Laurent series of $\frac{1}{(1-z^{a_1})\cdots(1-z^{a_d})z^n}$ expanded about $z = 0$ is naturally equivalent to "shifting" this function by one exponent and computing the residue at $z = 0$ of the function

$$f(z) := \frac{1}{(1 - z^{a_1})(1 - z^{a_2}) \cdots (1 - z^{a_d}) z^{n+1}}.$$

Now let C_r be a positively oriented circle of radius $r > 1$, centered at the origin. The residue $\operatorname{Res}(z = 0) = p_A(n)$ is one of the residues that are picked up by the integral

$$\frac{1}{2\pi i} \int_{C_r} f = \operatorname{Res}(z = 0) + \sum_w \operatorname{Res}(z = w),$$

where the sum is over all nonzero poles w of f that lie inside C_r. These poles are at the $a_1^{\text{th}}, a_2^{\text{th}}, \ldots, a_d^{\text{th}}$ roots of unity. Moreover, with the help of Exercise 14.1, we can show that

$$0 = \lim_{r \to \infty} \frac{1}{2\pi i} \int_{C_r} f$$

$$= \lim_{r \to \infty} \left(\operatorname{Res}(z = 0) + \sum_w \operatorname{Res}(z = w) \right)$$

$$= \operatorname{Res}(z = 0) + \sum_w \operatorname{Res}(z = w),$$

where the sum extends over all $a_1^{\text{th}}, a_2^{\text{th}}, \ldots, a_d^{\text{th}}$ roots of unity. In other words,

$$p_A(n) = \text{Res}(z = 0) = -\sum_w \text{Res}(z = w).$$

To obtain the restricted partition function p_A, it remains to compute the residues at the roots of unity, and we invite the reader to realize that this computation is equivalent to the partial fraction expansion of Chapter 1 (Exercise 14.2). □

Analogous residue computations could replace any of the constant-term calculations that we performed in the earlier chapters.

14.2 The Weierstraß $\wp$- and ζ-Functions

The main character in our play is the **Weierstraß ζ-function**, defined by

$$\zeta(z) = \frac{1}{z} + \sum_{(m,n)\in\mathbb{Z}^2\backslash(0,0)} \left(\frac{1}{z - (m + ni)} + \frac{1}{m + ni} + \frac{z}{(m + ni)^2} \right). \quad (14.1)$$

This infinite sum converges uniformly on compact subsets of the lattice-punctured plane $\mathbb{C} \setminus \mathbb{Z}^2$ (Exercise 14.5), and hence forms a meromorphic function of z.

The Weierstraß ζ-function possesses the following salient properties, which follow immediately from (14.1):

(1) ζ has a simple pole at every integer point $m + ni$ and is analytic elsewhere.
(2) The residue of ζ at each integer point $m + ni$ equals 1.

We can easily check (Exercise 14.6) that

$$\wp(z) := -\zeta'(z) = \frac{1}{z^2} + \sum_{(m,n)\in\mathbb{Z}^2\backslash(0,0)} \left(\frac{1}{(z - (m + ni))^2} - \frac{1}{(m + ni)^2} \right),$$
$$(14.2)$$

the **Weierstraß $\wp$-function**. The $\wp$-function has a pole of order 2 at each integer point $m + ni$ and is analytic elsewhere, but has residue equal to zero at each integer point $m + ni$. However, $\wp$ possesses a very pleasant property that ζ does not: $\wp$ is **doubly periodic** on $\mathbb{C}$. We may state this more concretely:

Lemma 14.3. $\wp(z + 1) = \wp(z + i) = \wp(z).$

Proof. We first invite the reader to prove the following two properties of $\wp'$ (Exercises 14.7 and 14.8):

$$\wp'(z+1) = \wp'(z), \tag{14.3}$$

$$\int_{z_0}^{z_1} \wp'(z)\,dz \quad \text{is path-independent.} \tag{14.4}$$

By (14.3),

$$\frac{d}{dz}\left(\wp(z+1) - \wp(z)\right) = \wp'(z+1) - \wp'(z) = 0,$$

so $\wp(z+1) - \wp(z) = c$ for some constant c. On the other hand, $\wp$ is an even function (Exercise 14.9), and so $z = -\frac{1}{2}$ gives us

$$c = \wp\left(\tfrac{1}{2}\right) - \wp\left(-\tfrac{1}{2}\right) = 0.$$

This shows that $\wp(z+1) = \wp(z)$ for all $z \in \mathbb{C} \setminus \mathbb{Z}^2$. An analogous proof, which we invite the reader to construct in Exercise 14.10, shows that $\wp(z+i) = \wp(z)$. $\qquad\square$

Lemma 14.3 implies that $\wp(z + m + ni) = \wp(z)$ for all $m, n \in \mathbb{Z}$. The following lemma shows that the Weierstraß ζ-function is only a conjugate-analytic term away from being doubly periodic.

Lemma 14.4. *There is a constant α such that the function $\zeta(z) + \alpha\bar{z}$ is doubly periodic with periods 1 and i.*

Proof. We begin with $w = m + ni$:

$$\zeta(z + m + ni) - \zeta(z) = -\int_{w=0}^{m+ni} \wp(z+w)\,dw, \tag{14.5}$$

by definition of $\wp(z) = -\zeta'(z)$. To make sure that (14.5) makes sense, we should also check that the definite integral in (14.5) is path-independent (Exercise 14.11).

Due to the double periodicity of $\wp$,

$$\int_{w=0}^{m+ni} \wp(z+w)\,dw = m\int_0^1 \wp(z+t)\,dt + ni\int_0^1 \wp(z+it)\,dt$$
$$= m\alpha(z) + ni\beta(z),$$

where

$$\alpha(z) := \int_0^1 \wp(z+t)\,dt \quad \text{and} \quad \beta(z) := \int_0^1 \wp(z+it)\,dt.$$

Now we observe that $\alpha(z + x_0) = \alpha(z)$ for every $x_0 \in \mathbb{R}$, so that $\alpha(x+iy)$ depends only on y. Similarly, $\beta(x+iy)$ depends only on x. But

$$\zeta(z + m + in) - \zeta(z) = -\left(m\alpha(y) + in\beta(x)\right)$$

must be analytic for all $z \in \mathbb{C} \setminus \mathbb{Z}^2$. If we now set $m = 0$, we conclude that $\beta(x)$ must be analytic in $\mathbb{C} \setminus \mathbb{Z}^2$, so that $\beta(x)$ must be a constant by the Cauchy–Riemann equations for analytic functions. Similarly, setting $n = 0$ implies that $\alpha(y)$ is constant. Thus

$$\zeta(z + m + in) - \zeta(z) = -(m\alpha + in\beta)$$

with constants α and β. Returning to the Weierstraß $\wp$-function, we can integrate the identity (Exercise 14.12)

$$\wp(iz) = -\wp(z) \tag{14.6}$$

to obtain the relationship $\beta = -\alpha$, since

$$\beta = \int_0^1 \wp(z + it)\, dt = \int_0^1 \wp(it)\, dt = -\int_0^1 \wp(t)\, dt = -\alpha.$$

To summarize,

$$\zeta(z + m + in) - \zeta(z) = -m\alpha + in\alpha = -\alpha\left(\overline{z + m + in} - \overline{z}\right),$$

so that $\zeta(z) + \alpha\overline{z}$ is doubly periodic. $\qquad\square$

14.3 A Contour-Integral Extension of Pick's Theorem

For the remainder of this chapter, let C be any piecewise-differentiable simple closed curve in the plane, with a counterclockwise parameterization. We let D denote the region that C contains in its interior.

Theorem 14.5. *Let C avoid every integer point, that is, $C \cap \mathbb{Z}^2 = \varnothing$. Let I denote the number of integer points interior to C, and A the area of the region D enclosed by the curve C. Then*

$$\frac{1}{2\pi i} \int_C \left(\zeta(z) - \pi\overline{z}\right) dz = I - A.$$

Proof. We have

$$\int_C \left(\zeta(z) + \alpha\overline{z}\right) dz = \int_C \zeta(z)\, dz + \alpha \int_C (x - iy)\, (dx + idy),$$

where α is as in Lemma 14.4. By Theorem 14.1, $\int_C \zeta(z)\, dz$ is equal to the sum of the residues of ζ at all of its interior poles. There are I such poles, and each pole of ζ has residue 1. Thus

$$\frac{1}{2\pi i} \int_C \zeta(z)\,dz = I\,. \tag{14.7}$$

On the other hand, Green's theorem tells us that

$$
\begin{aligned}
\int_C (x - iy)\,(dx + idy) &= \int_C (x - iy)\,dx + (y + ix)\,dy \\
&= \int_D \frac{\partial}{\partial x}(y + ix) - \frac{\partial}{\partial y}(x - iy) \\
&= \iint_D 2i \\
&= 2iA\,.
\end{aligned}
$$

Returning to (14.7), we get

$$\int_C (\zeta(z) + \alpha\bar{z})\,dz = 2\pi i I + \alpha\,(2iA)\,. \tag{14.8}$$

We have only to show that $\alpha = -\pi$. Consider the particular curve C that is a square path, centered at the origin, traversing the origin counterclockwise and bounding a square of area 1. Thus $I = 1$ for this path. Since $\zeta(z) + \alpha\bar{z}$ is doubly periodic by Lemma 14.4, the integral in (14.8) vanishes. We can conclude that

$$0 = 2\pi i \cdot 1 + \alpha\,(2i \cdot 1)\,,$$

so that $\alpha = -\pi$. □

Notice that Theorem 14.5 has given us information about the Weierstraß ζ-function, namely that $\alpha = -\pi$.

This chapter offers a detour into an infinite landscape of discrete results that meet their continuous counterparts. Equipped with the modest tools offered in this book, we hope that we have motivated the reader to explore this landscape further.

Notes

1. The Weierstraß $\wp$-function, named after Karl Theodor Wilhelm Weierstraß (1815–1897),[1] can be extended to every 2-dimensional lattice $\mathcal{L} = \{kw_1 + jw_2 : k, j \in \mathbb{Z}\}$ for some $w_1, w_2 \in \mathbb{C}$ that are linearly independent over $\mathbb{R}$:

$$\wp_{\mathcal{L}}(z) = \frac{1}{z^2} + \sum_{m \in \mathcal{L}\setminus\{0\}} \left(\frac{1}{(z-m)^2} - \frac{1}{m^2}\right).$$

---

[1] For more information about Weierstraß, see
http://www-history.mcs.st-andrews.ac.uk/Biographies/Weierstrass.html.

The Weierstraß $\wp_{\mathcal{L}}$-function and its derivative $\wp_{\mathcal{L}}'$ satisfy a polynomial relationship, namely, $\left(\wp_{\mathcal{L}}'\right)^2 = 4\left(\wp_{\mathcal{L}}\right)^3 - g_2\,\wp_{\mathcal{L}} - g_3$ for some constants g_2 and g_3 that depend on $\mathcal{L}$. This is the beginning of a wonderful friendship between complex analysis and elliptic curves.

2. Theorem 14.5 appeared in [105]. There it is also shown that one can recover Pick's theorem (Theorem 2.8) from Theorem 14.5.

Exercises

14.1. ♣ Show that for positive integers $a_1, a_d, \ldots, a_d, n$,

$$\lim_{r\to\infty} \int_{C_r} \frac{1}{(1 - z^{a_1})\cdots(1 - z^{a_d})\,z^{n+1}} = 0\,.$$

This computation shows that the integrand above "has no pole at infinity."

14.2. ♣ Compute the residues at the nontrivial roots of unity of

$$f(z) = \frac{1}{(1 - z^{a_1})\cdots(1 - z^{a_d})\,z^{n+1}}\,.$$

For simplicity, you may assume that $a_1, a_2, \ldots, a_d$ are pairwise relatively prime.

14.3. This exercise yields a proof of (8.4), namely, that the Bernoulli–Barnes polynomials $B_k^A(x)$, defined in Chapter 8 through the generating function

$$\frac{z^d e^{xz}}{(e^{a_1 z} - 1)(e^{a_2 z} - 1)\cdots(e^{a_d z} - 1)} = \sum_{k\geq 0} B_k^A(x)\,\frac{z^k}{k!}\,,$$

allow us to express the polynomial part of $p_A(n)$ as

$$\operatorname{poly}_A(n) = \frac{(-1)^d}{(d-1)!}\,B_d^A(-t)\,.$$

As above, we let

$$f(z) = \frac{1}{(1 - z^{a_1})\cdots(1 - z^{a_d})\,z^{n+1}}$$

and assume that $a_1, a_2, \ldots, a_d$ are pairwise relatively prime.

(a) Show that $\operatorname{poly}_A(n)$ equals the negative of the residue of $f(z)$ at $z = 1$.
(b) Deduce the above formula for $\operatorname{poly}_A(n)$ in terms of a Bernoulli–Barnes polynomial. (*Hint:* convince yourself that the residue of $f(z)$ at $z = 1$ equals the residue of $e^z f(e^z)$ at $z = 0$.)

14.4. Give an integral version of Theorem 2.13.

14.5. ♣ Show that

$$\zeta(z) = \frac{1}{z} + \sum_{(m,n)\in\mathbb{Z}^2\setminus(0,0)} \left(\frac{1}{z - (m+ni)} + \frac{1}{m+ni} + \frac{z}{(m+ni)^2} \right)$$

converges absolutely for z belonging to compact subsets of $\mathbb{C} \setminus \mathbb{Z}^2$.

14.6. ♣ Prove (14.2), that is,

$$\zeta'(z) = -\frac{1}{z^2} - \sum_{(m,n)\in\mathbb{Z}^2\setminus(0,0)} \left(\frac{1}{(z - (m+ni))^2} - \frac{1}{(m+ni)^2} \right).$$

14.7. ♣ Prove (14.3), that is, show that $\wp'(z+1) = \wp'(z)$.

14.8. ♣ Prove (14.4), that is, show that for all $z_0, z_1 \in \mathbb{C} \setminus \mathbb{Z}^2$, the integral $\int_{z_0}^{z_1} \wp'(w)\,dw$ is path-independent.

14.9. ♣ Show that $\wp$ is even, that is, $\wp(-z) = \wp(z)$.

14.10. ♣ Finish the proof of Lemma 14.3 by showing that $\wp(z+i) = \wp(z)$.

14.11. ♣ Prove that the integral in (14.5),

$$\zeta(z+m+ni) - \zeta(z) = -\int_{w=0}^{w=m+ni} \wp(z+w)\,dw,$$

is path-independent.

14.12. ♣ Prove (14.6), that is, $\wp(iz) = -\wp(z)$.

Open Problems

14.13. Can we get even more information about the Weierstraß $\wp$- and ζ-functions using more detailed knowledge of the discrepancy between I and A for special curves C?

14.14. Find a complex-analytic extension of Theorem 14.5 to higher dimensions.

Appendix A
Vertex and Hyperplane Descriptions of Polytopes

Everything should be made as simple as possible, but not simpler.

Albert Einstein

In this appendix, we prove that every convex polytope has both a vertex and a hyperplane description. This appendix owes everything to Günter Ziegler's beautiful exposition in [258]; in fact, these pages contain merely a few cherries picked from [258, Lecture 1].

As in Chapter 3, it is easier to move to the world of cones. To be as concrete as possible, let's call $\mathcal{K} \subseteq \mathbb{R}^d$ an **h-cone** if

$$\mathcal{K} = \left\{ \mathbf{x} \in \mathbb{R}^d : \mathbf{A}\,\mathbf{x} \leq \mathbf{0} \right\}$$

for some $\mathbf{A} \in \mathbb{R}^{m \times d}$; in this case, $\mathcal{K}$ is given as the intersection of m half-spaces determined by the rows of $\mathbf{A}$. We use the notation $\mathcal{K} = \mathrm{hcone}(\mathbf{A})$.

On the other hand, we call $\mathcal{K} \subseteq \mathbb{R}^d$ a **v-cone** if

$$\mathcal{K} = \left\{ \mathbf{B}\,\mathbf{y} : \mathbf{y} \geq \mathbf{0} \right\}$$

for some $\mathbf{B} \in \mathbb{R}^{d \times n}$, that is, $\mathcal{K}$ is a pointed cone with the column vectors of $\mathbf{B}$ as generators. In this case, we use the notation $\mathcal{K} = \mathrm{vcone}(\mathbf{B})$.

Note that according to our definitions, every h-cone and every v-cone contains the origin in its apex. We will prove that every h-cone is a v-cone and vice versa. More precisely:

Theorem A.1. *For every* $\mathbf{A} \in \mathbb{R}^{m \times d}$*, there exists* $\mathbf{B} \in \mathbb{R}^{d \times n}$ *(for some n) such that* $\mathrm{hcone}(\mathbf{A}) = \mathrm{vcone}(\mathbf{B})$*. Conversely, for every* $\mathbf{B} \in \mathbb{R}^{d \times n}$*, there exists* $\mathbf{A} \in \mathbb{R}^{m \times d}$ *(for some m) such that* $\mathrm{vcone}(\mathbf{B}) = \mathrm{hcone}(\mathbf{A})$*.*

We will prove the two halves of Theorem A.1 in Sections A.1 and A.2. For now, let's record that Theorem A.1 implies our goal, that is, the equivalence of the vertex and half-space description of a polytope:

Corollary A.2. *If $\mathcal{P}$ is the convex hull of finitely many points in $\mathbb{R}^d$, then $\mathcal{P}$ is the intersection of finitely many half-spaces in $\mathbb{R}^d$. Conversely, if $\mathcal{P}$ is given as the bounded intersection of finitely many half-spaces in $\mathbb{R}^d$, then $\mathcal{P}$ is the convex hull of finitely many points in $\mathbb{R}^d$.*

Proof. If $\mathcal{P} = \text{conv}\,\{\mathbf{v}_1, \mathbf{v}_2, \ldots, \mathbf{v}_n\}$ for some $\mathbf{v}_1, \mathbf{v}_2, \ldots, \mathbf{v}_n \in \mathbb{R}^d$, then coning over $\mathcal{P}$ (as defined in Chapter 3) gives

$$\text{cone}(\mathcal{P}) = \text{vcone}\begin{pmatrix} \mathbf{v}_1 & \mathbf{v}_2 & \cdots & \mathbf{v}_n \\ 1 & 1 & & 1 \end{pmatrix}.$$

By Theorem A.1, we can find a matrix $(\mathbf{A}, \mathbf{b}) \in \mathbb{R}^{m \times (d+1)}$ such that

$$\text{cone}(\mathcal{P}) = \text{hcone}(\mathbf{A}, \mathbf{b}) = \left\{ \mathbf{x} \in \mathbb{R}^{d+1} : (\mathbf{A}, \mathbf{b})\,\mathbf{x} \le \mathbf{0} \right\}.$$

We recover the polytope $\mathcal{P}$ on setting $x_{d+1} = 1$, that is,

$$\mathcal{P} = \left\{ \mathbf{x} \in \mathbb{R}^d : \mathbf{A}\mathbf{x} \le -\mathbf{b} \right\},$$

which is a hyperplane description of $\mathcal{P}$.

These steps can be reversed: Suppose the polytope $\mathcal{P}$ is given as

$$\mathcal{P} = \left\{ \mathbf{x} \in \mathbb{R}^d : \mathbf{A}\mathbf{x} \le -\mathbf{b} \right\}$$

for some $\mathbf{A} \in \mathbb{R}^{m \times d}$ and $\mathbf{b} \in \mathbb{R}^m$. Then $\mathcal{P}$ can be obtained from

$$\text{hcone}(\mathbf{A}, \mathbf{b}) = \left\{ \mathbf{x} \in \mathbb{R}^{d+1} : (\mathbf{A}, \mathbf{b})\,\mathbf{x} \le \mathbf{0} \right\}$$

by setting $x_{d+1} = 1$. By Theorem A.1, we can construct a matrix $\mathbf{B} \in \mathbb{R}^{(d+1) \times n}$ such that

$$\text{hcone}(\mathbf{A}, \mathbf{b}) = \text{vcone}(\mathbf{B}).$$

We may normalize the generators of $\text{vcone}(\mathbf{B})$, that is, the columns of $\mathbf{B}$, such that they all have their $(d+1)^{\text{st}}$ variable equal to 1:

$$\mathbf{B} = \begin{pmatrix} \mathbf{v}_1 & \mathbf{v}_2 & \cdots & \mathbf{v}_n \\ 1 & 1 & & 1 \end{pmatrix}.$$

Since $\mathcal{P}$ can be recovered from $\text{vcone}(\mathbf{B})$ by setting $x_{d+1} = 1$, we conclude that $\mathcal{P} = \text{conv}\,\{\mathbf{v}_1, \mathbf{v}_2, \ldots, \mathbf{v}_n\}$. $\square$

A.1 Every h-Cone Is a v-Cone

Suppose

$$\mathcal{K} = \text{hcone}(\mathbf{A}) = \left\{ \mathbf{x} \in \mathbb{R}^d : \mathbf{A}\mathbf{x} \le \mathbf{0} \right\}$$

for some $\mathbf{A} \in \mathbb{R}^{m \times d}$. We introduce an auxiliary m-dimensional variable $\mathbf{y}$ and write

$$\mathcal{K} = \left\{ \begin{pmatrix} \mathbf{x} \\ \mathbf{y} \end{pmatrix} \in \mathbb{R}^{d+m} : \mathbf{A}\mathbf{x} \leq \mathbf{y} \right\} \cap \left\{ \begin{pmatrix} \mathbf{x} \\ \mathbf{y} \end{pmatrix} \in \mathbb{R}^{d+m} : \mathbf{y} = \mathbf{0} \right\}. \qquad (A.1)$$

(Strictly speaking, this is $\mathcal{K}$ lifted into a d-dimensional subspace of $\mathbb{R}^{d+m}$.) Our goal in this section is to prove the following two lemmas.

Lemma A.3. *The h-cone* $\left\{ \begin{pmatrix} \mathbf{x} \\ \mathbf{y} \end{pmatrix} \in \mathbb{R}^{d+m} : \mathbf{A}\mathbf{x} \leq \mathbf{y} \right\}$ *is a v-cone.*

Lemma A.4. *If* $\mathcal{K} \subseteq \mathbb{R}^d$ *is a v-cone, then so is* $\mathcal{K} \cap \{\mathbf{x} \in \mathbb{R}^d : x_k = 0\}$, *for every* k.

The first half of Theorem A.1 follows with these two lemmas, since we can start with (A.1) and intersect with one hyperplane $y_k = 0$ at a time.

Proof of Lemma A.3. We begin by noting that

$$\mathcal{K} = \left\{ \begin{pmatrix} \mathbf{x} \\ \mathbf{y} \end{pmatrix} \in \mathbb{R}^{d+m} : \mathbf{A}\mathbf{x} \leq \mathbf{y} \right\}$$

$$= \left\{ \begin{pmatrix} \mathbf{x} \\ \mathbf{y} \end{pmatrix} \in \mathbb{R}^{d+m} : (\mathbf{A}, -\mathbf{I}) \begin{pmatrix} \mathbf{x} \\ \mathbf{y} \end{pmatrix} \leq \mathbf{0} \right\}$$

is an h-cone; here $\mathbf{I}$ represents the $m \times m$ identity matrix. Let's denote the kth unit vector by $\mathbf{e}_k$. Then we can decompose

$$\begin{pmatrix} \mathbf{x} \\ \mathbf{y} \end{pmatrix} = \sum_{j=1}^{d} x_j \begin{pmatrix} \mathbf{e}_j \\ \mathbf{A}\mathbf{e}_j \end{pmatrix} + \sum_{k=1}^{m} (y_k - (\mathbf{A}\mathbf{x})_k) \begin{pmatrix} \mathbf{0} \\ \mathbf{e}_k \end{pmatrix}$$

$$= \sum_{j=1}^{d} |x_j| \operatorname{sign}(x_j) \begin{pmatrix} \mathbf{e}_j \\ \mathbf{A}\mathbf{e}_j \end{pmatrix} + \sum_{k=1}^{m} (y_k - (\mathbf{A}\mathbf{x})_k) \begin{pmatrix} \mathbf{0} \\ \mathbf{e}_k \end{pmatrix}.$$

Note that if $\begin{pmatrix} \mathbf{x} \\ \mathbf{y} \end{pmatrix} \in \mathcal{K}$, then $y_k - (\mathbf{A}\mathbf{x})_k \geq 0$ for all k, and so $\begin{pmatrix} \mathbf{x} \\ \mathbf{y} \end{pmatrix}$ can be written as a nonnegative linear combination of the vectors $\operatorname{sign}(x_j) \begin{pmatrix} \mathbf{e}_j \\ \mathbf{A}\mathbf{e}_j \end{pmatrix}$, $1 \leq j \leq d$, and $\begin{pmatrix} \mathbf{0} \\ \mathbf{e}_k \end{pmatrix}$, $1 \leq k \leq m$. But this means that $\mathcal{K}$ is a v-cone. $\square$

Proof of Lemma A.4. Suppose $\mathcal{K} = \operatorname{vcone}(\mathbf{B})$, where $\mathbf{B}$ has the column vectors $\mathbf{b}_1, \mathbf{b}_2, \ldots, \mathbf{b}_n \in \mathbb{R}^d$; that is, $\mathbf{b}_1, \mathbf{b}_2, \ldots, \mathbf{b}_n$ are the generators of $\mathcal{K}$. Fix $k \leq d$ and construct a new matrix $\mathbf{B}_k$ whose column vectors are all the $\mathbf{b}_j$ for which $b_{jk} = 0$ together with the combinations $b_{ik}\mathbf{b}_j - b_{jk}\mathbf{b}_i$ whenever $b_{ik} > 0$ and $b_{jk} < 0$. We claim that

$$\mathcal{K} \cap \{\mathbf{x} \in \mathbb{R}^d : x_k = 0\} = \operatorname{vcone}(\mathbf{B}_k).$$

Every $\mathbf{x} \in \text{vcone}(\mathbf{B}_k)$ satisfies $x_k = 0$ by construction of $\mathbf{B}_k$, and so $\text{vcone}(\mathbf{B}_k) \subseteq \mathcal{K} \cap \{\mathbf{x} \in \mathbb{R}^d : x_k = 0\}$ follows immediately. We need to do some more work to prove the reverse containment.

Suppose $\mathbf{x} \in \mathcal{K} \cap \{\mathbf{x} \in \mathbb{R}^d : x_k = 0\}$, that is, $\mathbf{x} = \lambda_1 \mathbf{b}_1 + \lambda_2 \mathbf{b}_2 + \cdots + \lambda_n \mathbf{b}_n$ for some $\lambda_1, \lambda_2, \ldots, \lambda_n \geq 0$ and $x_k = \lambda_1 b_{1k} + \lambda_2 b_{2k} + \cdots + \lambda_n b_{nk} = 0$. This allows us to define

$$\Lambda = \sum_{i:\, b_{ik}>0} \lambda_i b_{ik} = -\sum_{j:\, b_{jk}<0} \lambda_j b_{jk}.$$

Note that $\Lambda \geq 0$. Now consider the decomposition

$$\mathbf{x} = \sum_{j:\, b_{jk}=0} \lambda_j \mathbf{b}_j + \sum_{i:\, b_{ik}>0} \lambda_i \mathbf{b}_i + \sum_{j:\, b_{jk}<0} \lambda_j \mathbf{b}_j. \tag{A.2}$$

If $\Lambda = 0$, then $\lambda_i b_{ik} = 0$ for all i such that $b_{ik} > 0$, and so $\lambda_i = 0$ for those i. Similarly, $\lambda_j = 0$ for all j such that $b_{jk} < 0$. Thus we conclude from $\Lambda = 0$ that

$$\mathbf{x} = \sum_{j:\, b_{jk}=0} \lambda_j \mathbf{b}_j \in \text{vcone}(\mathbf{B}_k).$$

Now assume $\Lambda > 0$. Then we can expand the decomposition (A.2) into

$$\mathbf{x} = \sum_{j:\, b_{jk}=0} \lambda_j \mathbf{b}_j + \frac{1}{\Lambda}\left(-\sum_{j:\, b_{jk}<0} \lambda_j b_{jk}\right)\left(\sum_{i:\, b_{ik}>0} \lambda_i \mathbf{b}_i\right)$$

$$+ \frac{1}{\Lambda}\left(\sum_{i:\, b_{ik}>0} \lambda_i b_{ik}\right)\left(\sum_{j:\, b_{jk}<0} \lambda_j \mathbf{b}_j\right)$$

$$= \sum_{j:\, b_{jk}=0} \lambda_j \mathbf{b}_j + \frac{1}{\Lambda} \sum_{\substack{i:\, b_{ik}>0 \\ j:\, b_{jk}<0}} \lambda_i \lambda_j \left(b_{ik}\mathbf{b}_j - b_{jk}\mathbf{b}_i\right),$$

which is by construction in $\text{vcone}(\mathbf{B}_k)$. $\square$

A.2 Every v-Cone Is an h-Cone

Suppose

$$\mathcal{K} = \text{vcone}(\mathbf{B}) = \{\mathbf{B}\mathbf{y} : \mathbf{y} \geq \mathbf{0}\}$$

for some $\mathbf{B} \in \mathbb{R}^{d \times n}$. Then $\mathcal{K}$ is the projection of

$$\left\{\begin{pmatrix} \mathbf{x} \\ \mathbf{y} \end{pmatrix} \in \mathbb{R}^{d+n} : \mathbf{y} \geq \mathbf{0},\ \mathbf{x} = \mathbf{B}\mathbf{y}\right\} \tag{A.3}$$

to the subspace $\left\{ \begin{pmatrix} \mathbf{x} \\ \mathbf{y} \end{pmatrix} \in \mathbb{R}^{d+n} : \mathbf{y} = \mathbf{0} \right\}$. The constraints for (A.3) can be written as

$$\mathbf{y} \geq \mathbf{0} \qquad \text{and} \qquad (\mathbf{I}, -\mathbf{B}) \begin{pmatrix} \mathbf{x} \\ \mathbf{y} \end{pmatrix} = \mathbf{0}.$$

Thus the set (A.3) is an h-cone, for which we can project one component of $\mathbf{y}$ at a time to obtain $\mathcal{K}$. This means that it suffices to prove the following lemma to finish the second half of Theorem A.1.

Lemma A.5. *If $\mathcal{K}$ is an h-cone, then the projection $\{\mathbf{x} - x_k\mathbf{e}_k : \mathbf{x} \in \mathcal{K}\}$ is also an h-cone, for every k.*

Proof. Suppose $\mathcal{K} = \mathrm{hcone}(\mathbf{A})$ for some $\mathbf{A} \in \mathbb{R}^{m \times d}$. Fix k and consider

$$\mathcal{P}_k = \{\mathbf{x} + \lambda\mathbf{e}_k : \mathbf{x} \in \mathcal{K}, \ \lambda \in \mathbb{R}\}.$$

The projection we are after can be constructed from this set as

$$\{\mathbf{x} - x_k\mathbf{e}_k : \mathbf{x} \in \mathcal{K}\} = \mathcal{P}_k \cap \{\mathbf{x} \in \mathbb{R}^d : x_k = 0\},$$

so that it suffices to prove that $\mathcal{P}_k$ is an h-cone.

Suppose $\mathbf{a}_1, \mathbf{a}_2, \ldots, \mathbf{a}_m$ are the row vectors of $\mathbf{A}$. We construct a new matrix $\mathbf{A}_k$ whose row vectors are all $\mathbf{a}_j$ for which $a_{jk} = 0$, and the combinations $a_{ik}\mathbf{a}_j - a_{jk}\mathbf{a}_i$ whenever $a_{ik} > 0$ and $a_{jk} < 0$. We claim that $\mathcal{P}_k = \mathrm{hcone}\,(\mathbf{A}_k)$.

If $\mathbf{x} \in \mathcal{K}$, then $\mathbf{A}\,\mathbf{x} \leq \mathbf{0}$, which implies $\mathbf{A}_k\,\mathbf{x} \leq \mathbf{0}$ because each row of $\mathbf{A}_k$ is a nonnegative linear combination of rows of $\mathbf{A}$; that is, $\mathcal{K} \subseteq \mathrm{hcone}\,(\mathbf{A}_k)$. However, the kth component of $\mathbf{A}_k$ is zero by construction, and so $\mathcal{K} \subseteq \mathrm{hcone}\,(\mathbf{A}_k)$ implies $\mathcal{P}_k \subseteq \mathrm{hcone}\,(\mathbf{A}_k)$.

Conversely, suppose $\mathbf{x} \in \mathrm{hcone}\,(\mathbf{A}_k)$. We need to find $\lambda \in \mathbb{R}$ such that $\mathbf{A}\,(\mathbf{x} - \lambda\mathbf{e}_k) \leq \mathbf{0}$, that is,

$$a_{11}x_1 + \cdots + a_{1k}(x_k - \lambda) + \cdots + a_{1d}x_d \leq 0,$$

$$\vdots$$

$$a_{m1}x_1 + \cdots + a_{mk}(x_k - \lambda) + \cdots + a_{md}x_d \leq 0.$$

The jth constraint is $\mathbf{a}_j \cdot \mathbf{x} - a_{jk}\lambda \leq 0$, that is, $\mathbf{a}_j \cdot \mathbf{x} \leq a_{jk}\lambda$. This gives the following conditions on λ:

$$\lambda \geq \frac{\mathbf{a}_i \cdot \mathbf{x}}{a_{ik}} \qquad \text{if } a_{ik} > 0,$$

$$\lambda \leq \frac{\mathbf{a}_j \cdot \mathbf{x}}{a_{jk}} \qquad \text{if } a_{jk} < 0.$$

Such a λ exists, because if $a_{ik} > 0$ and $a_{jk} < 0$, then (since $\mathbf{x} \in \mathrm{hcone}\,(\mathbf{A}_k)$)

$$(a_{ik}\mathbf{a}_j - a_{jk}\mathbf{a}_i) \cdot \mathbf{x} \leq 0,$$

which is equivalent to

$$\frac{\mathbf{a}_i \cdot \mathbf{x}}{a_{ik}} \leq \frac{\mathbf{a}_j \cdot \mathbf{x}}{a_{jk}} .$$

Thus we can find λ that satisfies

$$\frac{\mathbf{a}_i \cdot \mathbf{x}}{a_{ik}} \leq \lambda \leq \frac{\mathbf{a}_j \cdot \mathbf{x}}{a_{jk}} ,$$

which proves $\mathrm{hcone}\,(\mathbf{A}_k) \subseteq \mathcal{P}_k$. $\qquad\qquad\square$

Hints for ♣ Exercises

Well here's another clue for you all.

John Lennon & Paul McCartney ("Glass Onion," *The White Album*)

Chapter 1

1.1 Set up the partial fraction expansion as

$$\frac{z}{1 - z - z^2} = \frac{A}{1 - \frac{1+\sqrt{5}}{2} z} + \frac{B}{1 - \frac{1-\sqrt{5}}{2} z}$$

and clear denominators to compute A and B; one can do so, for example, by specializing z.

1.2 Multiply out $(1 - z)\left(1 + z + z^2 + \cdots + z^n\right)$. For the infinite sum, note that $\lim_{k \to \infty} z^k = 0$ if $|z| < 1$.

1.3 Start with the observation that there are $\lfloor x \rfloor + 1$ lattice points in the interval $[0, x]$.

1.4 (i) & (j) Write $n = qm + r$ for some integers q, r such that $0 \le r < m$. Distinguish the cases $r = 0$ and $r > 0$.

1.9 Use the fact that for m and n relatively prime and $a \in \mathbb{Z}$, there exists $b \in \mathbb{Z}$ (which is unique modulo n) such that $mb \equiv a \pmod{n}$. For the second equality of sets, think about the case $a = 0$.

1.12 First translate the line segment to the origin and explain why this translation leaves the integer-point enumeration invariant. For the case $(a, b) = (0, 0)$, first study the problem under the restriction that $\gcd(c, d) = 1$.

1.17 Given a triangle $\mathcal{T}$ with vertices on the integer lattice, consider the parallelogram $\mathcal{P}$ formed by two fixed edges of $\mathcal{T}$. Use integral translates of

$\mathcal{P}$ to tile the plane $\mathbb{R}^2$. Conclude from this tiling that $\mathcal{P}$ contains only its vertices as lattice points if and only if the area of $\mathcal{P}$ is 1.

1.20 Given an integer b, the Euclidean algorithm asserts the existence of $m_1, m_2, \ldots, m_d \in \mathbb{Z}$ such that b can be represented as $b = m_1 a_1 + m_2 a_2 + \cdots + m_d a_d$. Convince yourself that we can demand that in this representation, $0 \le m_2, m_3, \ldots, m_d < a_1$. Conclude that all integers beyond $(a_1 - 1)(a_2 + a_3 + \cdots + a_d)$ are representable in terms of $a_1, a_2, \ldots, a_d$. (This argument can be refined to yield another proof of Theorem 1.2.)

1.21 Use the setup

$$f(z) = \frac{A_1}{z} + \frac{A_2}{z^2} + \cdots + \frac{A_n}{z^n} + \frac{B_1}{z-1} + \frac{B_2}{(z-1)^2} + \sum_{k=1}^{a-1} \frac{C_k}{z - \xi_a^k} + \sum_{j=1}^{b-1} \frac{D_j}{z - \xi_b^j}.$$

To compute C_k, multiply both sides by $\left(z - \xi_a^k\right)$ and calculate the limit as $z \to \xi_a^k$. The coefficients D_j can be computed in a similar fashion.

1.22 Use Exercise 1.9 (with $m = b^{-1}$) on the left-hand side of the equation.

1.24 Suppose $a > b$. The integer $a + b$ certainly has a representation in terms of a and b, namely, $1 \cdot a + 1 \cdot b$. Think about how the coefficient of b would change if we changed the coefficient of a.

1.31 Use the partial fraction setup (1.11), multiply both sides by $\left(z - \xi_{a_1}^k\right)$, and take the limit as $z \to \xi_{a_1}^k$.

1.33 Convince yourself of the generating-function setup

$$\sum_{n \ge 1} p_A^\circ(n) \, z^n = \left(\frac{z^{a_1}}{1 - z^{a_1}} \right) \left(\frac{z^{a_2}}{1 - z^{a_2}} \right) \cdots \left(\frac{z^{a_d}}{1 - z^{a_d}} \right).$$

Now use the machinery of Section 1.5.

Chapter 2

2.1 Use Exercise 1.3 for the closed interval. For open intervals, you can use Exercise 1.4(j) or the $\lceil \ldots \rceil$ notation of Exercise 1.4(e). To show the quasipolynomial character, rewrite the greatest-integer function in terms of the fractional-part function.

2.2 Write $\mathcal{R}$ as a direct product of two intervals and use Exercise 1.3.

2.6 Start by showing that the convex hull of a d-element subset W of V is a face of Δ. This allows you to prove the first statement by induction (using Exercise 2.5). For the converse statement, given a supporting hyperplane H that defines the face $\mathcal{F}$ of Δ, let $W \subseteq V$ consist of those vertices of Δ that are in H. Now prove that every point

$$\mathbf{x} = \lambda_1 \mathbf{v}_1 + \lambda_2 \mathbf{v}_2 + \cdots + \lambda_{d+1} \mathbf{v}_{d+1}$$

in $\mathcal{F}$ has to satisfy $\lambda_k = 0$ for all $\mathbf{v}_k \notin W$.

2.7 First show that the linear inequalities and equations describing a rational polytope can be chosen with rational coefficients, and then clear denominators.

2.8

(a) Prove that as rational functions, $\sum_{j \geq 0} j^d z^j = (-1)^{d+1} \sum_{j \geq 0} j^d \left(\frac{1}{z} \right)^j$.

(b) Use the fact that $\frac{\sum_{k=0}^{d} A(d,k) z^k}{(1-z)^{d+1}} = z \frac{d}{dz} \left(\frac{\sum_{k=0}^{d-1} A(d-1,k) z^k}{(1-z)^d} \right)$.

(c) Begin by proving one of the remarks in the notes of Chapter 2, namely, that $A(d,k)$ counts the permutations of $\{1, 2, \ldots, d\}$ with $k - 1$ ascents.

(d) Use the fact that $\sum_{k=0}^{d} A(d,k) z^k = (1-z)^{d+1} \sum_{j \geq 0} j^d z^j$.

2.9 Write $\frac{1}{(1-z)^{d+1}} = \left(\sum_{k_1 \geq 0} z^{k_1} \right) \left(\sum_{k_2 \geq 0} z^{k_2} \right) \cdots \left(\sum_{k_{d+1} \geq 0} z^{k_{d+1}} \right)$ and come up with a combinatorial enumeration scheme to compute the coefficients of this power series.

2.10 Write $\binom{t+k}{d} = \frac{(t+k)(t+k-1) \cdots (t+k-d+1)}{d!}$ and switch t to $-t$.

2.14 Think about the poles of the function $\frac{z}{e^z - 1}$ and use a theorem from complex analysis.

2.15 Compute the generating function of $B_d(1-x)$ and rewrite it as $\frac{z e^{-xz}}{1 - e^{-z}}$.

2.16 Show that $\frac{z}{e^z - 1} + \frac{1}{2} z$ is an even function of z.

2.23 Follow the steps of the proof of Theorem 2.4.

2.25 Extend $\mathcal{T}$ to a rectangle whose diagonal is the hypotenuse of $\mathcal{T}$, and consider the lattice points on this diagonal separately.

2.26 For the area use elementary calculus. For the number of boundary points on $t\mathcal{P}$, extend Exercise 1.12 to a set of line segments whose union forms a simple closed curve.

2.33 Rewrite the inequality as $\left(\left\lceil \frac{ta}{d} \right\rceil - 1 \right) e + \left(\left\lceil \frac{tb}{d} \right\rceil - 1 \right) f \leq tr$ and compare this with the definition of $\mathcal{T}$.

2.34 To compute C_3, multiply both sides of (2.20) by $(z-3)^2$ and compute the limit as $z \to 1$. The coefficients A_j and B_l can be computed in a similar fashion. To compute C_2, first move $\frac{C_3}{(z-1)^3}$ in (2.20) to the left-hand side, then multiply by $(z-1)^2$ and take the limit as $z \to 1$. A similar, even more elaborate, computation gives C_1. (Alternatively, compute the Laurent series of the function in (2.20) at $z = 1$ with a computer algebra system such as Maple, Mathematica, or Sage.)

2.35 Show that $\lim_{z \to \xi_a^k} \frac{1}{1-z^{ab}} + \frac{\xi_a^k}{ab} \left(z - \xi_a^k\right)^{-1} = \frac{ab-1}{2ab}$.

2.36 Follow the proof of Theorem 2.10. Use Exercise 2.35 to compute the additional coefficients in the partial fraction expansion of the generating function corresponding to this lattice-point count.

2.38 Start with computing the constant term of

$$\frac{1}{(1 - z_1 z_2)(1 - z_1^2 z_2)(1 - z_1)(1 - z_2) z_1^{3t} z_2^{2t}}$$

with respect to z_2 by treating z_1 as a constant and setting up a partial fraction expansion of this function with respect to z_2.

Chapter 3

3.3

(a) Convince yourself that if the supporting hyperplane of $\mathcal{F}$ is horizontal, then $h_1, h_2, \ldots, h_n$ in (3.1) were not chosen randomly.
(b) Prove that if $a_{d+1} > 0$, then there cannot exist a point $(x_1, x_2, \ldots, x_{d+1}) \in \mathcal{F}$ and an $\epsilon > 0$ such that $(x_1, x_2, \ldots, x_{d+1} - \epsilon) \in \mathcal{Q}$, and if $a_{d+1} < 0$ and $(x_1, x_2, \ldots, x_{d+1}) \in \mathcal{F}$, then there exists $\epsilon > 0$ such that $(x_1, x_2, \ldots, x_{d+1} - \epsilon) \in \mathcal{Q}$.
(c) Use the fact that $\mathcal{F}_1 \cap \mathcal{F}_2$ is a face common to both $\mathcal{F}_1$ and $\mathcal{F}_2$.

3.5 Write the simplicial cones as cones over simplices and use Exercise 2.6.

3.8 Write down a typical term of the product

$$\sigma_S (z_1, z_2, \ldots, z_m) \, \sigma_T (z_{m+1}, z_{m+2}, \ldots, z_{m+n}) \,.$$

3.9 Multiply out $\mathbf{z}^\mathbf{m} \sigma_{\mathcal{K}}(\mathbf{z})$.

3.10 Write a typical term in $\sigma_S \left(\frac{1}{z_1}, \frac{1}{z_2}, \ldots, \frac{1}{z_d}\right) = \sigma_S \left(z_1^{-1}, z_2^{-1}, \ldots, z_d^{-1}\right)$.

3.13 Given the polynomial f, split up the generating function on the left-hand side according to the terms of f and use (2.2). Conversely, if the polynomial g is given, use (2.6).

3.18 Show that $H \cap \mathbb{Z}^d$ is a $\mathbb{Z}$-module. Therefore, it has a basis; extend this basis to a basis of $\mathbb{Z}^d$.

3.19 Think about a small (even irrational) perturbation of all hyperplanes in the right direction.

3.21 View

$$
\det\begin{pmatrix}
x_1^d & x_1^{d-1} & \cdots & x_1 & 1 \\
x_2^d & x_2^{d-1} & \cdots & x_2 & 1 \\
\vdots & \vdots & & \vdots & \vdots \\
x_{d+1}^d & x_{d+1}^{d-1} & \cdots & x_{d+1} & 1
\end{pmatrix}
$$

as a polynomial in x_1, considering $x_2, x_3, \ldots, x_{d+1}$ constants. Show that this polynomial has roots at $x_1 = x_2$, $x_1 = x_3$, $\ldots$, $x_1 = x_{d+1}$, and compute its leading coefficient.

3.24 Given f, split up the generating function on the left-hand side according to the constituents of f; then use Exercise 3.13. Conversely, given g and h, multiply both by a polynomial to get the denominator on the right-hand side into the form $(1 - z^p)^{d+1}$; then use (2.6).

3.25 Start with the setup on page 80, and closely orient yourself along the proof of Theorem 3.8.

3.32 Convince yourself that if $\mathcal{P}$ has denominator p, then $L_{\mathcal{P}}(0)$ equals the constant term of the Ehrhart *polynomial* of $p\mathcal{P}$.

3.34 Use Lemma 3.19.

Chapter 4

4.1 Use Exercise 2.1.

4.2 Use the explicit description of Π given by (4.3).

4.4 Consider each simplicial cone $\mathcal{K}_j$ separately, and look at the arrangement of its bounding hyperplanes. For each hyperplane, use Exercise 3.18.

4.7 For (a), convince yourself that $Q(-t)$ is also a quasipolynomial. For (b), use (1.3). For (c), differentiate (1.3). For (d), think about one constituent of the quasipolynomial at a time.

4.8 In the generating function for $L_{\mathcal{P}}(t - k)$, make a change in the summation variable; then use Theorem 4.4.

4.13 Use the fact that $\mathbf{A}$ has only integral entries. For the second part, write down the explicit hyperplane descriptions of $(t + 1)\mathcal{P}^\circ$ and $t\mathcal{P}$.

4.14 Assume that there exist $t \in \mathbb{Z}$ and a facet hyperplane H of $\mathcal{P}$ such that there is a lattice point between tH and $(t + 1)H$. Translate this lattice point to a lattice point that violates (4.12).

Chapter 5

5.3 To prove the inclusion $\subseteq$ (the other inclusion is clear), you need to show that every point in $\mathcal{P}$ lies in the interior of some face $\mathcal{F}$ of $\mathcal{P}$, and that the relative interiors of two different faces are disjoint. One can prove both facts by looking for a minimal face that contains a given point in $\mathcal{P}$ in its relative interior.

5.5 Consider an **interval** $[\mathcal{F}, \mathcal{P}]$ in the face lattice of $\mathcal{P}$; namely, $[\mathcal{F}, \mathcal{P}]$ contains all faces $\mathcal{G}$ such that $\mathcal{F} \subseteq \mathcal{G} \subseteq \mathcal{P}$. Prove that if $\mathcal{P}$ is simple, then every such interval is isomorphic to a Boolean lattice.

5.6 Use Exercise 2.6 to show that the face lattice of a simplex is isomorphic to a Boolean lattice.

5.7 Orient yourself along the proof of Theorem 5.3, but start with the Euler relation (Theorem 5.2) for a given face $\mathcal{F}$ instead of (5.2).

5.9 Assume that the origin is in the interior of $\mathcal{P}$.

(a) Given a face $\mathcal{F}$ of $\mathcal{P}$, show that

$$\mathcal{F}^* := \{\mathbf{x} \in \mathbb{R}^d : \mathbf{x} \cdot \mathbf{y} \leq 1 \text{ for all } \mathbf{y} \in \mathcal{P}, \ \mathbf{x} \cdot \mathbf{y} = 1 \text{ for all } \mathbf{y} \in \mathcal{F}\}$$

is a face of $\mathcal{P}^*$, and that $\mathcal{G} \subseteq \mathcal{F}$ if and only if $\mathcal{G}^* \supseteq \mathcal{F}^*$.
(b) Figure out a prominent property of the face lattice of every simple/simplicial polytope; then use (a).
(c) Try a proof that is dual to the one hinted at in Exercise 5.7.

5.14

(a) First choose any two of the five points on the sphere, and draw a great circle through them. This great circle partitions the sphere into two hemispheres, and the union of these two hemispheres now contains three of the five given points. Now the pigeonhole principle tells us that (at least) two of the five given points must lie in one of these two hemispheres.
(b) Normalize the five given vectors to have unit length and use part (a) of this exercise.

Chapter 6

6.1 Think permutation matrices.

6.3 Show that the rank of (6.5) is $2n - 1$.

6.5 Start by showing that all permutation matrices are indeed vertices. Then use Exercise 6.4 to show that there are no other vertices.

6.6 Establish a bijection between semimagic squares with line sum $t - n$ and semimagic squares with *positive* entries and line sum t.

6.7 Think about the smallest possible line sum if the entries of the square are positive integers.

6.8 Follow the computation on page 120 that led to the formula for H_2.

6.9 Multiply both sides of (6.7) by $\left(w - \frac{1}{z_k} \right)$ and take the limit as $w \to \frac{1}{z_k}$.

6.10 Orient yourself along the computation in (6.10).

6.16 Compute the matrix equivalent to (6.5) for the polytope describing all magic squares of a given size. Show that this matrix has rank $2n + 1$.

6.18 Orient yourself along the computation on page 120.

Chapter 7

7.2 Use Exercise 7.1.

7.5 Differentiate (1.3).

7.6 Use (1.3).

7.7 Write an arbitrary function on $\mathbb{Z}$ with period b in terms of $\delta_m(x)$, $1 \le m \le b$.

7.8 Use the definition (7.6) of the inner product and the properties $z\bar{z} = |z|^2$ and $\overline{(zw)} = \bar{z} \cdot \bar{w}$ for complex numbers z and w.

7.15 Use the definition (7.4) and simplify the fractional-part function in the sum on the right-hand side.

7.23 Use the definition of **F**.

Chapter 8

8.7 Use Exercise 1.9.

8.9 Use the methods outlined in the hints for Exercises 1.21 and 2.34 to compute the partial fraction coefficients for $z = 1$ in (8.3).

8.12 Multiply out all the terms on the left-hand side and make use of Exercises 1.9 and 7.15.

8.15 Use the methods outlined in the hints for Exercises 1.21 and 2.34 to compute a partial fraction expansion of (8.8).

Chapter 9

9.1 Write out a typical term in $p(z_1, z_2, \ldots, z_d)\, q(z_1, z_2, \ldots, z_d)$.

9.5 Review the definition of $\mathcal{Z}(\mathbf{u}_1, \mathbf{u}_2, \ldots \mathbf{u}_n)$ to show that

$$\Pi_1 \cup \Pi_2 \cup \cdots \cup \Pi_k \cup \mathcal{P}_1 \cup \mathcal{P}_2 \cup \cdots \cup \mathcal{P}_m = \mathcal{Z}(\mathbf{u}_1, \mathbf{u}_2, \ldots \mathbf{u}_n)$$

and then show that this union is disjoint, using the precise definition of the $\mathcal{P}_j$'s.

9.9 Convince yourself that $\mathcal{P}_d$ lies in the hyperplane given by the equation $x_1 + x_2 + \cdots + x_d = \binom{d}{2}$, and find $d - 1$ linearly independent vectors among those defining $\mathcal{P}_d$. To show that $\mathbf{v} = (\pi(1) - 1, \pi(2) - 1, \ldots, \pi(d) - 1)$ is a vertex of $\mathcal{P}_d$, consider the hyperplane through $\mathbf{v}$ with normal vector $\mathbf{v}$.

Chapter 10

10.1 Expand the powers of $(1 - z)$ in the definition of $h_{\mathcal{P}}(z)$.

10.2 Assume that $\mathcal{P}$ is full-dimensional and consider a facet of $\mathcal{S}$ by constructing one of its supporting hyperplanes. Show that this hyperplane can contain no more than d vertices of $\mathcal{S}$. To show that the face numbers of $\mathcal{S}$ equal those of the triangulation of $\partial\mathcal{P}$, try to construct a bijection.

10.4 Given the unimodular k-simplex $\Delta \subset \mathbb{R}^d$, construct a bijection $\operatorname{span}(\Delta) \to \mathbb{R}^k$ that maps Δ to $\{\mathbf{x} \in \mathbb{R}^k_{\geq 0} : x_1 + x_2 + \cdots + x_k \leq 1\}$ and $\operatorname{span}(\Delta) \cap \mathbb{Z}^d$ to $\mathbb{Z}^k$.

10.7 Start by showing that if Δ is a k-simplex and

$$\Pi := \{\lambda_1 \mathbf{w}_1 + \lambda_2 \mathbf{w}_2 + \cdots + \lambda_{k+1} \mathbf{w}_{k+1} : 0 < \lambda_1, \lambda_2, \ldots, \lambda_{k+1} < 1\}$$

is the open fundamental parallelepiped of $\operatorname{cone}(\Delta)$, then

$$\Pi = -\Pi + \mathbf{w}_1 + \mathbf{w}_2 + \cdots + \mathbf{w}_{k+1}.$$

10.8 Carefully review the definition of $\operatorname{link}(\Delta)$.

10.9 Begin by rewriting (10.3), collecting terms stemming from simplices Φ of the same dimension.

10.10 Begin by proving that

$$\Pi(\Delta) = \{\mathbf{0}\} \cup \bigcup_{\Omega \subseteq \Delta} \Pi(\Omega)^\circ,$$

where the union is over all nonempty faces of Δ, and that this union is disjoint.

10.11 Build an almost-one-to-one correspondence between the simplices in T and those in $T_0 \setminus T$, and see what this correspondence gives about the respective h-vectors of links.

10.12 Establish a one-to-one correspondence between $\mathrm{link}(\Delta)$ and the boundary faces of a polytope of dimension $d - \dim(\Delta) - 1$ respecting the face relations.

10.13 Writing $a(z) = a_d z^d + a_{d-1} z^{d-1} + \cdots + a_0$ and $b(z) = b_{d-1} z^{d-1} + b_{d-2} z^{d-2} + \cdots + b_0$, give a concrete rule how to compute a_0, a_d, b_{d-1}, b_0, a_1, a_{d-1}, b_{d-2}, $\ldots$ from the coefficients of $p(z)$.

10.14 See what your rule from Exercise 10.13 says about the coefficients of $h_{\mathcal{P}}^*(z)$ when one takes into account that the a_j's and b_j's are nonnegative.

Chapter 11

11.4 Consider the hyperplanes $H_1, H_2, \ldots, H_{d+1}$ that bound Δ. For each hyperplane H_k, denote by H_k^+ the closed half-space bounded by H_k that contains Δ, and by H_k^- the open half-space bounded by H_k that does not contain Δ. Show that every tangent cone of Δ is the intersection of some of the H_k^+'s, and conversely, that every intersection of some of the H_k^+'s, except for $\Delta = \bigcap_{k=1}^{d+1} H_k^+$, is a tangent cone of Δ. Since $H_k^+ \cup H_k^- = \mathbb{R}^d$ as a disjoint union, for each k, the point $\mathbf{x}$ is either in H_k^+ or H_k^-. Prove that the intersection of those H_k^+ that contain $\mathbf{x}$ is the sought-after tangent cone.

11.6 As in Exercise 5.6, show that the face lattice of a simplex is a Boolean lattice. Note that every sublattice of a Boolean lattice is again Boolean.

11.8 One approach to this problem is first to dilate $\mathcal{P}$ and the corresponding hyperplanes in H by a small factor. To avoid subtleties, first translate $\mathcal{P}$ by an integer vector, if necessary, to ensure that none of the hyperplanes in H contains the origin. Use Exercise 3.18.

11.9 Adjust the steps in Section 11.4 to open polytopes. Start by proving a Brianchon–Gram identity for open simplices, by analogy with Theorem 11.5. This implies a Brion-type identity for open simplices, as in Corollary 11.6. Finally, adjust the proof of Theorem 11.7 to open polytopes.

Chapter 12

12.1 Use (12.3), Exercise 2.18, and (2.11).

12.2 Review the proof of Theorem 3.5.

12.3 Use the definition of unimodularity to show that the only integer point in the fundamental parallelepiped of $\mathcal{K}$ is $\mathbf{v}$.

12.6 Orient yourself along the proof of Theorem 12.3; instead of a sum over vertex cones, just consider one simple cone $\mathcal{K}$.

Chapter 13

13.5 Multiply out $\mathbf{z}^m \alpha_{\mathcal{K}}(\mathbf{z})$.

13.6 Orient yourself along the proof of Theorem 4.2. Note that for solid angles, we do not require the condition that the boundary of $\mathcal{K}$ contains no lattice point.

13.7 As a warmup exercise, show that

$$\sum_{\substack{\mathcal{F} \subseteq \Delta \\ \dim \mathcal{F} > 0}} \sum_{\mathbf{v} \text{ a vertex of } \mathcal{F}} \sigma_{\mathcal{K}_{\mathbf{v}}(\mathcal{F})^\circ}(\mathbf{z}) = \sum_{\mathbf{v} \text{ a vertex of } \Delta} \sum_{\substack{\mathcal{F} \subseteq \mathcal{K}_{\mathbf{v}} \\ \dim \mathcal{F} > 0}} \sigma_{\mathcal{F}^\circ}(\mathbf{z}).$$

13.8 Start with the setup of our second proof of Ehrhart's theorem in Section 11.5; that is, it suffices to prove that if p is the denominator of $\mathcal{P}$, then $A_{\mathcal{P}}(-r - pt) = (-1)^{\dim \mathcal{P}} A_{\mathcal{P}}(r + pt)$ for all integers r and t with $0 \le r < p$ and $t > 0$. (Think of r as fixed and t as variable.) Now orient yourself along the proof on page 232.

Chapter 14

14.1 Bound the integral from above, using the length of C_r and an upper bound for the absolute value of the integrand.

14.2 The nontrivial roots of unity are simple poles of f, for which the residue computation boils down to a simple limit.

14.5 Start by combining the terms $\frac{1}{z-(m+ni)}$ and $\frac{1}{m+ni}$ into one fraction.

14.6 Differentiate (14.1) term by term.

14.7 Compute $\wp'$ explicitly.

14.8 Use a famous theorem from complex analysis.

14.9 Compute $\wp(-z)$ and use the fact that $(-(m+in))^2 = (m+in)^2$.

14.10 Repeat the proof of Lemma 14.3, but now starting with the proof of $\wp'(z+i) = \wp'(z)$.

14.11 Use a famous theorem from complex analysis.

14.12 Use the definition of the Weierstraß $\wp$-function.

References

1. *The On-Line Encyclopedia of Integer Sequences*, published electronically at http://oeis.org, 2014.
2. Maya Ahmed, Jesús A. De Loera, and Raymond Hemmecke, *Polyhedral cones of magic cubes and squares*, Discrete and Computational Geometry, Algorithms Combin., vol. 25, Springer, Berlin, 2003, pp. 25–41, `arXiv:math.CO/0201108`.
3. Maya Mohsin Ahmed, *How many squares are there, Mr. Franklin? constructing and enumerating Franklin squares*, Amer. Math. Monthly **111** (2004), no. 5, 394–410.
4. Iskander Aliev, Martin Henk, and Aicke Hinrichs, *Expected Frobenius numbers*, J. Combin. Theory Ser. A **118** (2011), no. 2, 525–531.
5. Iskander M. Aliev and Peter M. Gruber, *An optimal lower bound for the Frobenius problem*, J. Number Theory **123** (2007), no. 1, 71–79.
6. Harsh Anand, Vishwa Chander Dumir, and Hansraj Gupta, *A combinatorial distribution problem*, Duke Math. J. **33** (1966), 757–769.
7. W. S. Andrews, *Magic Squares and Cubes*, Dover Publications Inc., New York, 1960.
8. Tom M. Apostol, *Generalized Dedekind sums and transformation formulae of certain Lambert series*, Duke Math. J. **17** (1950), 147–157.
9. Tom M. Apostol and Thiennu H. Vu, *Identities for sums of Dedekind type*, J. Number Theory **14** (1982), no. 3, 391–396.
10. Federico Ardila, Federico Castillo, and Michael Henley, *The arithmetic Tutte polynomials of the classical root systems*, Int. Math. Res. Not. (to appear), `arXiv:1305.6621`.
11. Vladimir I. Arnold, *Arnold's problems*, Springer-Verlag, Berlin, 2004, Translated and revised edition of the 2000 Russian original, with a preface by V. Philippov, A. Yakivchik, and M. Peters.
12. Christos A. Athanasiadis, *h^*-vectors, Eulerian polynomials and stable polytopes of graphs*, Electron. J. Combin. **11** (2004/06), no. 2, Research Paper 6, 13 pp.
13. ———, *Ehrhart polynomials, simplicial polytopes, magic squares and a conjecture of Stanley*, J. Reine Angew. Math. **583** (2005), 163–174, `arXiv:math/0312031`.
14. Velleda Baldoni, Nicole Berline, Matthias Köppe, and Michèle Vergne, *Intermediate sums on polyhedra: computation and real Ehrhart theory*, Mathematika **59** (2013), no. 1, 1–22, `arXiv:math/1011.6002`.
15. Velleda Baldoni, Nicole Berline, and Michèle Vergne, *Local Euler–Maclaurin expansion of Barvinok valuations and Ehrhart coefficients of a rational polytope*, Integer points in polyhedra—geometry, number theory, representation theory, algebra, optimization, statistics, Contemp. Math., vol. 452, Amer. Math. Soc., Providence, RI, 2008, pp. 15–33.
16. Velleda Baldoni and Michèle Vergne, *Residue formulae for volumes and Ehrhart polynomials of convex polytopes*, Preprint (`arXiv:math.CO/0103097`).

17. Philippe Barkan, *Sur les sommes de Dedekind et les fractions continues finies*, C. R. Acad. Sci. Paris Sér. A-B **284** (1977), no. 16, A923–A926.

18. Peter Barlow, *An Elementary Investigation of the Theory of Numbers*, J. Johnson & Co., London, 1811.

19. Alexander Barvinok, *Exponential integrals and sums over convex polyhedra*, Funktsional. Anal. i Prilozhen. **26** (1992), no. 2, 64–66.

20. _____, *A Course in Convexity*, Graduate Studies in Mathematics, vol. 54, American Mathematical Society, Providence, RI, 2002.

21. _____, *Integer points in polyhedra*, Zurich Lectures in Advanced Mathematics, European Mathematical Society (EMS), Zurich, 2008.

22. Alexander Barvinok and James E. Pommersheim, *An algorithmic theory of lattice points in polyhedra*, New Perspectives in Algebraic Combinatorics (Berkeley, CA, 1996–97), Math. Sci. Res. Inst. Publ., vol. 38, Cambridge Univ. Press, Cambridge, 1999, pp. 91–147.

23. Alexander Barvinok and Kevin Woods, *Short rational generating functions for lattice point problems*, J. Amer. Math. Soc. **16** (2003), no. 4, 957–979, arXiv:math.CO/0211146.

24. Alexander I. Barvinok, *A polynomial time algorithm for counting integral points in polyhedra when the dimension is fixed*, Math. Oper. Res. **19** (1994), no. 4, 769–779.

25. Victor V. Batyrev, *Dual polyhedra and mirror symmetry for Calabi–Yau hypersurfaces in toric varieties*, J. Algebraic Geom. **3** (1994), no. 3, 493–535, arXiv:alg-geom/9310003.

26. Victor V. Batyrev and Dimitrios I. Dais, *Strong McKay correspondence, string-theoretic Hodge numbers and mirror symmetry*, Topology **35** (1996), no. 4, 901–929.

27. Abdelmejid Bayad and Matthias Beck, *Relations for Bernoulli–Barnes numbers and Barnes zeta functions*, Int. J. Number Theory **10** (2014), no. 5, 1321–1335, arXiv:1301.7097.

28. Abdelmejid Bayad and Yilmaz Simsek, *Dedekind sums involving Jacobi modular forms and special values of Barnes zeta functions*, Ann. Inst. Fourier (Grenoble) **61** (2011), no. 5, 1977–1993 (2012).

29. Matthias Beck, *Counting lattice points by means of the residue theorem*, Ramanujan J. **4** (2000), no. 3, 299–310, arXiv:math.CO/0306035.

30. _____, *Multidimensional Ehrhart reciprocity*, J. Combin. Theory Ser. A **97** (2002), no. 1, 187–194, arXiv:math.CO/0111331.

31. _____, *Dedekind cotangent sums*, Acta Arith. **109** (2003), no. 2, 109–130, arXiv:math.NT/0112077.

32. Matthias Beck and Benjamin Braun, *Euler–Mahonian statistics via polyhedral geometry*, Adv. Math. **244** (2013), 925–954, arXiv:1109.3353.

33. Matthias Beck and Anastasia Chavez, *Bernoulli-Dedekind sums*, Acta Arith. **149** (2011), no. 1, 65–82, arXiv:1008.0038.

34. Matthias Beck, Beifang Chen, Lenny Fukshansky, Christian Haase, Allen Knutson, Bruce Reznick, Sinai Robins, and Achill Schürmann, *Problems from the Cottonwood Room*, Integer Points in Polyhedra—Geometry, Number Theory, Algebra, Optimization, Contemp. Math., vol. 374, Amer. Math. Soc., Providence, RI, 2005, pp. 179–191.

35. Matthias Beck, Moshe Cohen, Jessica Cuomo, and Paul Gribelyuk, *The number of "magic" squares, cubes, and hypercubes*, Amer. Math. Monthly **110** (2003), no. 8, 707–717, arXiv:math.CO/0201013.

36. Matthias Beck, Jesús A. De Loera, Mike Develin, Julian Pfeifle, and Richard P. Stanley, *Coefficients and roots of Ehrhart polynomials*, Integer Points in Polyhedra—Geometry, Number Theory, Algebra, Optimization, Contemp. Math., vol. 374, Amer. Math. Soc., Providence, RI, pp. 15–36, arXiv:math.CO/0402148.

37. Matthias Beck, Ricardo Diaz, and Sinai Robins, *The Frobenius problem, rational polytopes, and Fourier–Dedekind sums*, J. Number Theory **96** (2002), no. 1, 1–21, arXiv:math.NT/0204035.

38. Matthias Beck and Richard Ehrenborg, *Ehrhart–Macdonald reciprocity extended*, Preprint (arXiv:math.CO/0504230).

39. Matthias Beck, Christian Haase, and Asia R. Matthews, *Dedekind–Carlitz polynomials as lattice-point enumerators in rational polyhedra*, Math. Ann. **341** (2008), no. 4, 945–961, arXiv:0710.1323.

40. Matthias Beck, Christian Haase, and Frank Sottile, *Formulas of Brion, Lawrence, and Varchenko on rational generating functions for cones*, Math. Intelligencer **31** (2009), no. 1, 9–17, arXiv:math.CO/0506466.

41. Matthias Beck and Curtis Kifer, *An extreme family of generalized Frobenius numbers*, Integers **11** (2011), Paper A24, 6 pp.

42. Matthias Beck and Dennis Pixton, *The Ehrhart polynomial of the Birkhoff polytope*, Discrete Comput. Geom. **30** (2003), no. 4, 623–637, arXiv:math.CO/0202267.

43. Matthias Beck and Sinai Robins, *Explicit and efficient formulas for the lattice point count in rational polygons using Dedekind–Rademacher sums*, Discrete Comput. Geom. **27** (2002), no. 4, 443–459, arXiv:math.CO/0111329.

44. _____, *A formula related to the Frobenius problem in two dimensions*, Number theory (New York, 2003), Springer, New York, 2004, pp. 17–23, arXiv:math.NT/0204037.

45. Matthias Beck, Sinai Robins, and Steven V Sam, *Positivity theorems for solid-angle polynomials*, Beiträge Algebra Geom. **51** (2010), no. 2, 493–507, arXiv:0906.4031.

46. Matthias Beck and Frank Sottile, *Irrational proofs for three theorems of Stanley*, European J. Combin. **28** (2007), no. 1, 403–409, arXiv:math.CO/0506315.

47. Matthias Beck and Andrew van Herick, *Enumeration of* 4×4 *magic squares*, Math. Comp. **80** (2011), no. 273, 617–621, arXiv:0907.3188.

48. Matthias Beck and Thomas Zaslavsky, *An enumerative geometry for magic and magilatin labellings*, Ann. Comb. **10** (2006), no. 4, 395–413, arXiv:math.CO/0506315.

49. Dale Beihoffer, Jemimah Hendry, Albert Nijenhuis, and Stan Wagon, *Faster algorithms for Frobenius numbers*, Electron. J. Combin. **12** (2005), no. 1, Research Paper 27, 38 pp.

50. Nicole Berline and Michèle Vergne, *Local Euler–Maclaurin formula for polytopes*, Mosc. Math. J. **7** (2007), no. 3, 355–386, 573, arXiv:math.CO/0507256.

51. Bruce C. Berndt, *Reciprocity theorems for Dedekind sums and generalizations*, Advances in Math. **23** (1977), no. 3, 285–316.

52. Bruce C. Berndt and Ulrich Dieter, *Sums involving the greatest integer function and Riemann–Stieltjes integration*, J. Reine Angew. Math. **337** (1982), 208–220.

53. Bruce C. Berndt and Boon Pin Yeap, *Explicit evaluations and reciprocity theorems for finite trigonometric sums*, Adv. in Appl. Math. **29** (2002), no. 3, 358–385.

54. Ulrich Betke and Peter McMullen, *Lattice points in lattice polytopes*, Monatsh. Math. **99** (1985), no. 4, 253–265.

55. Christian Bey, Martin Henk, and Jörg M. Wills, *Notes on the roots of Ehrhart polynomials*, Discrete Comput. Geom. **38** (2007), no. 1, 81–98, arXiv:math.MG/0606089.

56. Louis J. Billera and A. Sarangarajan, *The combinatorics of permutation polytopes*, Formal Power Series and Algebraic Combinatorics (New Brunswick, NJ, 1994), Amer. Math. Soc., Providence, RI, 1996, pp. 1–23.

57. Garrett Birkhoff, *Tres observaciones sobre el álgebra lineal*, Revista Facultad de Ciencias Exactas, Puras y Aplicadas Universidad Nacional de Tucumán, Serie A (Matemáticas y Física Teórica) **5** (1946), 147–151.

58. G. R. Blakley, *Combinatorial remarks on partitions of a multipartite number*, Duke Math. J. **31** (1964), 335–340.

59. Sebastian Böcker and Zsuzsanna Lipták, *The money changing problem revisited: computing the Frobenius number in time* $O(ka_1)$, Computing and combinatorics, Lecture Notes in Comput. Sci., vol. 3595, Springer, Berlin, 2005, pp. 965–974.

60. Alfred Brauer, *On a problem of partitions*, Amer. J. Math. **64** (1942), 299–312.

61. Benjamin Braun, *An Ehrhart series formula for reflexive polytopes*, Electron. J. Combin. **13** (2006), no. 1, Note 15, 5 pp.

62. ———, *Norm bounds for Ehrhart polynomial roots*, Discrete Comput. Geom. **39** (2008), no. 1–3, 191–193, arXiv:math.CO/0602464.

63. Benjamin Braun and Robert Davis, *Ehrhart series, unimodality, and integrally closed reflexive polytopes*, to appear in Ann. Comb., arXiv:math/1403.5378.

64. Henrik Bresinsky, *Symmetric semigroups of integers generated by 4 elements*, Manuscripta Math. **17** (1975), no. 3, 205–219.

65. Charles J. Brianchon, *Théorème nouveau sur les polyèdres*, J. Ecole (Royale) Polytechnique **15** (1837), 317–319.

66. Michel Brion, *Points entiers dans les polyèdres convexes*, Ann. Sci. École Norm. Sup. (4) **21** (1988), no. 4, 653–663.

67. Michel Brion and Michèle Vergne, *Residue formulae, vector partition functions and lattice points in rational polytopes*, J. Amer. Math. Soc. **10** (1997), no. 4, 797–833.

68. Arne Brøndsted, *An Introduction to Convex Polytopes*, Graduate Texts in Mathematics, vol. 90, Springer-Verlag, New York, 1983.

69. Richard A. Brualdi and Peter M. Gibson, *Convex polyhedra of doubly stochastic matrices. I. Applications of the permanent function*, J. Combinatorial Theory Ser. A **22** (1977), no. 2, 194–230.

70. ———, *Convex polyhedra of doubly stochastic matrices. III. Affine and combinatorial properties of U_n*, J. Combinatorial Theory Ser. A **22** (1977), no. 3, 338–351.

71. Heinz Bruggesser and Peter Mani, *Shellable decompositions of cells and spheres*, Math. Scand. **29** (1971), 197–205 (1972).

72. Winfried Bruns, *Commutative algebra arising from the Anand–Dumir–Gupta conjectures*, Commutative algebra and combinatorics, Ramanujan Math. Soc. Lect. Notes Ser., vol. 4, Ramanujan Math. Soc., Mysore, 2007, pp. 1–38.

73. Winfried Bruns and Joseph Gubeladze, *Polytopes, rings, and K-theory*, Springer Monographs in Mathematics, Springer, Dordrecht, 2009.

74. Winfried Bruns and Tim Römer, *h-vectors of Gorenstein polytopes*, J. Combin. Theory Ser. A **114** (2007), no. 1, 65–76.

75. Daniel Bump, Kwok-Kwong Choi, Pär Kurlberg, and Jeffrey Vaaler, *A local Riemann hypothesis. I*, Math. Z. **233** (2000), no. 1, 1–19.

76. Kristin A. Camenga, *Vector spaces spanned by the angle sums of polytopes*, Beiträge Algebra Geom. **47** (2006), no. 2, 447–462, arXiv:math.MG/0508629.

77. Schuyler Cammann, *Old Chinese magic squares*, Sinologica **7** (1962), 14–53.

78. ———, *Islamic and Indian magic squares, Parts I and II*, History of Religions **8** (1969), 181–209; 271–299.

79. Sylvain E. Cappell and Julius L. Shaneson, *Euler–Maclaurin expansions for lattices above dimension one*, C. R. Acad. Sci. Paris Sér. I Math. **321** (1995), no. 7, 885–890.

80. Leonard Carlitz, *Some theorems on generalized Dedekind sums*, Pacific J. Math. **3** (1953), 513–522.

81. ———, *Some polynomials associated with Dedekind sums*, Acta Math. Acad. Sci. Hungar. **26** (1975), no. 3–4, 311–319.

82. John W. S. Cassels, *An Introduction to the Geometry of Numbers*, Classics in Mathematics, Springer-Verlag, Berlin, 1997, Corrected reprint of the 1971 edition.

83. Clara S. Chan, David P. Robbins, and David S. Yuen, *On the volume of a certain polytope*, Experiment. Math. **9** (2000), no. 1, 91–99, arXiv:math.CO/9810154.

84. Beifang Chen, *Weight functions, double reciprocity laws, and volume formulas for lattice polyhedra*, Proc. Natl. Acad. Sci. USA **95** (1998), no. 16, 9093–9098.

85. ———, *Lattice points, Dedekind sums, and Ehrhart polynomials of lattice polyhedra*, Discrete Comput. Geom. **28** (2002), no. 2, 175–199.

86. Beifang Chen and Vladimir Turaev, *Counting lattice points of rational polyhedra*, Adv. Math. **155** (2000), no. 1, 84–97.

87. Louis Comtet, *Advanced Combinatorics*, enlarged ed., D. Reidel Publishing Co., Dordrecht, 1974.

88. David Cox, Christian Haase, Takayuki Hibi, and Akihiro Higashitani, *Integer decomposition property of dilated polytopes*, Electron. J. Combin. **21** (2014), no. 4, P4.28, 17 pp.

89. Michele D'Adderio and Luca Moci, *Ehrhart polynomial and arithmetic Tutte polynomial*, European J. Combin. **33** (2012), no. 7, 1479–1483, arXiv:math/1102.0135.

90. Wolfgang Dahmen and Charles A. Micchelli, *The number of solutions to linear Diophantine equations and multivariate splines*, Trans. Amer. Math. Soc. **308** (1988), no. 2, 509–532.

91. Vladimir I. Danilov, *The geometry of toric varieties*, Uspekhi Mat. Nauk **33** (1978), 85–134, 247.

92. Robert Davis, *Ehrhart series of polytopes related to symmetric doubly-stochastic matrices*, Preprint (arXiv:math/1409.2742).

93. J. Leslie Davison, *On the linear Diophantine problem of Frobenius*, J. Number Theory **48** (1994), no. 3, 353–363.

94. Jesús A. De Loera, David Haws, Raymond Hemmecke, Peter Huggins, and Ruriko Yoshida, *A user's guide for LattE v1.1, software package LattE*, 2004. Electronically available at https://www.math.ucdavis.edu/ latte/.

95. Jesús A. De Loera, Raymond Hemmecke, and Matthias Köppe, *Algebraic and Geometric Ideas in the Theory of Discrete Optimization*, MOS-SIAM Series on Optimization, vol. 14, Society for Industrial and Applied Mathematics (SIAM), Philadelphia, PA; Mathematical Optimization Society, Philadelphia, PA, 2013.

96. Jesús A. De Loera, Raymond Hemmecke, Jeremiah Tauzer, and Ruriko Yoshida, *Effective lattice point counting in rational convex polytopes*, J. Symbolic Comput. **38** (2004), no. 4, 1273–1302.

97. Jesús A. De Loera and Shmuel Onn, *The complexity of three-way statistical tables*, SIAM J. Comput. **33** (2004), no. 4, 819–836, arXiv:math.CO/0207200.

98. Jesús A. De Loera, Jörg Rambau, and Francisco Santos, *Triangulations*, Algorithms and Computation in Mathematics, vol. 25, Springer-Verlag, Berlin, 2010.

99. Richard Dedekind, *Erläuterungen zu den Fragmenten XXVIII*, Collected Works of Bernhard Riemann, Dover Publ., New York, 1953, pp. 466–478.

100. Max Dehn, *Die Eulersche Formel im Zusammenhang mit dem Inhalt in der nichteuklidischen Geometrie*, Math. Ann. **61** (1905), 279–298.

101. József Dénes and Anthony D. Keedwell, *Latin squares and their applications*, Academic Press, New York, 1974.

102. Graham Denham, *Short generating functions for some semigroup algebras*, Electron. J. Combin. **10** (2003), Research Paper 36, 7 pp.

103. David Desario and Sinai Robins, *Generalized solid-angle theory for real polytopes*, Q. J. Math. **62** (2011), no. 4, 1003–1015, arXiv:0708.0042.

104. Persi Diaconis and Anil Gangolli, *Rectangular arrays with fixed margins*, Discrete Probability and Algorithms (Minneapolis, MN, 1993), Springer, New York, 1995, pp. 15–41.

105. Ricardo Diaz and Sinai Robins, *Pick's formula via the Weierstrass ℘-function*, Amer. Math. Monthly **102** (1995), no. 5, 431–437.

106. ———, *The Ehrhart polynomial of a lattice polytope*, Ann. of Math. (2) **145** (1997), no. 3, 503–518.

107. Ulrich Dieter, *Das Verhalten der Kleinschen Funktionen* $\log \sigma_{g,h}(\omega_1, \omega_2)$ *gegenüber Modultransformationen und verallgemeinerte Dedekindsche Summen*, J. Reine Angew. Math. **201** (1959), 37–70.

108. ———, *Cotangent sums, a further generalization of Dedekind sums*, J. Number Theory **18** (1984), no. 3, 289–305.

109. Eugène Ehrhart, *Sur les polyèdres rationnels homothétiques à n dimensions*, C. R. Acad. Sci. Paris **254** (1962), 616–618.

110. ———, *Sur les carrés magiques*, C. R. Acad. Sci. Paris Sér. A-B **277** (1973), A651–A654.

111. ———, *Polynômes arithmétiques et méthode des polyèdres en combinatoire*, Birkhäuser Verlag, Basel, 1977, International Series of Numerical Mathematics, Vol. 35.

112. Günter Ewald, *Combinatorial convexity and algebraic geometry*, Graduate Texts in Mathematics, vol. 168, Springer-Verlag, New York, 1996.

113. Leonid G. Fel and Boris Y. Rubinstein, *Restricted partition function as Bernoulli and Euler polynomials of higher order*, Ramanujan J. **11** (2006), no. 3, 331–348, arXiv:math.NT/0304356.

114. Matthew H. J. Fiset and Alexander M. Kasprzyk, *A note on palindromic δ-vectors for certain rational polytopes*, Electron. J. Combin. **15** (2008), no. 1, Note 18, 4 pp.

115. Lenny Fukshansky and Sinai Robins, *Bounds for solid angles of lattices of rank three*, J. Combin. Theory Ser. A **118** (2011), no. 2, 690–701, arXiv:1006.0743.

116. William Fulton, *Introduction to Toric Varieties*, Annals of Mathematics Studies, vol. 131, Princeton University Press, Princeton, NJ, 1993.

117. Stavros Garoufalidis and James E. Pommersheim, *Values of zeta functions at negative integers, Dedekind sums and toric geometry*, J. Amer. Math. Soc. **14** (2001), no. 1, 1–23.

118. Ewgenij Gawrilow and Michael Joswig, *polymake: a framework for analyzing convex polytopes*, Polytopes—combinatorics and computation (Oberwolfach, 1997), DMV Sem., vol. 29, Birkhäuser, Basel, 2000, pp. 43–73, Software polymake available at www.polymake.org/.

119. Ira M. Gessel, *Generating functions and generalized Dedekind sums*, Electron. J. Combin. **4** (1997), no. 2, Research Paper 11, 17 pp.

120. Kurt Girstmair, *On the fractional parts of Dedekind sums*, Int. J. Number Theory **11** (2015), no. 1, 29–38.

121. Jørgen P. Gram, *Om rumvinklerne i et polyeder*, Tidsskrift for Math. (Copenhagen) **4** (1874), no. 3, 161–163.

122. Nick Gravin, Jean B. Lasserre, Dmitrii V. Pasechnik, and Sinai Robins, *The inverse moment problem for convex polytopes*, Discrete Comput. Geom. **48** (2012), no. 3, 596–621.

123. Nick Gravin, Sinai Robins, and Dmitry Shiryaev, *Translational tilings by a polytope, with multiplicity*, Combinatorica **32** (2012), no. 6, 629–649, arXiv:1103.3163.

124. Harold Greenberg, *An algorithm for a linear Diophantine equation and a problem of Frobenius*, Numer. Math. **34** (1980), no. 4, 349–352.

125. Peter M. Gruber, *Convex and discrete geometry*, Grundlehren der Mathematischen Wissenschaften [Fundamental Principles of Mathematical Sciences], vol. 336, Springer, Berlin, 2007.

126. Branko Grünbaum, *Convex Polytopes*, second ed., Graduate Texts in Mathematics, vol. 221, Springer-Verlag, New York, 2003, Prepared and with a preface by Volker Kaibel, Victor Klee, and Günter M. Ziegler.

127. Victor Guillemin, *Riemann–Roch for toric orbifolds*, J. Differential Geom. **45** (1997), no. 1, 53–73.

128. Ulrich Halbritter, *Some new reciprocity formulas for generalized Dedekind sums*, Results Math. **8** (1985), no. 1, 21–46.

129. Richard R. Hall and Martin N. Huxley, *Dedekind sums and continued fractions*, Acta Arith. **63** (1993), no. 1, 79–90.

130. Richard R. Hall, J. C. Wilson, and Don Zagier, *Reciprocity formulae for general Dedekind–Rademacher sums*, Acta Arith. **73** (1995), no. 4, 389–396.

131. Martin Henk, Achill Schürmann, and Jörg M. Wills, *Ehrhart polynomials and successive minima*, Mathematika **52** (2005), no. 1–2, 1–16 (2006), arXiv:math.MG/0507528.

132. Jürgen Herzog, *Generators and relations of abelian semigroups and semigroup rings.*, Manuscripta Math. **3** (1970), 175–193.

133. Takayuki Hibi, *Some results on Ehrhart polynomials of convex polytopes*, Discrete Math. **83** (1990), no. 1, 119–121.

134. ———, *Algebraic Combinatorics on Convex Polytopes*, Carslaw, 1992.

135. ———, *Dual polytopes of rational convex polytopes*, Combinatorica **12** (1992), no. 2, 237–240.

136. Dean Hickerson, *Continued fractions and density results for Dedekind sums*, J. Reine Angew. Math. **290** (1977), 113–116.

137. Lutz Hille and Harald Skarke, *Reflexive polytopes in dimension 2 and certain relations in* $SL_2(\mathbb{Z})$, J. Algebra Appl. **1** (2002), no. 2, 159–173.

138. Friedrich Hirzebruch, *Neue topologische Methoden in der algebraischen Geometrie*, Ergebnisse der Mathematik und ihrer Grenzgebiete (N.F.), Heft 9, Springer-Verlag, Berlin, 1956.

139. ———, *Eulerian polynomials*, Münster J. Math. **1** (2008), 9–14.

140. Friedrich Hirzebruch and Don Zagier, *The Atiyah–Singer Theorem and Elementary Number Theory*, Publish or Perish Inc., Boston, Mass., 1974.

141. Jeffrey Hood and David Perkinson, *Some facets of the polytope of even permutation matrices*, Linear Algebra Appl. **381** (2004), 237–244.

142. Masa-Nori Ishida, *Polyhedral Laurent series and Brion's equalities*, Internat. J. Math. **1** (1990), no. 3, 251–265.

143. Maruf Israilovich Israilov, *Determination of the number of solutions of linear Diophantine equations and their applications in the theory of invariant cubature formulas*, Sibirsk. Mat. Zh. **22** (1981), no. 2, 121–136, 237.

144. Ravi Kannan, *Lattice translates of a polytope and the Frobenius problem*, Combinatorica **12** (1992), no. 2, 161–177.

145. Jean-Michel Kantor and Askold G. Khovanskiĭ, *Une application du théorème de Riemann–Roch combinatoire au polynôme d'Ehrhart des polytopes entiers de* $\mathbf{R}^d$, C. R. Acad. Sci. Paris Sér. I Math. **317** (1993), no. 5, 501–507.

146. Yael Karshon, Shlomo Sternberg, and Jonathan Weitsman, *The Euler–Maclaurin formula for simple integral polytopes*, Proc. Natl. Acad. Sci. USA **100** (2003), no. 2, 426–433.

147. Askold G. Khovanskiĭ and Aleksandr V. Pukhlikov, *The Riemann–Roch theorem for integrals and sums of quasipolynomials on virtual polytopes*, Algebra i Analiz **4** (1992), no. 4, 188–216.

148. Peter Kirschenhofer, Attila Pethő, and Robert F. Tichy, *On analytical and Diophantine properties of a family of counting polynomials*, Acta Sci. Math. (Szeged) **65** (1999), no. 1–2, 47–59.

149. Daniel A. Klain and Gian-Carlo Rota, *Introduction to Geometric Probability*, Lezioni Lincee, Cambridge University Press, Cambridge, 1997.

150. Victor Klee, *A combinatorial analogue of Poincaré's duality theorem*, Canad. J. Math. **16** (1964), 517–531.

151. Donald E. Knuth, *Permutations, matrices, and generalized Young tableaux*, Pacific J. Math. **34** (1970), 709–727.

152. ———, *Notes on generalized Dedekind sums*, Acta Arith. **33** (1977), no. 4, 297–325.

153. ———, *The Art of Computer Programming. Vol. 2*, second ed., Addison-Wesley Publishing Co., Reading, Mass., 1981.

154. Matthias Köppe, *A primal Barvinok algorithm based on irrational decompositions*, SIAM J. Discrete Math. **21** (2007), no. 1, 220–236, `arXiv:math.CO/0603308`. Software `LattE macchiato` available at http://www.math.ucdavis.edu/ mkoeppe/latte/.

155. Thomas W. Körner, *Fourier Analysis*, Cambridge University Press, Cambridge, 1988.

156. Maximilian Kreuzer and Harald Skarke, *Classification of reflexive polyhedra in three dimensions*, Adv. Theor. Math. Phys. **2** (1998), no. 4, 853–871, `arXiv:hep-th/9805190`.

157. ———, *Complete classification of reflexive polyhedra in four dimensions*, Adv. Theor. Math. Phys. **4** (2000), no. 6, 1209–1230, `arXiv:hep-th/0002240`.

158. Jean B. Lasserre, *Moments, positive polynomials and their applications*, Imperial College Press Optimization Series, vol. 1, Imperial College Press, London, 2010.

159. Jean B. Lasserre and Eduardo S. Zeron, *On counting integral points in a convex rational polytope*, Math. Oper. Res. **28** (2003), no. 4, 853–870.

160. Jim Lawrence, *Valuations and polarity*, Discrete Comput. Geom. **3** (1988), no. 4, 307–324.

161. _____, *Polytope volume computation*, Math. Comp. **57** (1991), no. 195, 259–271.

162. _____, *A short proof of Euler's relation for convex polytopes*, Canad. Math. Bull. **40** (1997), no. 4, 471–474.

163. Arjen K. Lenstra, Hendrik W. Lenstra, Jr., and László Lovász, *Factoring polynomials with rational coefficients*, Math. Ann. **261** (1982), no. 4, 515–534.

164. Eva Linke, *Rational Ehrhart quasi-polynomials*, J. Combin. Theory Ser. A **118** (2011), no. 7, 1966–1978, arXiv:1006.5612.

165. László Lovász, *Combinatorial Problems and Exercises*, second ed., North-Holland Publishing Co., Amsterdam, 1993.

166. Ian G. Macdonald, *Polynomials associated with finite cell-complexes*, J. London Math. Soc. (2) **4** (1971), 181–192.

167. Percy A. MacMahon, *Combinatory Analysis*, Chelsea Publishing Co., New York, 1960, reprint of the 1915 original.

168. Evgeny N. Materov, *The Bott formula for toric varieties*, Mosc. Math. J. **2** (2002), no. 1, 161–182, arXiv:math.AG/9904110.

169. Tyrrell B. McAllister and Kevin M. Woods, *The minimum period of the Ehrhart quasi-polynomial of a rational polytope*, J. Combin. Theory Ser. A **109** (2005), no. 2, 345–352, arXiv:math.CO/0310255.

170. Peter McMullen, *On zonotopes*, Trans. Amer. Math. Soc. **159** (1971), 91–109.

171. _____, *Valuations and Euler-type relations on certain classes of convex polytopes*, Proc. London Math. Soc. (3) **35** (1977), no. 1, 113–135.

172. _____, *Lattice invariant valuations on rational polytopes*, Arch. Math. (Basel) **31** (1978/79), no. 5, 509–516.

173. Peter McMullen and Geoffrey C. Shephard, *Diagrams for centrally symmetric polytopes*, Mathematika **15** (1968), 123–138.

174. Curt Meyer, *Über einige Anwendungen Dedekindscher Summen*, J. Reine Angew. Math. **198** (1957), 143–203.

175. Jeffrey L. Meyer, *Character analogues of Dedekind sums and transformations of analytic Eisenstein series*, Pacific J. Math. **194** (2000), no. 1, 137–164.

176. Werner Meyer and Robert Sczech, *Über eine topologische und zahlentheoretische Anwendung von Hirzebruchs Spitzenauflösung*, Math. Ann. **240** (1979), no. 1, 69–96.

177. Ezra Miller and Bernd Sturmfels, *Combinatorial Commutative Algebra*, Graduate Texts in Mathematics, vol. 227, Springer-Verlag, New York, 2005.

178. Hermann Minkowski, *Geometrie der Zahlen*, Bibliotheca Mathematica Teubneriana, Band 40, Johnson Reprint Corp., New York, 1968.

179. Marcel Morales, *Syzygies of monomial curves and a linear Diophantine problem of Frobenius*, Preprint, Max-Planck-Institut für Mathematik, Bonn, 1986.

180. _____, *Noetherian symbolic blow-ups*, J. Algebra **140** (1991), no. 1, 12–25.

181. Louis J. Mordell, *Lattice points in a tetrahedron and generalized Dedekind sums*, J. Indian Math. Soc. (N.S.) **15** (1951), 41–46.

182. Robert Morelli, *Pick's theorem and the Todd class of a toric variety*, Adv. Math. **100** (1993), no. 2, 183–231.

183. Gerald Myerson, *On semiregular finite continued fractions*, Arch. Math. (Basel) **48** (1987), no. 5, 420–425.

184. Walter Nef, *Zur Einführung der Eulerschen Charakteristik*, Monatsh. Math. **92** (1981), no. 1, 41–46.

185. Benjamin Nill and Andreas Paffenholz, *On the equality case in Ehrhart's volume conjecture*, Adv. Geom. **14** (2014), no. 4, 579–586, arXiv:1205.1270.

186. C. D. Olds, Anneli Lax, and Giuliana P. Davidoff, *The Geometry of Numbers*, Anneli Lax New Mathematical Library, vol. 41, Mathematical Association of America, Washington, DC, 2000.

187. Paul C. Pasles, *The lost squares of Dr. Franklin: Ben Franklin's missing squares and the secret of the magic circle*, Amer. Math. Monthly **108** (2001), no. 6, 489–511.

188. Sam Payne, *Ehrhart series and lattice triangulations*, Discrete Comput. Geom. **40** (2008), no. 3, 365–376, arXiv:math/0702052.

189. Micha A. Perles and Geoffrey C. Shephard, *Angle sums of convex polytopes*, Math. Scand. **21** (1967), 199–218.

190. Georg Alexander Pick, *Geometrisches zur Zahlenlehre*, Sitzenber. Lotos (Prague) **19** (1899), 311–319.

191. Clifford A. Pickover, *The Zen of Magic Squares, Circles, and Stars*, Princeton University Press, Princeton, NJ, 2002.

192. Christopher Polis, *Pick's theorem extended and generalized*, Math. Teacher (1991), 399–401.

193. James E. Pommersheim, *Toric varieties, lattice points and Dedekind sums*, Math. Ann. **295** (1993), no. 1, 1–24.

194. Bjorn Poonen and Fernando Rodriguez-Villegas, *Lattice polygons and the number 12*, Amer. Math. Monthly **107** (2000), no. 3, 238–250.

195. Tiberiu Popoviciu, *Asupra unei probleme de patitie a numerelor*, Acad. Republicii Populare Romane, Filiala Cluj, Studii si cercetari stiintifice **4** (1953), 7–58.

196. Alexander Postnikov, *Permutohedra, associahedra, and beyond*, Int. Math. Res. Not. (2009), no. 6, 1026–1106, arXiv:math/0507163.

197. Hans Rademacher, *Generalization of the reciprocity formula for Dedekind sums*, Duke Math. J. **21** (1954), 391–397.

198. _____, *Some remarks on certain generalized Dedekind sums*, Acta Arith. **9** (1964), 97–105.

199. Hans Rademacher and Emil Grosswald, *Dedekind Sums*, The Mathematical Association of America, Washington, D.C., 1972.

200. Jorge L. Ramírez Alfonsín, *Complexity of the Frobenius problem*, Combinatorica **16** (1996), no. 1, 143–147.

201. _____, *The Diophantine Frobenius Problem*, Oxford Lecture Series in Mathematics and its Applications, vol. 30, Oxford University Press, Oxford, 2005.

202. J. E. Reeve, *On the volume of lattice polyhedra*, Proc. London Math. Soc. (3) **7** (1957), 378–395.

203. Les Reid and Leslie G. Roberts, *Monomial subrings in arbitrary dimension*, J. Algebra **236** (2001), no. 2, 703–730.

204. Jason M. Ribando, *Measuring solid angles beyond dimension three*, Discrete Comput. Geom. **36** (2006), no. 3, 479–487.

205. Adrian Rice, *Brought to book: the curious story of Guglielmo Libri (1803–69)*, European Mathematical Society Newsletter **48** (2003), 12–14.

206. Jürgen Richter-Gebert, *Realization spaces of polytopes*, Lecture Notes in Mathematics, vol. 1643, Springer-Verlag, Berlin Heidelberg, 1996.

207. Bjarke Hammersholt Roune, *Solving thousand-digit Frobenius problems using Gröbner bases*, J. Symbolic Comput. **43** (2008), no. 1, 1–7, arXiv:math/0702040.

208. Boris Y. Rubinstein, *Expression for restricted partition function through Bernoulli polynomials*, Ramanujan J. **15** (2008), no. 2, 177–185.

209. Steven V Sam, *A bijective proof for a theorem of Ehrhart*, Amer. Math. Monthly **116** (2009), no. 8, 688–701, arXiv:0801.4432v5.

210. Jan Schepers and Leen Van Langenhoven, *Unimodality questions for integrally closed lattice polytopes*, Ann. Comb. **17** (2013), no. 3, 571–589, arXiv:math/1110.3724.

211. Ludwig Schläfli, *Theorie der vielfachen Kontinuität*, Ludwig Schläfli, 1814–1895, Gesammelte Mathematische Abhandlungen, Vol. I, Birkhäuser, Basel, 1950, pp. 167–387.

212. Pieter Hendrik Schoute, *Analytic treatment of the polytopes regularly derived from the regular polytopes*, Verhandelingen der Koninklijke Akademie von Wetenschappen te Amsterdam **11** (1911), no. 3.

213. Alexander Schrijver, *Combinatorial Optimization. Polyhedra and Efficiency. Vol. A–C*, Algorithms and Combinatorics, vol. 24, Springer-Verlag, Berlin, 2003.

214. Paul R. Scott, *On convex lattice polygons*, Bull. Austral. Math. Soc. **15** (1976), no. 3, 395–399.

215. Sinan Sertöz, *On the number of solutions of the Diophantine equation of Frobenius*, Diskret. Mat. **10** (1998), no. 2, 62–71.

216. Jeffrey Shallit, *The Frobenius problem and its generalizations*, Developments in language theory, Lecture Notes in Comput. Sci., vol. 5257, Springer, Berlin, 2008, pp. 72–83.

217. Jeffrey Shallit and James Stankewicz, *Unbounded discrepancy in Frobenius numbers*, Integers **11** (2011), A2, 8 pp.

218. Geoffrey C. Shephard, *An elementary proof of Gram's theorem for convex polytopes*, Canad. J. Math. **19** (1967), 1214–1217.

219. ———, *Combinatorial properties of associated zonotopes*, Canad. J. Math. **26** (1974), 302–321.

220. Carl Ludwig Siegel, *Lectures on the Geometry of Numbers*, Springer-Verlag, Berlin, 1989, Notes by B. Friedman, rewritten by Komaravolu Chandrasekharan with the assistance of Rudolf Suter, with a preface by Chandrasekharan.

221. R. Jamie Simpson and Robert Tijdeman, *Multi-dimensional versions of a theorem of Fine and Wilf and a formula of Sylvester*, Proc. Amer. Math. Soc. **131** (2003), no. 6, 1661–1671.

222. David Solomon, *Algebraic properties of Shintani's generating functions: Dedekind sums and cocycles on* $\mathrm{PGL}_2(\mathbf{Q})$, Compositio Math. **112** (1998), no. 3, 333–362.

223. Duncan M. Y. Sommerville, *The relation connecting the angle-sums and volume of a polytope in space of n dimensions*, Proc. Roy. Soc. London, Ser. A **115** (1927), 103–119.

224. Richard P. Stanley, *Linear homogeneous Diophantine equations and magic labelings of graphs*, Duke Math. J. **40** (1973), 607–632.

225. ———, *Combinatorial reciprocity theorems*, Advances in Math. **14** (1974), 194–253.

226. ———, *Decompositions of rational convex polytopes*, Ann. Discrete Math. **6** (1980), 333–342.

227. ———, *On the Hilbert function of a graded Cohen–Macaulay domain*, J. Pure Appl. Algebra **73** (1991), no. 3, 307–314.

228. ———, *A zonotope associated with graphical degree sequences*, Applied geometry and discrete mathematics, DIMACS Ser. Discrete Math. Theoret. Comput. Sci., vol. 4, Amer. Math. Soc., Providence, RI, 1991, pp. 555–570.

229. ———, *Combinatorics and Commutative Algebra*, second ed., Progress in Mathematics, vol. 41, Birkhäuser Boston Inc., Boston, MA, 1996.

230. ———, *Enumerative Combinatorics. Volume 1*, second ed., Cambridge Studies in Advanced Mathematics, vol. 49, Cambridge University Press, Cambridge, 2012.

231. ———, *How the Upper Bound Conjecture Was Proved*, Ann. Comb. **18** (2014), no. 3, 533–539.

232. Alan Stapledon, *Additive number theory and inequalities in Ehrhart theory*, Preprint (arXiv:0904.3035v2).

233. ———, *Weighted Ehrhart theory and orbifold cohomology*, Adv. Math. **219** (2008), no. 1, 63–88, arXiv:math/0711.4382.

234. ———, *Inequalities and Ehrhart δ-vectors*, Trans. Amer. Math. Soc. **361** (2009), no. 10, 5615–5626, arXiv:math/0801.0873.

235. Bernd Sturmfels, *On vector partition functions*, J. Combin. Theory Ser. A **72** (1995), no. 2, 302–309.

236. ———, *Gröbner Bases and Convex Polytopes*, University Lecture Series, vol. 8, American Mathematical Society, Providence, RI, 1996.

237. James J. Sylvester, *On the partition of numbers*, Quaterly J. Math. **1** (1857), 141–152.

238. ———, *On subinvariants, i.e. semi-invariants to binary quantics of an unlimited order*, Amer. J. Math. **5** (1882), 119–136.

239. _____, *Mathematical questions with their solutions*, Educational Times **41** (1884), 171–178.

240. András Szenes and Michèle Vergne, *Residue formulae for vector partitions and Euler–Maclaurin sums*, Adv. in Appl. Math. **30** (2003), no. 1–2, 295–342, arXiv:math.CO/0202253.

241. Lajos Takács, *On generalized Dedekind sums*, J. Number Theory **11** (1979), no. 2, 264–272.

242. Audrey Terras, *Fourier Analysis on Finite Groups and Applications*, London Mathematical Society Student Texts, vol. 43, Cambridge University Press, Cambridge, 1999.

243. John A. Todd, *The geometrical invariants of algebraic loci*, Proc. London Math. Soc. **43** (1937), 127–141.

244. _____, *The geometrical invariants of algebraic loci (second paper)*, Proc. London Math. Soc. **45** (1939), 410–424.

245. Amitabha Tripathi, *The number of solutions to $ax + by = n$*, Fibonacci Quart. **38** (2000), no. 4, 290–293.

246. _____, *On the Frobenius problem for geometric sequences*, Integers **8** (2008), A43, 5 pp.

247. _____, *On the Frobenius problem for $\{a^k, a^k + 1, a^k + a, \ldots, a^k + a^{k-1}\}$*, Integers **10** (2010), A44, 523–529.

248. Helge Tverberg, *How to cut a convex polytope into simplices*, Geometriae Dedicata **3** (1974), 239–240.

249. Alexander N. Varchenko, *Combinatorics and topology of the arrangement of affine hyperplanes in the real space*, Funktsional. Anal. i Prilozhen. **21** (1987), no. 1, 11–22.

250. Sven Verdoolaege, *Software package* barvinok, (2004), electronically available at http://freshmeat.net/projects/barvinok/.

251. John von Neumann, *A certain zero-sum two-person game equivalent to the optimal assignment problem*, Contributions to the Theory of Games, vol. 2, Annals of Mathematics Studies, no. 28, Princeton University Press, Princeton, N. J., 1953, pp. 5–12.

252. Herbert S. Wilf, *generatingfunctionology*, second ed., Academic Press Inc., Boston, MA, 1994, electronically available at http://www.math.upenn.edu/~wilf/DownldGF.html.

253. Kevin Woods, *Computing the period of an Ehrhart quasi-polynomial*, Electron. J. Combin. **12** (2005), no. 1, Research Paper 34, 12 pp.

254. Guoce Xin, *Constructing all magic squares of order three*, Discrete Math. **308** (2008), no. 15, 3393–3398, arXiv:math.CO/0409468.

255. Don Zagier, *Higher dimensional Dedekind sums*, Math. Ann. **202** (1973), 149–172.

256. Claudia Zaslavsky, *Africa Counts*, Prindle, Weber & Schmidt, Inc., Boston, Mass., 1973.

257. Doron Zeilberger, *Proof of a conjecture of Chan, Robbins, and Yuen*, Electron. Trans. Numer. Anal. **9** (1999), 147–148.

258. Günter M. Ziegler, *Lectures on Polytopes*, Springer-Verlag, New York, 1995.

List of Symbols

The following table contains a list of symbols that are frequently used throughout the book. The page numbers refer to the first appearance/definition of each symbol.

Symbol	Meaning	Page
$\hat{a}(m)$	Fourier coefficient of $a(n)$	137
$A(d,k)$	Eulerian number	30
$\mathcal{A}^{\perp}$	orthogonal complement of $\mathcal{A}$	210
$A_{\mathcal{P}}(t)$	solid-angle sum of $\mathcal{P}$	229
$\alpha_{\mathcal{P}}(\mathbf{z})$	solid-angle generating function	231
$B_k(x)$	Bernoulli polynomial	34
B_k	Bernoulli number	34
$\mathcal{B}_n$	Birkhoff polytope	116
$\mathrm{BiPyr}(\mathcal{P})$	bipyramid over $\mathcal{P}$	39
$\mathrm{cone}\,\mathcal{P}$	cone over $\mathcal{P}$	63
$\mathrm{const}\,f$	constant term of the generating function f	14
$\mathrm{conv}\,S$	convex hull of S	27
d-cone	d-dimensional cone	62
d-polytope	d-dimensional polytope	28
$\dim\mathcal{P}$	dimension of $\mathcal{P}$	28
$\delta_m(x)$	delta function	140
$\mathrm{Ehr}_{\mathcal{P}}(z)$	Ehrhart series of $\mathcal{P}$	30
$\mathrm{Ehr}_{\mathcal{P}^{\circ}}(z)$	Ehrhart series of the interior of $\mathcal{P}$	93
$\mathbf{e}_a(x)$	root-of-unity function $e^{2\pi i a x/b}$	140
f_k	face number	101
$F_k(t)$	lattice-point enumerator of the k-skeleton	103
$\mathbf{F}(f)$	Fourier transform of f	139
$g(a_1, a_2, \dots, a_d)$	Frobenius number	6
$h_{\mathcal{P}}^*(z)$	h^*-polynomial of $\mathcal{P}$	72
$H_n(t)$	number of semimagic $n \times n$ squares with line sum t	115

Symbol	Meaning	Page
$\mathcal{K}_\mathcal{F}$	tangent cone of $\mathcal{F} \subseteq \mathcal{P}$	204
$L_\mathcal{P}(t)$	lattice-point enumerator of $\mathcal{P}$	29
$L_{\mathcal{P}^\circ}(t)$	lattice-point enumerator of the interior of $\mathcal{P}$	30
$M_n(t)$	number of magic $n \times n$ squares with line sum t	115
$\mathcal{N}(p(z_1, z_2, \ldots, z_d))$	Newton polytope of $p(z_1, z_2, \ldots, z_d)$	167
$\omega_\mathcal{P}(\mathbf{x})$	solid angle of $\mathbf{x}$ (with respect to $\mathcal{P}$)	227
$p_A(n)$	restricted partition function	6
$\text{poly}_A(n)$	polynomial part of $p_A(n)$	151
$\mathcal{P}$	a closed polytope	27
$\mathcal{P}^\circ$	interior of the polytope $\mathcal{P}$	30
$\mathcal{P}_1 + \mathcal{P}_2 + \cdots + \mathcal{P}_n$	Minkowski sum of $\mathcal{P}_1, \mathcal{P}_2, \ldots, \mathcal{P}_n$	167
$\mathcal{P}(\mathbf{h})$	perturbed polytope	221
$\text{Pyr}(\mathcal{P})$	pyramid over $\mathcal{P}$	36
$\wp(z)$	Weierstraß $\wp$-function	243
Π	fundamental parallelepiped of a cone	66
$r_n(a, b)$	Dedekind–Rademacher sum	155
$s(a, b)$	Dedekind sum	138
$s_n(a_1, a_2, \ldots, a_m; b)$	Fourier–Dedekind sum	15
$\text{Solid}_\mathcal{P}(x)$	solid-angle series	236
$\text{span}\,\mathcal{P}$	affine space spanned by $\mathcal{P}$	28
$\sigma_S(\mathbf{z})$	integer-point transform of S	64
$t\mathcal{P}$	t^{th} dilation of $\mathcal{P}$	29
Todd_h	Todd operator	214
$\text{vol}\,\mathcal{P}$	(continuous) volume of $\mathcal{P}$	76
V_G	vector space of all complex-valued functions on $G = \{0, 1, 2, \ldots, b-1\}$	139
ξ_a	root of unity $e^{2\pi i/a}$	8
$\zeta(z)$	Weierstraß ζ-function	243
$\mathcal{Z}(\mathbf{u}_1, \mathbf{u}_2, \ldots, \mathbf{u}_n)$	zonotope spanned by $\mathbf{u}_1, \mathbf{u}_2, \ldots, \mathbf{u}_n$	168
$[\mathbf{x}, \mathbf{y}]$	line segment joining $\mathbf{x}$ and $\mathbf{y} \in \mathbb{R}^d$	12
$\lfloor x \rfloor$	greatest integer function	10
$\{x\}$	fraction-part function	10
$((x))$	sawtooth function	137
$\mathbf{z}^\mathbf{c}$	$z_1^{c_1} z_2^{c_2} \cdots z_m^{c_m}$	50
$\binom{m}{n}$	binomial coefficient	30
$\langle f, g \rangle$	inner product of f and g	140
$(f * g)(t)$	convolution of f and g	143
$1_S(\mathbf{x})$	characteristic function of S	205
$\#S$	number of elements in S	6
♣	an exercise that is used in the text	5

Index

Printed in the United States
By Bookmasters